T0324110

INTELLIGENT DATA MINING AND FUSION SYSTEMS IN AGRICULTURE

INTELLIGENT DATA MINING AND FUSION SYSTEMS IN AGRICULTURE

XANTHOULA EIRINI PANTAZI

DIMITRIOS MOSHOU

DIONYSIS BOCHTIS

Academic Press is an imprint of Elsevier
125 London Wall, London EC2Y 5AS, United Kingdom
525 B Street, Suite 1650, San Diego, CA 92101, United States
50 Hampshire Street, 5th Floor, Cambridge, MA 02139, United States
The Boulevard, Langford Lane, Kidlington, Oxford OX5 1GB, United Kingdom

Library of Congress Cataloging-in-Publication Data
A catalog record for this book is available from the Library of Congress

British Library Cataloguing-in-Publication Data
A catalogue record for this book is available from the British Library

ISBN 978-0-12-814391-9

For information on all Academic Press publications
visit our website at https://www.elsevier.com/books-and-journals

Publisher: Charlotte Cockle
Acquisition Editor: Nancy Maragioglio
Editorial Project Manager: Devlin Person
Production Project Manager: Nirmala Arumugam
Cover Designer: Greg Harris

Typeset by SPi Global, India

CONTENTS

CHAPTER 1
Sensors in agriculture

Contents

1.1 Milestones in agricultural automation and monitoring in agriculture

1.1.1 Introduction

By 2050, is expected that world population will reach 9.8 billion according to a new launched United Nations report (*www.un.org*). Consequently, as the global demand for food and agricultural crops elevates, novel and sustainable approaches are needed that employ agricultural technologies focusing not only on agricultural activities for crop production but also on the global impacts concerning the appropriate nitrogen fertilizer use, reduced GHG emissions and water footprints.

The intensification of agricultural activities, mechanization, and automation throughout the years are regarded the main reasons that contributed significantly to the rise of agricultural productivity. This led to the evolution of autonomous farming systems for field operations, livestock management systems and growth control systems oriented mostly on greenhouse monitoring, climate control and irrigation management systems. The efficient

Intelligent Data Mining and Fusion Systems in Agriculture
https://doi.org/10.1016/B978-0-12-814391-9.00001-7

agricultural production management enables the decrease of negative environmental impacts allowing both efficiency, as well agricultural products safety.

Agriculture has to meet significant challenges in the face of providing technical solutions for increasing production, while the environmental impact decreases by reduced application of agro-chemicals and increased use of environmental friendly management practices. A benefit of this is the reduction of production costs. Technologies of sensors produce tools to achieve the above-mentioned goals. The explosive technological advances and development in recent years enormously facilitates the attainment of these objectives by removing many barriers for their implementation, including reservations expressed by farmers themselves. Precision Agriculture is an emerging area, where sensor-based technologies play an important role. Farmers, researchers, and technical manufacturers, all together, are joining efforts to find efficient solutions and improvements in production and in to reductions in costs.

Precision agriculture (PA) is based on the idea of sensing for monitoring and deploying management actions according to spatial and temporal crop variability. Therefore, sensor technologies are regarded as fundamental components of PA systems. PA technologies combine sensors, farm management information systems, compatible machinery that can apply inputs according to map requirements, for production optimization. Through adaptation of PA production inputs, an efficient and precise resources management for protecting the environment while maintaining the sustainability of the food supply is achieved.

Precision agriculture comprises a powerful tool for monitoring food production chain and securing both agricultural production quantity and quality by employing novel technologies to optimize the agricultural operations for producing simultaneously hiring yield and quality with site-specific input. More specifically, sensor-based monitoring technologies give precise yield predictions and early alerts concerning the crop condition. PA makes farming more credible by improving registration of operations, by monitoring and documenting. Crop monitoring provides more precise predictions on agricultural products quality, making consequently the food chain easier to be monitored for stakeholders (producers, retailers, customers). It is also capable of giving vital information for plant health condition. Recent technologies are able to monitor plants and crops at different scales of resolution. The monitoring scales vary from field (ca. $30 \times 30\,m$) to plant monitoring (ca. $30 \times 30\,cm$). It is expected that the upcoming technologies will be capable of making leaf level (ca. 3×3 cm) and symptoms on leaves (ca. 0.5×0.5 cm) detectable by sensor based optical diagnostics.On the other hand, diseases untraceable by conventional means will be detected by automated optical sensing and optimal planning options.

1.1.2 Sensing systems for PA

Crop motoring is not a new tendency. The first approaches for crop monitoring are dated back in ancient Egypt, where attempts to examine the River Nile water level fluctuations effect on crop yield have been indicated (Luiz, Formaggio, & Epiphanio, 2011). The above measurements were employed to contribute not only to the tax management system but also to the prevention of famine. Nowadays, it is more than necessary to use the intelligent tools of agriculture to meet the multiple social needs that have arisen concerning trustworthy crop product information which consequently are capable of guaranteeing and assessing crop and food safety (Becker-Reshef et al., 2010).

It should be stressed that sensing monitoring systems estimations in agriculture are crucial to be provided at an early stage during the growing period and updated regularly until harvesting so as timely vital information for crop production, condition and yield productivity both at the (sub)regional to the national level are provided. Consequently, this information are combined by forming early warning systems, which are capable of providing homogeneous data sets. The statistically valid precision and accuracy data gives the opportunity to stakeholders, to recognise and discriminate areas, which vary a lot in terms of production and productivity, and to take early decisions. For example, a synergy between a satellite based remote sensing approach and modeling tools can offer timely useful and spatial detailed information for crop status and yield with reasonable costs (Roughgarden, Running, & Matson, 1991).

However, sensing monitoring of agricultural production often conveys some additional limiting-factors closely related to the ability of unfavorable growing conditions altering within short time periods (FAO, 2011) including the seasonal patterns associated with crops biological lifecycle, the physical landscape, the management practices and the climate variability. Food and Agriculture Organization (FAO) (2011), stresses the contribution of early and accurate information that concerns crop status and yield productivity to underlying agricultural statistics and to the decision making process of the associated monitoring systems. On the contrary, information is not considered of high value if it is available too late. Considering variables space and time sensitivity and crops perishable nature, it is concluded that agricultural monitoring systems are vital to be timely assessed.

1.1.3 Benefits through sensing for PA farming systems

The positive effects of PA farming systems adaptation in agricultural activities are reflected mainly in two areas including the farmers' profitability and the protection of the environment. For achieving profitability, it is important that the cost of investing in such promising sensing technologies is fully focused on some envisioned reasonable return. PA farming systems allows detailed agricultural production tracking and tuning by enabling farmers to modify and timely program the fertilization and agrochemicals application according the field's spatial and temporal variability. By adapting PA technologies, the farmers are given the opportunity to manage, their own agricultural machinery in a more precise and efficient way to meet the cultivated crop needs. Moreover, PA systems are capable of forming a database of many years, which enables the combination of both yield and weather data for adjusting crop management and predicting crop productivity. Agricultural machinery movements and their work recordings, also contribute to the creation of a useful database tool able to assess the duration of several farming activities. Based on the crop yield variability in a field and the inputs cost, farmers are enabled to assess the economic risk and consequently estimate the cash return over the costs per hectare in a more precise and trustworthy way. Moreover, parts within a field demonstrated low yield productivity, are isolated and subjected to site-specific management treatment (Goddard, 1997).

Nowadays, there are environmental legislations already established in some European countries including Denmark, Germany, and UK and in USA and Australia, More countries across Europe are expected to follow and introduce novel legislations focusing on the rational use of water resources and on the significant reduction of fertilizers and agro-chemicals usage. As PA farming systems are fully complied to the principles of those legislations, novel and effective solutions for accurate application, recording of farming activities in the field, operation to operation tracking, and transfer of recorded crop information after harvesting will be offered (Stafford, 2000).

Automation systems has contributed significantly to lowered production cost by reducing the manual labor drudgery, offering higher quality productivity, leading to a more effective environmental control. Automation in agriculture employs advanced technologies to face its complexity and its highly variable environment. The GNSS-based vehicle guidance is used extensively for several PA activities because it allows the operation of agricultural vehicles working on parallel tracks or on predefined paths, contributing to more comfortable driving, avoiding gaps and overlaps. Initially, navigation systems were used to help agricultural vehicles operators by utilizing visual feedback (e.g. light bars, graphical displays). Current auto-guidance systems do not need any direct input from operators.

Taking into account the high complexity and variability of agricultural environment and the extensive use of resources. Intelligent PA's sensing technologies are capable of offering beneficial solutions to difficult problems including crop variability and environment (size, shape, location, soil properties, and weather), delicate products, and hostile environmental conditions (dust, dirt, and extreme temperature and humidity).

1.1.4 Current trends in agricultural automation

PA systems utilizes a several heterogeneous technologies including:

- Automated steering systems. They control specific driving tasks (auto-steering, overhead turning) following field edges and overlapping of rows. Through this type of agricultural automation, human errors are avoided while soil and site management is achieved. For example, automated headland turns contribute to fuel consumption (from 2% up to 10%).
- Sensors and remote sensing. Parameters such as moisture content, soil nutrients, soil compaction, and possible crop infections are collected from distanced sensors usually installed on mobile machines to assess soil and crop status. To date, many farmers across Europe already utilizes several sensors (thermal, optical, mechanical and chemical) for capturing variations in climatic conditions, soil and crop properties and status so as to quantify accurately crop biomass, fertilization or pest application.
- Future autonomous agricultural robots. These robots will utilize electricity power produced at the field and will be able to reconfigure their own architecture to perform various tasks offering an enormous potential for sustainability: Their weight will minimise soil compaction due to their lighter weight. Future autonomous agricultural robots will be also permanently situated on the field area in order to function only where they are needed. Robots will optimize inputs used by farmers (fertilizers, pesticides, insecticides) and reduce the impact on soils and water tables.

1.1.5 Current challenges of PA sensing applications

Crop infestation, biotic and abiotic stresses and lack of the adequate monitoring tools, are responsible for affecting negatively yield prediction, consequently leading to low productivity, crop quality and safety. For this reason, a framework in which smart sensors are combined with intelligent decision support software should be developed in order to enable efficient agricultural production management through real-time error detection technology, crop condition monitoring, automation and product quality assessment. Moreover, the combination of sensor fusion and novelty detection provides situational awareness and automation in several types of monitoring.

The central milestones for automation and crop sensing monitoring include:

1. Determination of fault detection architecture, algorithms requirements and specifications for the development of monitoring systems in decision-making support and automation frameworks that meet the principles of PA.
2. Development of situational framework recognition based on information fusion for agricultural production.
3. Development of technical innovation in detecting abnormal situations in several agricultural processes.
4. Health oriented crop-monitoring application by using non- contact optical sensors for achieving online continuous operation.
5. Agricultural products monitoring through non-destructive detection for guaranteeing quality or measuring yield.
6. Implementation, evaluation and decision support systems applications for an efficient control at early stage.

To address the above challenge there have been introduced a number of sensor technologies specialized in aiding and contributing into several agricultural conditions which affect a variety of factors, including crop yield, soil properties and nutrients, crop nutrients, crop canopy volume and biomass, water content, and pest conditions (disease, weeds, and insects), consequently concerning crop status and management. These sensor technologies are given as follows:

1.1 Sensors for Soil Analysis and Characteristics
1.2 Yield Sensing
1.3 Sensors for crops and fruits assessment
1.4 Sensors for weed management
1.5 Sensors for disease detection and classification

The above mentioned sensor types and their use is presented and discussed in Sections 1.2–1.5. The main advantage of them lies on their ability of guiding decision support systems either autonomously or by integrating with other homogeneous or heterogeneous sensor technologies in sensor fusion architectures. This characteristic is regarded a great asset in reaching a trustworthy conclusion for efficient crop monitoring and control.

1.2 Sensors for soil analysis and characteristics

There are several sensing methods investigated for soil nutrients and other characteristics assessment such as Near Infrared Radiation (NIR), MIR and Raman spectroscopy, spectral libraries, electrodes, thermal imaging, fluorescence kinetics, and electromagnetic radiation (microwaves). The above-mentioned methods are individually presented below.

1.2.1 Visible (VIS)/NIR spectroscopy

Advances in the spectroscopy field have brought about novel approaches for determining the chemical elements concentration. Ultraviolet (UV), VIS and NIR reflectance spectroscopy belong to the most common techniques. The main advantage of these techniques lays on their ability to assess soil properties in a fast and non-destructive way.

Soil organic matter (SOM) is one of the initially investigated soil properties soil fertility due to its strong correlation to soil fertility and high affection to crop production. Bowers and Hanks (1965) have also reported that soil moisture is inversely proportional to the reflectance and can be possibly assessed through reflectance measuring.

Soil organic carbon (SOC) content is regarded also a prominent constituent of SOM. Linear and PLS regression models (Vasques, Grunwald, & Sickman, 2008), NIR and fluorescence approaches (Rinnan & Rinnan, 2007), and spectral features (Bartholomeus et al., 2008) belong to the most common techniques utilized for determining SOC. Soil mineral-N was also investigated. Ehsani, Upadhyaya, Slaughter, Shafii, and Pelletier (1999) utilized PLS and principal component regression (PCR) models together with soil NIR reflectance for assessing soil mineral-N content in the range of 1100–2500 nm. The calibration models were proven quite robust. However, a site-specific calibration was essential when the models had not included some other interfering factors. A soil moisture content determination approach using a NIR calibration equation has been reported by

Slaughter, Pelletier, and Upadhyaya (2001). High correlations between soil moisture content and NIR absorbance data were indicated when the calibration set and the unknown samples were of the same soil type and particle size. However, the model correspondence appeared to be low for unknown soil samples whose particle size differed.

Moisture content is another important soil property. NIR spectroscopy is considered suitable for estimating soil water content due to the fact that there are concrete water absorption bands (960, 1410, 1460, and 1910 nm). Mouazen, Karoui, De Baerdemaeker, and Ramon (2006) presented the soil water content effect on other soil properties estimation with VIS and NIR spectroscopy. Three different classifiers off limited texture and color variation were utilized in order to form the validation data sets for assessing water content on soil, reaching an accuracy of 95%. Shibusawa et al. (1999) used a handheld spectrophotometer for estimating in real-time the underground soil reflectance in a range from 400 to 1700 nm. After being revised, the same soil spectrophotometer was used for collecting soil reflectance data in a paddy field in order to predict soil moisture, SOM and NO_3-N content, EC, and pH. The R^2 values between 0.54 and 0.66 were reported to form the validation samples.

1.2.2 Soil sensing using airborne and satellite imaging

Airborne and satellite images have been utilized widely in several studies for the assessment of different soil properties. Baumgardner, Silva, Biehl, and Stoner (1985) managed to detect different soil properties via airborne and satellites based soil reflectance measurements. In this study, GIS has been surprisingly referred to a georeferenced information system instead of "geographic information system". At this time, GIS was considered a newly introduced technology for soil data recording and monitoring.

Soil P, soil organic matter and soil moisture (Muller & Decamps, 2000; Varvel, Schlemmer, & Schepers, 1999) have been indicated with the help of aerial imagery. Barnes and Baker (2002) utilized multispectral aerial and satellite imagery for soil textural class mapping development. The differentiation between field properties appeared to be responsible for worsening the spectral classification results. On the other hand, after spectral classification performed field-to-field, reasonable accuracy results were achieved. Later, Barnes et al. (2003) presented a soil properties assessment approach including electrical conductivity (EC), soil compaction, SOM and N levels by using both ground and remote sensing data. The synergy between physical and empirical models could form a highly sophisticated and effective method for collecting soil property information via airborne reflectance data. Moreover, the near infrared reflectance analysis was proven an useful approach not only for soil properties assessment but also for facing remote sensing challenges when different sensing methods are employed. It was also stressed that multispectral imagery integration and ground-based sensing data have the potential to provide soil maps of better accuracy. Ben-Dor (2002) demonstrated the soil quantitative remote sensing techniques, the sensors of high spectral resolution (HSR), the Soil-radiation interactions processes and the main factors that affect remote sensing.

1.2.2.1 Electrodes

Electrodes use is regarded a relatively newly introduced method for soil properties detection. 1990s, Adsett and Zoerb (1991) measured nitrate levels in soil by using a ion-selective electrode technology. It has been reported that the filed nitrate content can be automatically monitored but the calibration process fails in case that the operating environment or the electrodes is not fully balanced.

A different ion-selective electrodes (ISE) application for phosphate assessment was performed by Kim, Hummel, Sudduth, and Birrell (2007). Cobalt rod-based electrodes appeared to be more sensitive when the typical soil phosphorus concentration range increased. Kim, Hummel, Sudduth, and Motavalli (2007) expanded the previous study for measuring simultaneously soil macronutrients including N, K, and P. The NO_3 ISEs demonstrated almost the same results to those from standard laboratory analysis (R2 = 0.89). However, the standard laboratory analysis results demonstrated 50% than K and 64% higher concentrations than P ISEs.

1.2.2.2 Microwaves

Passive and active are two different techniques in microwave soil moisture sensing. Both methods remain unaffected by cloud cover and deliver highly trustworthy results when soil is barren or soil has low vegetation cover. The active technique is used for soil properties assessing, and has higher spatial resolution than the passive technique. A radar is required as a microwave source. The active technique is characterized by its high sensitivity to soil water content, surface, vegetation structure and soil geometry, making consequently soil water content hard to estimate. It is not affected by atmospheric conditions, solar radiation, cloud presence and rain spraying.

The passive technique, it is more often used than active microwave sensing technique, due to its lower sensitivity to soil surface roughness and geometry, and vegetation structure. There is no signal source needed as microwave source. Moreover, it provides higher temporal resolution than the active method. Njoku and Entekhabi (1996a, 1996b) described basic principles of soil moisture passive microwave remote sensing. It was proven that the soil thermal microwave radiation emission was closely connected to soil moisture content.

1.3 Yield sensing

Yield maps are useful for depicting the overall yield variability, consequently offering useful insights for improving management practices. The mapped variation in yield and crop quality depicted by yield maps envisages to the optimal profitability. For avoiding profit loss from unexpected events and optimizing inputs in intensive agricultural systems, the administration of inputs such as water, fertilizers, and pesticides for plant protection is required to be optimally controlled according to the crop needs. A useful tool for estimating the crop needs regarding nutrients and water content is Vis-NIR reflectance spectroscopy, which is used for mapping plant growth progress indicated by plant biomass, the levels of chlorophylls.

Crop yield is regarded the most valuable information in precision agriculture. It is a fusion of several factors with complex dynamics including soil properties, terrain features, crop intensity, and nutrient and health plant status. A yield map can be used for deriving site-specific management practices combined with crop stress information (Searcy, Schueller, Bae, Borgelt, & Stout, 1989). Remote sensing maps obtained during growth, enable both with-in and after-season management.

Vegetation indices (VIs) resulting from multispectral remote sensing are already used as a standard method for crop yield mapping (Yang & Everitt, 2002). These VIs result from ratios of visible and near-infrared (NIR) narrow wavebands. eNIR/Red ratio and normalized difference vegetation index (NDVI) are two standard VIs (Rouse, Haas, Shell, & Deering, 1973). A common index with wide applicability is soil-adjusted vegetation index (SAVI) (Huete, 1988). Other Vis such as NIR/Green and green NDVI (GNDVI) have been employed for yield prediction (Yang, Wang, Liu, Wang, & Zhu, 2006).

Without doubt, there is wide remote sensing research on several annual crops; however, in the case of specialty crops (fruits and vegetables) remote sensing finds limited use due to the discrete crop character and the complex crop geometry. Koller and Upadhyaya (2005a, 2005b) developed a model for predicting processing tomato yield utilizing. Leaf area index (LAI) and a modified NDVI. Their results demonstrated similar tendencies between the actual and the predicted yield maps.

Partial least squares (PLS) regression models were utilized by Ye, Sakai, Manago, Asada, and Sasao (2007) for yield prediction in citrus trees. In this study, canopy features were extracted from airborne hyperspectral imagery and they were compared with vegetation indices. The hyperspectral features, combined with PLS models successfully predicted citrus yield ($R^2=0.51$–0.90). Ye, Sakai, Asada, and Sasao (2008) also correlated particular canopy features obtained by airborne multispectral remote sensing to the yields of citrus trees. They found a significant correlation between the spectral responses of mature leaves and the yield of the current season. The spectral characteristics of the younger leaves were correlated to the yield of the next season.

Airborne images can be fused effectively with sampled ground truth information providing interpolated reference data and Machine Learning models providing a statistical an alternative for yield mapping and prediction. The timing of remote sensing for predicting yield is important because growth stages at which the estimation can bear a faithful and trustworthy model depends on the growth dynamics of each individual type of growth. Yield variability patterns and homogeneous management zones with a similar yield tendency as correlated to spectral features from airborne remote sensing are useful at different stages during growth and after harvesting. The yield prediction of specialty crops is still at an early stage of development and there is a need to carry research on new techniques based on remote and proximal sensing. Research results indicated that hyperspectral remote sensing features carry more information than multispectral features as related to yield prediction.

1.4 Sensors for weed management

Weed identification can be a result of proximal sensing obtained from ground vehicles and remote sensing from unmanned aerial vehicles (UAV) or satellites. Sensors combined with Machine Learning algorithms can provide a real-time platform detection for identifying and mapping crop type, land cover, and particularly weed patches in order to enable targeted chemical or mechanical treatment in a site-specific management regime.

Detection and identification of weed species via machine imaging systems is applied simultaneously with crop-health sensors based on fluorescence, thermal infrared cameras and ultrasonic mapping. Advances in imaging spectroscopy has been further expanded to the detection of fungal infection in arable crops, enabling selective harvesting, and contributing to the concentration of harmful mycotoxins reduction (Gebbers & Adamchuk, 2010).

1.5 Sensors for disease detection and classification

For sustainable intensive crop production, inputs are required to follow certain environmental regulations. Pesticides are applied in uniform dosages in fields, whilst most diseases appear in patches. Large ecological and financial benefits are expected to occur if site-specific management treatment would be applied only to the infected areas and adjusted to the specific plants nutritional need. Based on the effect of disease and nutrient stresses on spectral plant properties, an optical sensor would recognize and control disease symptoms and nutrient demands. There are sensor system including multispectral, hyperspectral sensing and chlorophyll fluorescence kinetics, capable of providing high-resolution data about crop condition (Mahlein, Oerke, Steiner, & Dehne, 2012). The effective combination of sensor equipment with advances in Machine Learning and GIS systems offer new prospects to precision agriculture, are the requirements for early stage recognition and identification of different types of pests (Moshou et al., 2004; Rumpf et al., 2010).

The data quality and quantity derived from crop sensing has radically increased. They are not predicting physiological responses directly, but obtain a spectral feature, which is combined of several biochemical and structural properties for the plant and environmental circumstances often related to lighting conditions and leaf geometry (Jensen, Humes, Vierling, & Hudak, 2008). These sensor data are highly intercorrelated, depend on several heterogeneous factors so their interpretation necessitates the utilization of advanced Machine Learning, and signal processing techniques.

The spatial distribution of disease symptoms and stresses can be obtained by UAVs. Current commercial satellite sensing is too coarse in terms of spatial resolution since most disease symptoms are very fine grained to allow early detection. On the other hand, satellite images can be useful by detecting relatively large infected or nutrient stressed areas, which can then be inspected by the farmer. However, several constraints related to availability of satellites at certain places and the invisibility due to meteorological conditions variability could lead to intermittent of remote sensing information. Airborne systems are more effective for obtaining remote sensing information compared to satellites due to customized availably.

Canopy spectral characteristics are affected by nitrogen status and other stress factor like diseases and water stress, which can cause a spatial variation of spectral characteristics. Crop diseases affect not only crop appearance and its nutritional content, but also is capable of decreasing crop quality in such an extent that can consequently induce health problems. There has been an increasing occurrence of fungal disease spread, degrading global food production and safety due to related mycotoxin risks (Fisher et al., 2012). Since mycotoxin presence depends on the species of fungus, the fungus mapping can be used to obtain a map of mycotoxin risk. By mapping the fungal presence in real-time based on symptoms detected by sensors it is possible to allow the construction of application map. More specifically, in arable crops, leaves or ears of grain show symptoms of senescent, which are symptoms of fungal infection.

These symptoms develop earlier on infected plants compared to healthy plants. However, the differences are clearer between an infected and a green canopy, while at later stage of disease progress the causal relation between disease and the senescent is not evident. This situation leads to the need to obtain more specific information related to the temporal development of symptoms in order to recognize the context of the health status in order to obtain custom multisensory or meteorological information to facilitate decision support for optimal crop handling according to real inflection risk.

References

Adsett, J. F., & Zoerb, G. C. (1991). Automated field monitoring of soil nitrate-levels. In (Vols. 11-91). *Automated agriculture for 21st century* (pp. 326–335). ASAE Pub.

Barnes, E. M., & Baker, M. G. (2002). Multispectral data for mapping soil texture: Possibility and limitations. *Applied Engineering in Agriculture*, *16*(6), 731–741.

Barnes, E. M., Sudduth, K. A., Hummel, J. W., Lesch, S. M., Corwin, D. L., Yang, C., et al. (2003). Remote- and ground-based sensor techniques to map soil properties. *Phtotogrammetric Engineering and Remote Sensing*, *69*(6), 619–630.

Bartholomeus, H. M., Schaepman, M. E., Kooistra, L., Stevens, A., Hoogmoed, W. B., & Spaargaren, O. S. P. (2008). Spectral reflectance based indices for soil organic carbon quantification. *Geoderma*, *145*(1–2), 28–36.

Baumgardner, M. F., Silva, L. F., Biehl, L. L., & Stoner, E. R. (1985). Reflectance properties of soils. *Advances in Agronomy*, *38*, 1–44.

Becker-Reshef, I., Justice, C. O., Sullivan, M., Vermote, E. F., Tucker, C., Anyamba, A., et al. (2010). Monitoring global croplands with coarse resolution Earth observation: The Global Agriculture Monitoring (GLAM) project. *Remote Sensing*, *2*, 1589–1609.

Ben-Dor, E. (2002). Quantitative remote sensing of soil properties. *Advances in Agronomy, 75*, 173–243.
Bowers, S. A., & Hanks, R. J. (1965). Reflection of radiant energy from soils. *Soil Science, 100*, 130–138.
Ehsani, M. R., Upadhyaya, S. K., Slaughter, D., Shafii, S., & Pelletier, M. (1999). A NIR technique for rapid determination of soil mineral nitrogen. *Precision Agriculture, 1*, 219–236.
Food and Agriculture Organization of the United Nations (FAO) (2011). *Global strategy to improveagricultural and rural statistics (Report No. 56719-GB)*. Rome, Italy: FAO.
Fisher, et al. (2012). Emerging fungal threats to animal, plant and ecosystems. *Nature, 484*, 186–194.
Gebbers, R., & Adamchuk, V. I. (2010). Precision agriculture and food security. *Science, 327* (5967), 828–831.
Goddard, T. (1997). *What is precision farming. Proceedings of Precision Farming Conference, January 20/21*. Alberta, Canada: Taber.
Huete, A. R. (1988). A soil-adjusted vegetation index (SAVI). *Remote Sensing of Environment, 25*(3), 295–309.
Jensen, J. L., Humes, K. S., Vierling, L. A., & Hudak, A. T. (2008). Discrete return lidar-based prediction of leaf area index in two conifer forests. *Remote Sensing of Environment, 112*(10), 3947–3957.
Kim, H. J., Hummel, J. W., Sudduth, K. A., & Birrell, S. J. (2007). Evaluation of phosphate ion-selective membranes and cobalt-based electrodes for soil nutrient sensing. *Transactions of the ASABE, 50*, 215–225.
Kim, H. J., Hummel, J. W., Sudduth, K. A., & Motavalli, P. P. (2007). Simultaneous analysis of soil macronutrients using ion-selective electrodes. *Soil Science Society of America Journal, 71*(6), 1867–1877.
Koller, M., & Upadhyaya, S. K. (2005a). Prediction of processing tomato yield using a crop growth model and remotely sensed aerial images. *Transactions of the ASABE, 48*(6), 2335–2341.
Koller, M., & Upadhyaya, S. K. (2005b). Relationship between modified normalized difference vegetation index and leaf area index for processing tomatoes. *Applied Engineering in Agriculture, 21*(5), 927–933.
Luiz, A. J. B., Formaggio, A. R., & Epiphanio, J. C. N. (2011). Objective sampling estimation of crop area based on remote sensing images. In H. A. Prado, A. J. B. Luiz, & H. Chaib Filho (Eds.), *Computational Methods for Agricultural Research. Advances and Applications* (pp. 73–95). Hershey, PA: IGI-Global-Global. Chapter 4.
Mahlein, A. K., Oerke, E. C., Steiner, U., & Dehne, H. W. (2012). Recent advances in sensing plant diseases for precision crop protection. *European Journal of Plant Pathology, 133*, 197–209.
Moshou, D., Bravo, C., West, J., Wahlen, S., McCartney, A., & Ramon, H. (2004). Automatic detection of 'yellow rust in wheat using reflectance measurements and neural networks. *Computers and Electronics in Agriculture, 44*(3), 173–188.
Mouazen, A. M., Karoui, R., De Baerdemaeker, J., & Ramon, H. (2006). Characterization of soil water content using measured visible and near infrared spectra. *Soil Science Society of America Journal, 70*, 1295–1302.
Muller, E., & Decamps, H. (2000). Modeling soil moisture—reflectance. *Remote Sensing of Environment, 76*, 173–180.
Njoku, E. G., & Entekhabi, D. (1996a). Passive microwave remote sensing of soil moisture. *Journal of Hydrology, 184*, 101–129.
Njoku, E. G., & Entekhabi, D. (1996b). Passive microwave remote sensing of soil moisture. *Journal of Hydrology, 184*, 101–129.

Rinnan, R., & Rinnan, A. (2007). Application of near infrared reflectance (NIR) and fluorescence spectroscopy to analysis of microbiological and chemical properties of arctic soil. *Soil Biology and Biochemistry*, *39*, 1664–1673.
Roughgarden, J., Running, S. W., & Matson, P. A. (1991). What does remote sensing do for ecology? *Ecology*, *72*, 1918–1922.
Rouse, J. W., Haas, R. H., Shell, J. A., & Deering, D. W. (1973). Monitoring vegetation systems in the Great Plains with ERTS-1. In *vol. 1. Proceedings of Third Earth Resources Technology Satellite Symposium, Goddard Space Flight Center, Washington, DC* (pp. 309–317).
Rumpf, T., Mahlein, A. K., Steiner, U., Oerke, E. C., Dehne, H. W., & Plümer, L. (2010). Early detection and classification of plant diseases with Support Vector Machines based on hyperspectral reflectance. *Computers and Electronics in Agriculture*, *74*, 91–99.
Searcy, S. W., Schueller, J. K., Bae, Y. H., Borgelt, S. C., & Stout, B. A. (1989). Mapping of spatially variable yield during grain combining. *Transactions of ASAE*, *32*, 826–829.
Shibusawa, S., Li, M. Z., Sakai, K., Saao, A., Sato, H., Hirako, S., et al. (1999). Spectrophotometer for real-time underground soil sensing. *ASAE Paper No. 993030*. St. Joseph, Mich: ASAE.
Slaughter, D. C., Pelletier, M. G., & Upadhyaya, S. K. (2001). Sensing soil moisture using NIR spectroscopy. *Applied Engineering in Agriculture*, *17*(12), 241–247.
Stafford, J. V. (2000). Implementing precision agriculture in the 21st century. *Journal of Agricultural Engineering Research*, *76*(3), 267–275.
Varvel, G. E., Schlemmer, M. R., & Schepers, J. S. (1999). Relationship between spectral data from an aerial image and soil organic matter and phosphorus levels. *Precision Agriculture*, *1*, 291–300.
Vasques, G. M., Grunwald, S., & Sickman, J. O. (2008). Comparison of multivariate methods for inferential modeling of soil carbon using visible/near-infrared spectra. *Geoderma*, *146*(1–2), 14–25.
Yang, C., & Everitt, J. H. (2002). Relationships between yield monitor data and airborne multidate multispectral digital imagery for grain sorghum. *Precision Agriculture*, *3*(4), 373–388.
Yang, J., Wang, P., Liu, L., Wang, Z., & Zhu, Q. (2006). Evolution characteristics of grain yield and plant type for mid-season indica rice cultivars. *Acta Agronomica Sinica*, *32*(7), 949.
Ye, X., Sakai, K., Asada, S., & Sasao, A. (2008). Inter-relationships between canopy features and fruit yield in citrus as detected by airborne multispectral imagery. *Transactions of the ASABE*, *51*, 739–751.
Ye, X., Sakai, K., Manago, M., Asada, S., & Sasao, A. (2007). Prediction of citrus yield from airborne hyperspectral imagery. *Precision Agriculture*, *8*, 111–125.

Further reading

Yang, C., Everitt, J. H., Bradford, J. M., & Escobar, D. E. (2000). Mapping grain sorghum growth and yield variations using airborne multispectral digital imagery. *Transactions of ASAE*, *43*(6), 1927–1938.

CHAPTER 2

Artificial intelligence in agriculture

Contents

Intelligent Data Mining and Fusion Systems in Agriculture
https://doi.org/10.1016/B978-0-12-814391-9.00002-9

2.1 Artificial intelligence and data mining basics

Artificial Intelligence (AI) is the field of engineering that attempts to reverse engineer functions of the human brain reproduced in silico. Abe (2005) defined two complementary branches, the so-called "Knowledge based" systems, and the creation of an autonomous agent that can exhibit an adaptive behavior. One of approach in the category of Knowledge-Based Systems is a simulation of human thinking through intelligent programming.

The term Artificial Intelligence has received many definitions during the past decades. The term 'intelligence' is followed by typical definitions including:

1. Ability to comprehend
2. The information acquisition
3. Innate ability for interpreting the environment
4. The ability to comprehend events, facts or situation awareness
5. The capacity for problem solving by planning and reasoning, deriving abstractions, capture ideas, language understanding, and adaptive behavior.

Stair and Reynolds enlarged this definition by including the use and the synthesis of symbolic information capturing experience in the form of heuristic rules. The primary focus of AI aims to developing intelligent machines by employing principles from phycology and philosophy. The objective with phycology based modeling concerned the simulation of 'intelligent' human behavior through constructing software reproducing human behavior (Pfeifer & Bongard, 2007) The goal of the philosophy rendered approach was to develop a simulation of intelligence as a computational entity (Russell & Norvig, 2001).

With evolution of sensing technologies, there are huge amounts of the gathered data in real time from the monitoring of various industrial and natural processes. The process steps that concern the analysis of large datasets for supporting decision-making require a framework that is able to extract fit for purpose knowledge for decision algorithms. Depending on the type of framework used for extracting the necessary knowledge for developing the decision framework the knowledge acquisition procedure appears with terms such as data mining process, knowledge extraction, pattern recognition and machine learning.

More specifically, data mining involves the exploratory analytics procedure of large datasets to reveal recurring patterns to obtain correlations and to construct association rules and inference mechanisms. The data mining process requires collaborative and synergistic activities of application specialists and information technology analysts who use relevant algorithms for capturing knowledge for decision rule formation. Interventions to the development of decision rules by the aforementioned experts during different phases are necessary for prohibiting automation of the current process. The acquired knowledge must be to repeatable and predictable results of measurable accuracy so as it to later verified by the future data and subsequently leads to trustworthy decisions.

Data Mining architecture is defined as the global structure of information processing pipe lying which includes information acquisition from different sources, extraction of meaningful patterns and statistical indicators for further analysis, derivation of abstraction models with predictive characteristics and actionable insights leading to adapted optimized practices and solutions. The term *mathematical learning theory* refers to the majority of mathematical models that constitute the data mining process that lead to knowledge extraction. The data mining procedure based on inferencing algorithms, aim to derive an inferential framework based on data regularities, expressed as patterns that lead to repeatable conclusions and on which we can base future predictions and decisions. The generalization of these models is the ultimate outcome of the data mining process since a limited set of examples that have formed these models can generalize in a very large number of future examples and draw real-time conclusions in an operational mode by feeding future novel data to these predictive models. The predictive models have been derived from data can take different shapes, such as linear operations, heuristic rules in inferential *form*, charts, decision trees, nonlinear models such as Neural Networks (NNs) and Support Vector Machines (SVMs).

The data collection process is considered mostly autonomous and oblivious of the data mining purposes, thus it does not appear to have any similarities with data collection practices that follow any classic statistical scheme. From this aspect, data mining is characterized as a *secondary* way of data analysis.

Taking into account the main aim of data analysis, the Data mining enterprises can be split to two very important research factors: *interpretation* and *prediction*.

- **Interpretation** aims to find ordinal motifs in the data, describe them with the help of novel rules and principles to provide specialists with the adequate information and knowledge about the system investigated. The interpretation is capable of finding associations between data sets that lead to the formulation of a rule that connects these two datasets embodying the knowledge that associates the physical phenomena that can have produced these data and apart from the association which is a statistical feature, it can illuminate the causality between the physical phenomena that have been associated and reveal hidden mechanisms behind the phenomena that are explained by physical principles.

- **Prediction.** The aim of prediction is to estimate the future value of a process at a certain time base on previous observations. There are miscellaneous calibration methods used to develop data mining models. A plethora of techniques had been proposed by computer science research such as classification and decision trees or heuristic rules, and more recent machine learning methods and neural networks.

The Data mining can be summarized in the following steps:

1st Step. The ANNs models are trained with time-series input data for achieving a trustworthy prediction;

2nd Step. Once input data collection is completed, the time-series data are filtered properly so as to be de-noised and alter their time series form to multidimensional before being fed in to the ANN model.

3rd Step. The employed ANN models are fine-tuned with historical data—The accuracy of the ANN models is evaluated with unseen data that are not included in the training set so that the ANN model accuracy is tested with an independent error criterion;

4th Step. The calibrated neural network outputs are used as predictors for predicting future values of a process and therefore can be used for preventive measures and to facilitate actions based on these predictions.

2.2 Artificial neural networks (ANNs)

Artificial Neural Networks (ANNs) are parallel calculation algorithms, which consist of many linked processing units. These processing units have a very simple structure and are able to convert an input signal to an output signal with an activation function in correspondence with the biological neurons. Parallel processing of a large number of inputs simultaneously in a hierarchical structure enables the simulation of several complex functions based on input data.

The Artificial Neural Networks (ANNs) term has been established earlier than the term of Artificial Intelligence (AI). First attempts on utilizing ANNs was in the form of simplified neuron models and date back to the early 1940s (McCulloch & Pitts, 1943). The concept of ANN learning

(Bergenti, Giezes, & Zambonelli, 2004,Wiener, 1948) and more specifically, the idea of developing different types of computing systems with intelligence ability comparable to that of biological organisms starts circa 1950 (Von Neumann, 1958).

The term "artificial" is used to indicate the effort of simulating the structure of biological neural networks. This trend is justified by the particular ability of Biological Neural Networks to process large amounts of data and to draw conclusions in a really short period of time. The benefits of applying those Biological Neural Networks make them ideal for predictive computing and data processing architectures. Particular features related to the structure of neural networks are their ability to learn, store information, generalize to new data, and group concepts.

The human brain comprises of at least one hundred billion neurons that are interconnected by dense network elongated structures (Fig. 2.1).

Stimulations emanating from specialized cell structures that can transduce environmental signals to electrical current (sensory cells) are intercepted by dendrites that transmit the information through induction of electrical impulses. The signals transmitted from one to another neuron in such a way that the information is passing quickly through the whole neural assembly allowing the attention to critical signals or deciding to overlook if it is not critical (Fig. 2.1).

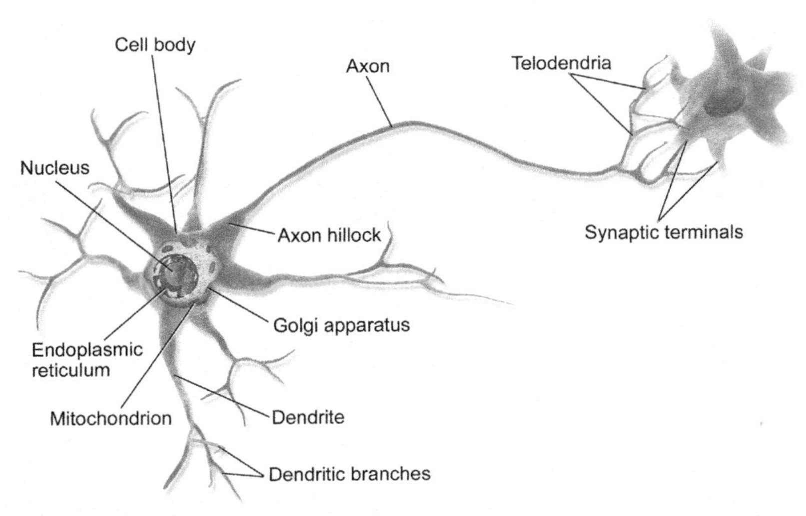

Fig. 2.1 Depiction of a typical neuron.

The term, ANN models refers often to the MultiLayer Perceptron (MLP), as a biomimetic model trying to reverse engineer biological network systems for constructing neural computing particles. The simplest structure of an ANN includes only one hidden layer, however, it demonstrates strong similarity to the network of neurons in a brain due to its vast interconnections between the neurons in consecutive layers. In Fig. 2.2, each circular node of the neural network is forming an artificial neuron, while the arrows denote the possible connection channels between neurons, modeling the paths from which the information passes, simulating the neuronal functions of several biological organisms including those of human brain.

This type of ANN model, is characterized by full connectivity and subsequent interaction between all neurons so that each neuron can receive input and perform elementary operations on the input data. The output activation of the neurons is transmitted to other neurons via connection which assigns a weight to a signal to modulate its strength. It has to be noted that the flow of information is one way, which means that the link will not receive any information feedback. The experience of the ANNs after it is trained, it is stored in a distributed way, in a different way connecting the neurons, similarly to the synapsis in a biological brain. This type of information storage is addressable from a point of view that is interrogated by an input that is similar to a trained one that can produce similar result. Therefore, ANNs have the capability to learn and generalize, and this capability is acquired by weight optimization base on the cost function of maximizing the overall prediction performance (Fig. 2.3).

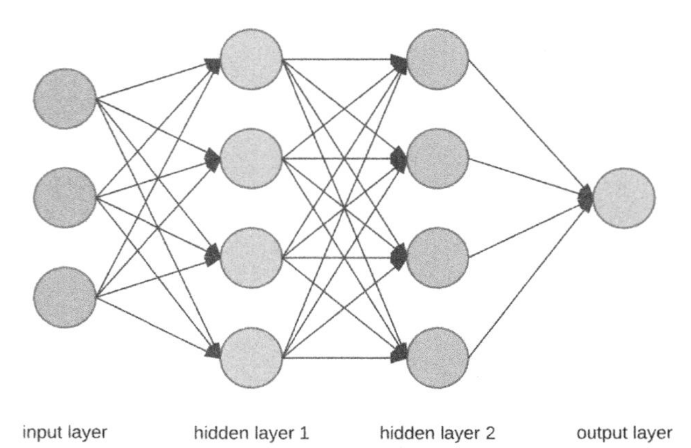

Fig. 2.2 A neural network architecture comprised of input units, hidden units, and output units.

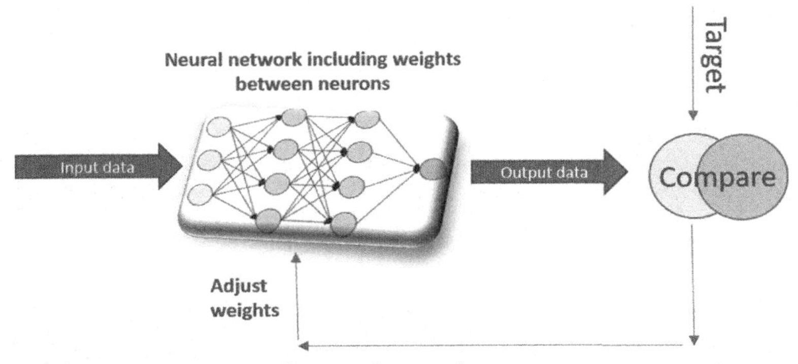

Fig. 2.3 The training process of a neural network.

An ANN comprises interlinked elementary operational units that communicate information through weighed linkages. These elementary units, defined as neurons or nodes, are simple processors that carry the basic operation of filtering an input signal through an activation function and produces and output value. The signals that are transmitted between neurons is weighted through a weight parameter, which indicates how strong the connection between two neurons is. The weight value is variable in a certain interval, for example in a range from −1 to 1, but in reality it depends on the situation.

The weight parameter carries a similar role to a chemical connecting two atoms that create a molecule. From a different point of view, a weight conveys information on the importance of the signal to the overall architecture of the network by connecting the two neurons. The size of the weight is proportional to the signal contribution. The neural network Connectivity graph is defined by the pattern described by the connections between the neurons. The associations between the neurons and the weight values between them. The connectivity pattern is defined by two processes: the first includes the expert's decision on neuron connections and the orientation, while the second concerns the learning procedure of the network which optimizes the weights. The weights can be defined without using a training procedure through some numeric algorithm with direct solutions. However, a unique characteristic of is their ability to train themselves in order to recognize patterns respond correctly to them in a similar way as they would do in an operational process. In reality, for many real life processes the procedure of learning from data from training is the only manner in which a neural network can be allowed to learn since they might be no other solution to tackle a problem by using a more traditional algorithm.

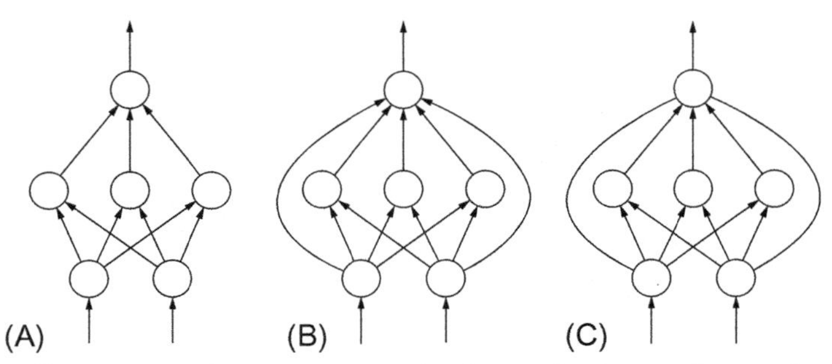

Fig. 2.4 Different ANN architectures emanating from a simple neural network structure. (A) A typical ANN architecture, (B) a feedforward ANN architecture, and (C) ANN architecture with feedback connections (Chang & Bai, 2018).

There are various NN topologies with different configuration of feedforward connections and comprising feedback loops with fixed architecture (Fig. 2.4).

Common applications of these topologies include: (a) pattern production, pattern identification, and categorization, (b) prediction, (c) clustering and classification, (d) non-linear regression, (e) optimization and content-based retrieval (Jain, Mao, & Mohiuddin, 1996). To attain this target, we rely on supervised learning in which the neural network is trained based on existing data samples, which are submitted by the users. The training procedure drives the adaptation of the neural networks range in order to produce the desired response for new input data, after the completion of the training procedure. In the architectures depicted in Fig. 2.4, the feedforward and feedback architectures may produce a "desired" output after calibrating the weights iteratively at the expense of the time needed to complete the operations.

The challenge that ANN theory has to face is to find appropriate network education algorithms and recall their information in order to simulate intelligent processes such as those mentioned above. This requires the definition of an appropriate learning environment, e.g., whether the network will be trained with **supervised learning**, by using some teacher-driver data, or whether the network will be self-organized alone (self-organizing network) and with what specific criterion and goal (unsupervised learning). Moreover, there are training rules so that the network is automatically programmed.

(A) Supervised training

A common form of training is supervised training, where t for each input that is fed into the network, there is also a target output. Typically, random values are selected as initial weights, which, using the input data, yield an output. The difference between the value calculated by the network and the actual value that it was supposed to compute is an error, used to determine the weights more accurately. In order correct this error; Delta (Widrow-Hoff) rule is applied. Its scope is to minimize the difference (error) between the desired output and the current output value after successive weight changes the so-called correction revolutions. Sometimes, this process may require large numbers of such corrections and hence large computing times.

(B) Unsupervised training

In unsupervised training, the information is provided to the network, but the targets are not matched, so there is no check or comparison error estimation. The network does not use an external parameter to change the weights. A specific process is followed and leads to network training. The network uses an internal control to seek for some input signals tendencies or regularity and tries to ensure that the outputs are of the same characteristics as the clusters. This is called self-supervised learning because the network checks itself and corrects data errors with a feedback mechanism. This approach is not as common as supervised training and is not fully understood, yet is very useful in the event there is no data provided. In this case, it is assumed that training is achieved when the network stops changing the weight values. This is because the exit error becomes zero or is too close (tending) to zero.

The commonly used models features are presented as follows:

(1) The processing units (neurons) have adjustable parameters so that they learn from new data mimicking the plasticity of the biological neurons of the brain.

(2) A large number of processing units (neurons) are used to achieve parallelization of the processing and distribution of information.

2.2.1 Back propagation artificial neural network

The backpropagation technique (Brooks, 1991) is commonly used to train MLPs. This technique could learn more than two layers of a network. The key idea of the back-propagation technique is that the error obtained from the output layer is propagated backwards to the hidden layer and is used to guide training of the weights between the hidden layer and the input layer. (Botros & Abdul-Aziz, 1994) showed that the BP algorithm is very sensitive to initial weight selection. Prototype patterns (Bai & Zhang, 2006) and the orthogonal least square algorithm (Boukerche, Juc'a, Sobral, & Notare, 2004) can be used to initialize the weights. The initialization of weights and biases has a great impact on both the network training (convergence) time and generalization performance. Usually, the weights and biases are initialized to small random values. If the random initial weights happen to be far from a good solution or they are near a poor local minimum, training may take a long time or become trapped there. Proper weight initialization will place the weights close to a good solution, which reduces training time and increases the possibility of reaching a good solution. As it is evident from Fig. 2.5, independent of the ANN model complexity, they are substantially simpler from the biological counterparts. However, these simplified models are reconstructed to gain knowledge of the capacity for performing simple functions but such simple units can tackle. The BPN learning process is a supervised procedure that is repeated at every cycle. If an ANN model encounters an input pattern at an early stage of training, it will produce as random estimation of weights for predicting the target. Then, it will assess the distance between the target value and the predicted one, and it will adjust its connection weights by utilizing gradient descent algorithm for calibrating its weights in order to achieve final convergence of the BPN algorithm. To achieve this, many iterations are necessary which consist of presenting the network of input data, that produces forward activation flow of outputs which internally activates a backward flow of error for weight calibration in order to reach a convergence state. In machine learning terminology, the iterative improvement is known as delta rule in the case of a mono-layer neural network, which constitutes a special restricted case of the more universally applicable error backpropagation algorithm.

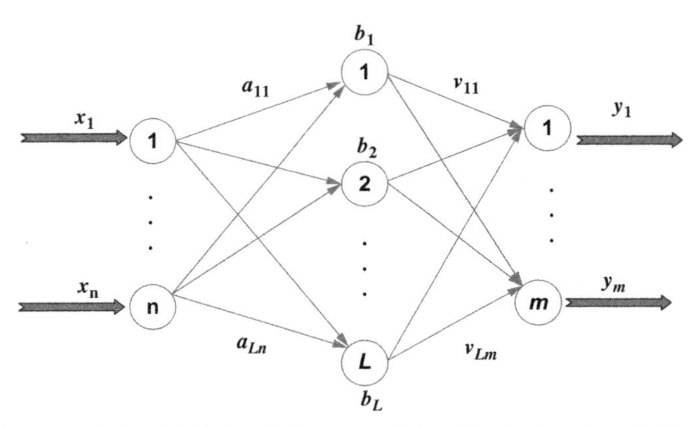

Fig. 2.5 Representative of SLFN and ELM comprising of n input units, L hidden neurons and m output units.

2.2.2 Single-layer feedforward neural network (SLFN) and extreme learning machine (ELM)

The Single-Layer Feedforward Neural Network (SLFN) training algorithm is characterized as adaptable, ample, and efficient in the occasion that high volume data pairs are available. The SLFN offer solutions to difficult problems and features by using easily formable behaviors, even with variable problem parameters. ELM algorithm is characterized by its ability to use examples for training in order to handle nonlinear functions, without altering the network architecture or activation operations. Instead, it modulates the weights connecting neurons. The operational graph of SLFN in provided in Fig. 2.6. For example, there is a known training set {(xi, yi) |xi ∈ Rn, yi ∈ Rm, $i=1, 2, 3,\ldots,$N}, comprising of N training data samples, denoting n-input and m-output for training the network in such a way to generalize in other datasets. In the occasion that there are L numbers of hidden neurons, the activation function g(x) is defined as:

$$\sum_{i=1}^{L} v_i g_i(x_j) = \sum_{i=1}^{L} v_i g(a_i \cdot x_j + b_i) \tag{2.1}$$

where $a_i=[a_{i1}, a_{i2}, \ldots, a_{in}]^T$ ($i=1, 2, 3, \ldots, L$) denotes the weight vector bridging the ith hidden neuron with the input data nodes, and $v_i=[v_{i1}, v_{i2}, \ldots, v_{in}]^T$ ($i=1, 2, 3, \ldots, L$) denotes the weight vector bridging the ith

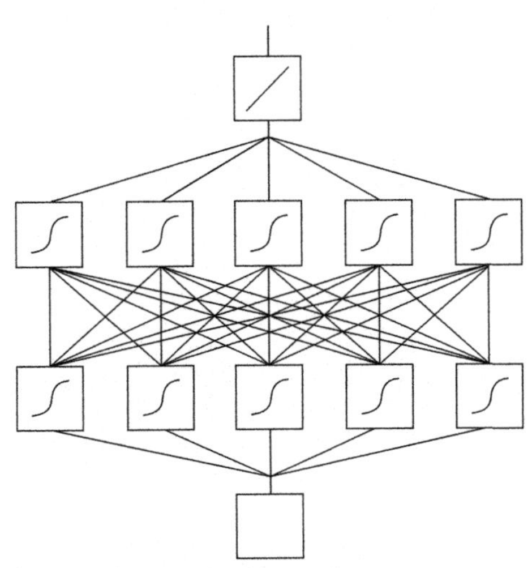

Fig. 2.6 Representative of an ANN model comprised of two hidden layers and sigmoidal activation function (Chang & Bai, 2018).

hidden neuron with output neurons. The b_i denotes the bias of the ith hidden neuron, while $g(x)$ denotes the activation function.

These activation functions can be in several forms depending on the formulation of the problem that has to be solved. These activation functions when they are applied in an MLP they improve their learning capability in the case a single layer of hidden neurons is not able to attain the learning goal. The addition of a layer of sigmoidal units (i.e., $f(x) = 1/(1 + e^x)$ in both hidden layers to the single layer is an example of such an improvement (Fig. 2.6).

With the use of these formulas, we can demonstrate the forward processing step of SLFN that generates the predictions matrices for N training data samples. The generic is summarized by the following equation (Eq. 2.2):

$$H \cdot V = Y \tag{2.2}$$

Here H symbolized the hidden layer activation matrix of the neural network, which is depicted as follows

$$H = \begin{bmatrix} g(a_1 \cdot x_1 + b_1) & g(a_L \cdot x_1 + b_L) \\ \vdots \quad \ddots & \vdots \\ g(a_1 \cdot x_N + b_1) & g(a_L \cdot x_N + b_L) \end{bmatrix}_{N \times L} \tag{2.3}$$

$$V = \begin{bmatrix} {v_1}^T \\ \vdots \\ {v_L}^T \end{bmatrix}_{L \times m}, Y = \begin{bmatrix} {y_1}^T \\ \vdots \\ {y_L}^T \end{bmatrix}_{N \times m} \tag{2.4}$$

where g denotes the activation function (Fig. 2.7). When this process comes to an end, then the SLFN is trained, seeking for specific sets of $\{V^{**}, a_i, b_i^*, i = 1, 2, 3, \ldots, L\}$ in order to lower the difference between approximations and targets (Eqs. 2.3 and 2.4) (Chang, Han, Yao, & Chen, 2012).

$$\|H({a_1}^*, \ldots, {a_L}^*, \ldots, {b_1}^*, \ldots, {b_L}^*) V^* - Y\|$$
$$= \min(\|H(a_1, \ldots, a_L, b_1, \ldots, b_L) V - Y\|)_{a_i, b_i, V} \tag{2.5}$$

The optimization process tries to find the gradient minimization algorithms that adapt the weights and biases through iteratively by backward propagation error according to the delta rule. This procedure is expressed mathematically as in Eq. (2.6):

$$W_k = W_{k-1} - \eta \frac{\partial E(W)}{\partial W} \tag{2.6}$$

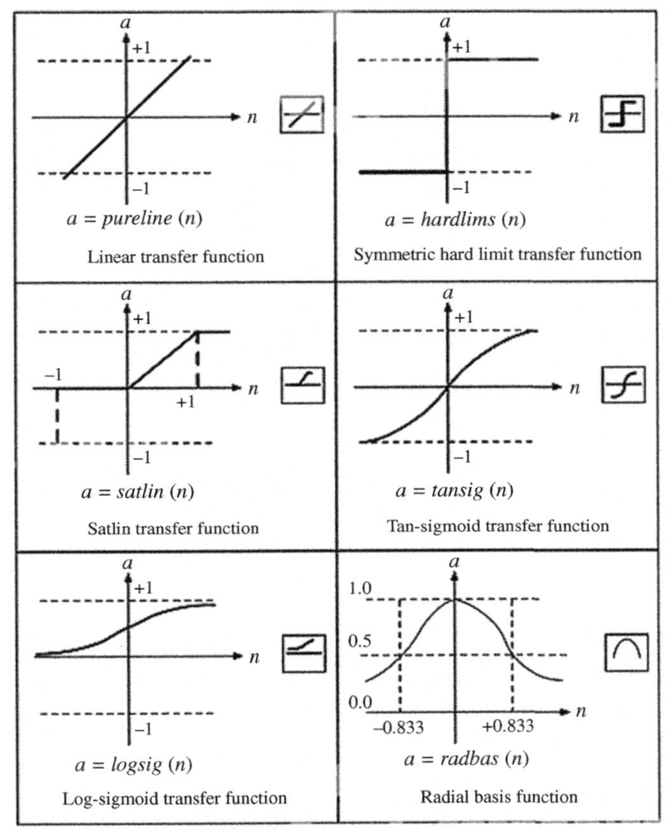

Fig. 2.7 Four types of activation functions with application in ANNs (Chang & Bai, 2018).

where η denotes the learning step and E denotes the error that occurs in each predictive step. These gradient descent-based are effective for a multitude of problems and situations. The only handicap related to such kind of algorithms is that the learning procedure has to be applied in a stepwise fashion, which forces them to run in a lower speed compared to batch algorithms. Additionally, they experience the problems that are related to local minima in comparison to global minima and they face overfitting problems.

ELM is a learning algorithm that is used for training feedforward networks with a very high speed that achieves the global optimum. Experience suggests that ELM are more effective and performant compared to other modern learning algorithms. The ELM algorithm was introduced by

Huang, Zhu, and Siew (2006a, 2006b), with objective of enhancing learning speed and prediction performance by using SLFNs (Huang et al., 2006a, 2006b). ELM differs from other feedforward algorithms whose parameters need to be calibrated with iterative techniques from nonlinear optimization. In ELM, the input unit and hidden neurons can be defined in a random way, while keeping the hidden output layer the same, in the case that the initiation function are widely differentiable (Huang et al., 2006a, 2006b). Consequently, the preparation process is similar to finding a base standard arrangement $\boldsymbol{HV}=\boldsymbol{Y}$ since the yield weights, $\boldsymbol{V}$, are the basic parameters of SLFN. So, for ELM, the equation is modified as follows:

$$\begin{aligned} &\|H(a_1, \ldots.., a_L, \ldots.., b_1, \ldots.b_L)V^* - Y\| \\ &= \min\left(\|H(a_1, \ldots., a_L, b_1, \ldots.., \mathrm{b}_L)V - Y\|\right)_V \end{aligned} \tag{2.7}$$

The linear system shown has an efficient numerical solution by using a least squares framework, and subsequently the output weights can be produced by the following equation:

$$V^* = H^\dagger \tag{2.8}$$

in which H† denotes the Moore-Penrose Generalized inverse of H (Huang et al., 2006a, 2006b). Since ELM does not rely on weight adaptation and bias adjustment for the neurons in the hidden layer, ELM requires minimal training time. By resolving efficiently the SLFN's problems, the ELM has found applications in a variety of fields such as image processing (Tang, Deng, & Huang, 2016), automatic control (Yu, Yang, & Zhang, 2012), system identification and prediction (Zhao, Li, & Xu, 2013), image recognition and feature extraction (Bai, Huang, Wang, Wang, & Westover, 2014), time series modeling (Butcher, Verstraeten, Schrauwen, Day, & Haycock, 2013), and earth observation (Chang, Bai, & Chen, 2015).

Because of the random way that the input weights and hidden layer are allocated, ELM is able to produce a direct reaction to the inputs and estimating a solution with extremely fast speed. However, in the case of big data with substantial variability, like in the case of environmental sensing or remote sensing, the existence of an abundance of outliers might affect heavily the prediction performance of ELM since the single time and instant determination of random weights is not sufficient for capturing the necessity of bias correction during prediction phase even by normalizing the raw data samples.

This couple of techniques for enhancing the accuracy of ELM when facing the presence of outliers in the training datasets by employing reinforcing learning methods. The first methods calculates the weights by using the training errors from ELM. The Regularized ELM (RELM) takes advantage of the regularization term to equalize the weight between the training error and the output, in order to enhance the final accuracy. Then, the weighted ELM has been introduced by Huynh and Won (2008) (WELM) through penalizing the training errors. Following the RELM's objective function, Deng, Zheng, and Chen (2009) introduced the weighted regularized ELM (WRELM) by putting further weighting parameters on training errors. The training error statistics deriving from RELM, are calculated by these regularization parameters. All the three of them (RELM, WELM, WRELM) are regarded efficient for offering solutions when dealing with outlier presence, consuming the same amount time as ELM. ELM-based algorithms have been introduced by Horata, Chiewchanwattana, and Sunat (2013) aiming to minimize the outlier impacts, but are regarded more computational time consuming than the traditional ones.

2.2.3 Radial basis function neural network

RBF model is a variation of the basic ANN modeling architecture. Similar to the general structure shown in Fig. 2.9, an RBF model comprises of three layers including an input layer, a hidden layer with a nonlinear RBF activation function, and a linear output layer. The justification for using a nonlinear RBF activation function is based on the fact that the nervous system is comprised of a large amount of neurons with locally tuned receptive fields (Fig. 2.8).

Commonly, RBF networks are calibrated by using a three-step procedure. The initial step concerns the choice of center vectors for the RBF neurons in the hidden layer. RBF centers can be selected in random fashion from a set of data by using supervised learning, or they can be defined by using k-means clusters by using unsupervised learning. The following step, includes a linear model determination whose weights associated with the outputs of the layers by using a dot product and objective function for target value prediction and error term production emanating from the subtraction of predicted and target values. The third step concerns the adjustment of all the RBF network parameters by using backpropagation iteration (Schwenker, Kestler, & Palm, 2001). Practically, this is an example of reinforcement learning algorithm.

RBFs can be characterized ideal for modeling local receptive fields in high dimensional spaces, representing the input vectors as point in

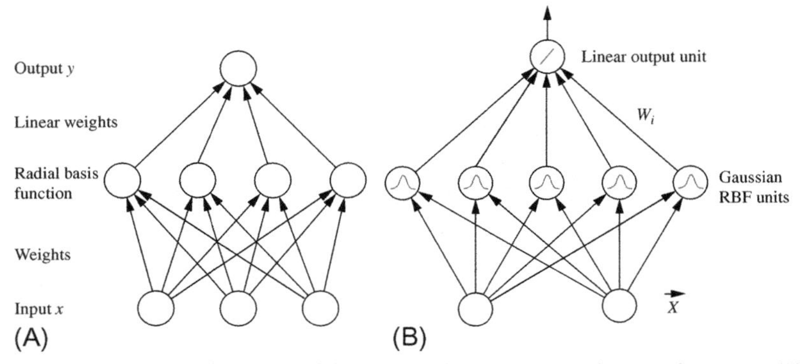

Fig. 2.8 The schematic depiction of the RBF model. (A) A general RBF architecture, (B) an RBF depiction with Gaussian activation function (Chang & Bai, 2018).

two-dimensional manifolds. In low-dimensional manifolds, we can estimate probability density functions by sampling a limited number of proximal data points to provide a local estimate for estimate density function that characterize the local receptive field by defining a Parzen window used for classification.

The problem that occurs in high-dimension vector spaces concerns that the processing of the inputs cannot be accomplished with high performance because of the redundant information introduced by the high dimension vectors. A way to overcome this to position the RBF senders according to learning scheme that allocates the senders to high density areas due to data information content. A common example when using unsupervised learning, is to find data clusters through the k-means algorithm in order to describe the topology of high dimensional data in terms of spatial distances. Another method concerns the use of a Kohonen Self-Organizing Map (SOM) that determines the centers of RBF on based on its own prototype vectors and performing a second training step when the RBFs are trained based in local data selected from the SOM Voronoi regions (Wehrens & Buydens, 2007).

2.2.4 Deep learning

From the early presence of ANNs till the development of deep learning algorithms, a variety of AI models dominate the last decade of developments oriented to automatic decision and classification systems. The main paradigms of deep learning models are:

- The deep autoencoder (DAE)

  An efficient method to perform nonlinear dimension reduction (Hinton & Salakhutdinov, 2006)

- Convolutional Neural Networks (CNNs)

  Method inspired from biological vision systems mimicking the multilayer convolutions that take place in the visual cortex.
- The Deep Belief Networks (DBNs)

  Approach based on physics, which is an evolution of the Hopfield Network. It proposes a method of overcoming the local optima, or avoiding the oscillating between states by employing a weight-energy-probability approach. The objective of this approach is to transfer the energy from the visible to the hidden layer, estimating the probability for altering the state (Krizhevsky & Hinton, 2010).
- The Recurrent Neural networks (RNNs)

  A neural network model which is based on connections of neurons that use a feedback process aiming to modulate the learning procedure of the whole network (Grossberg, 2013). The current architecture assures the capability to learn from time series data (Lipton, Berkowitz, & Elkan, 2015).

All the aforementioned deep learning approaches have different conceptual structures but share the concept of ANNs, the learning capacity, the dimension reduction, prediction performance and resource efficiency. As an insight it can be concluded that the representation is enhanced when the model sophistication increases, while on the other hand the model selection is dependent on the balance between representation capability and resource efficiency (Fig. 2.10) (Wang & Raj, 2017).

Natural images usually have stationarity property. Given the salience characteristic of the phenomena depicted in natural images, the statistics of one part of the image are the same throughout the full image. This two-dimensional structure facilitates a more effective image processing based on simpler feature extraction by utilizing the features that are typical of an image like edges, corners, endpoints, etc. By following this approach, while neurons of a typical ANN model have a full connectivity between layers in the case of CNNs, a neuron accepts as input a small set of inputs based on features that have mutual spatio-temporal proximity. This type of processing enforces invariance of translation of the input image by forming a receptive filed between layers that is specialized in the processing of low dimensional structure and then in involves hierarchically in the first layers of the CNNs. The CNN design is accomplishing its purpose by grouping pixels from the same neighborhood that are highly correlated and demonstrate common local associations. Similarly, the distant pixels do not share common features. The 2-D planes of neurons depicted in Fig. 2.9 belonging to downstream layers after resulting from dimensionality

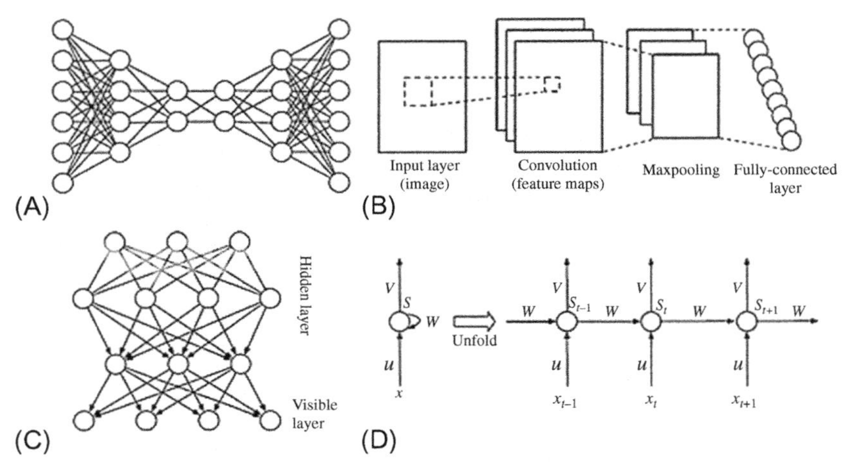

Fig. 2.9 Block diagrams depicting a of four typical deep learning approaches: (A) AE, (B) CNN, (C) DBN, and (D) RNN (Chang & Bai, 2018).

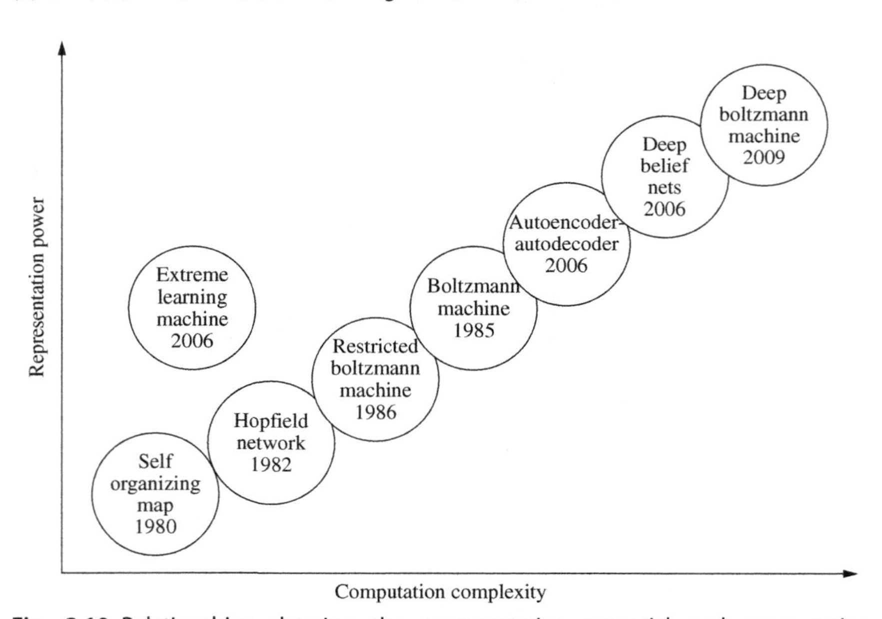

Fig. 2.10 Relationships showing the representative potential and computation complexity of different ANNs, which can be used for the development of more complex and effective ones (Chang & Bai, 2018).

reduction operation in a CNN, the so called feature maps, which share common characteristics that each neuron is interconnected to a specific receptive field that belongs to the previous neuron. Simultaneously, each neuron that belongs to the feature map links to the same weights connected to the target

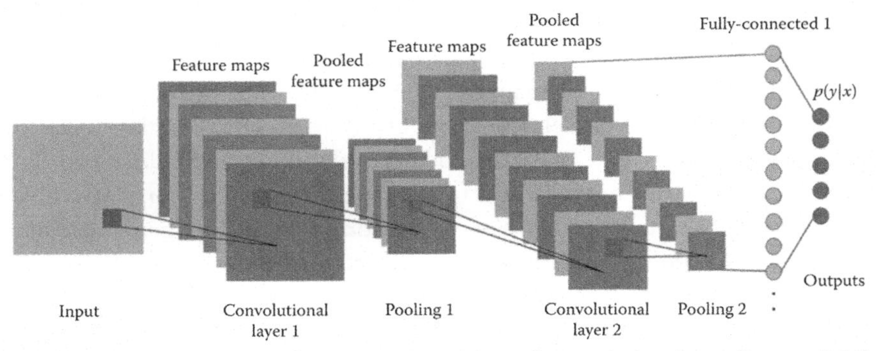

Fig. 2.11 A typical CNN architecture, comprising of convolutional, pooling, and full connectivity layers (Albelwi & Mahmood, 2017).

feature, each feature map scans the previous layer, examining the presence and frequency of each feature occurrence.

A common CNN model is depicted in Fig. 2.11, which comprises of 5 layers, without considering the input layer, in which all the layers comprise of weights which are tunable in a supervised manner by performing a iterative convolutional pooling operations to process important features and localizing the focal points iteratively for achieving classification convergence.

Calibrating a standard MLP demands a high volume data in order to tune the weights for a specific spatial or temporal points of interest. TheCNNs are considered a special version of MLPs employed for exploiting spatio-temporal invariance by learning more abstract high level features in a feed-forward iteration. The CNN manages to minimize efficiently the amount of the network's weights because the input in CNN is two dimensional image pixel data or 3-D volume data.

2.2.4.1 Generalization accuracy of ANNs

There is a variety of application issues in the calibration of neural networks that need special attention due to the trade off between the bias-variance. The first issue has to do with model learning and selection of hyperparameters. In the occasion that, the ANN is validated only with the same dataset that was used to calibrate it, no satisfactory results are expected to be obtained when testing with new data because of the overfitting. Hence, the hyperparameters are calibrated on a separated set that does not interfere with a procedure during which the neural weights are trained.

When provided with an annotated dataset, it is crucial to utilize these information for calibration validation and testing the performance of the ANN. Obviously, nobody manages to take advantage of the entirety of annotated data for the model development. More precisely, in the occasion that the same data is set for both model development and validation, the more likely is for the model to fail to estimate the accuracy correctly. This occurs due to the main objective of classification is to achieve generalization of a model constructed with annotated data, to unseen test data. Moreover, the subset of the data utilized for model choice and parameter learning is necessary to be different from the one used to calibrate the model. A usual mistake is to utilize the same data for learning and final testing. By this way, the calibration and validation data are partially mixed and the final performance is regarded optimistic. A provided data set needs to be split in three sets that are formed based on the way in which the data are utilized:

1. **Calibration data**: This is a subset of the data that is utilized for constructing the model during the training process, for example for calibrating the weights of the neural network during the learning procedure. A variety of designing parameters has to be decided during the model development. It is possible that the ANN can take advantage of various hyperparameters related to the learning rate or used for regularization. The same calibration dataset can be used several times by selecting different values for the hyperparameters or with different training algorithms for comparing the results in multiple ways. This procedure facilitates the comparison of different configurations of ANNs in terms of performance with different hyperparameter settings. Through this procedure, the best model is picked out as the most performant. It is crucial to notice that the selection of the best model is not performed on the training data set, but a separate unseen validation data set which does not contain any of the training data. This procedure is followed for assuring generalization, by avoiding overfitting models.
2. **Validation data**: This is a subset of the data utilized for picking the most appropriate NN architecture and complexity level together with hyperparameter tuning. As an example, the learning rate may be chosen by calibrating the model on the calibration data set and subsequently validate these models in terms of accuracy by using the validation set. Different sets of parameters are chosen from a range of values and validated in terms of performance by using the validation set. The optimal set of parameters is obtained by using performance criterion as it is attained by

using this validation set. From this aspect, the validation data set can be considered as a special case of data set that is used to select the parameters of the architecture and training regime like the learning step, the number of hidden layers and neurons in each of the layer or the selection of the best activation function.

3. **Testing data**: This is subset of part of the data utilized for evaluating the performance of the fined tuned model. It is crucial that the testing dataset is not involved in the parameter tuning and architectural choices that take place to avoid overfitting. The test data set is used only for operational testing so it appears at the last step of the ANN calibration process (independent validation). Additionally, in the occasion that the NN expert perform model adjustment based on the test data, then the outcome will be corrupted with prior information obtained with the training data set. The requirement of the test data to be used only at final stage is an important restriction since related to the operational use of the model in the real life with real data. However, it is regularly overlooked in benchmark with real data sets.

The segmentation of the annotated data set is depicted in Fig. 2.12. More specifically, the validation data belongs to the training data set, due to the impact it has on the final model (though the reference of the term training data usually describes the data used for the basic weight calibration of the ANN model. The segmentation according to the rule of 2:1:1 is a heuristic

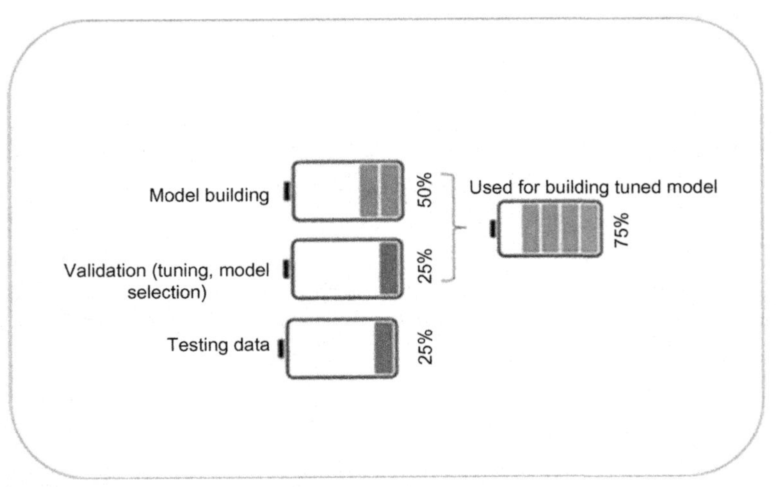

Fig. 2.12 Segmenting an annotated dataset for evaluation purposes.

choice that has been habitually used since 1990. On the other hand, this rule is not mandatory. For large size annotated datasets, there is the need for limited size of testing dataset for performance assessment.

In the case of a very large data available for calibration, it is sensible to take advantage of its size for model calibration, since the variance that will present due to the validation testing data sets will be quite low. A fixed size validation and testing data sets a fixed size (a few thousand samples) for the validation training datasets is sufficient for providing accurate results. Hence, the 2:1:1 rule was transferred from a period in which the datasets had limited size (Fig. 2.13). In more recent times, data sets are usually large, the majority of data points are devoted to model calibration while a lower quantity, is reserved for testing. It is not rare to utilize ratios like 98:1:1.

2.2.4.2 Validation by using partial datasets to alleviate bias

The already mentioned method of segmenting the annotated data into three partitions is an indirect demonstration of a method that is known for partitioning the annotated data into several partitions. In fact, the partitioning into three parts is not implemented in one phase. On the contrary, the calibration data set is initially partitioned into two sets, one for calibration and one for testing. The testing part is kept away and not subjected to any subsequent analysis, before the end of the process that is used once. Then, the rest of the data set is partitioned again in calibration and validation partitions. This type of hierarchical state segmentation is depicted in Fig. 2.12. A key aspect is that the partition categories at both sides of the hierarchy are of the same concept.

Hold-out

In the hold-out strategy, a small amount of the occurrences are utilized to manufacture the preparation model. The rest of the occurrences, which are likewise alluded to as the held-out examples, are utilized for testing. The performance of anticipating the labels of the held-out occasions is then revealed as the general general performance. Such a methodology guarantees, that the detailed performance isn't an aftereffect of overfitting to the particular data set, on the grounds that various instances are utilized for preparing and testing. The methodology, nonetheless, has a negatively bias view of the genuine precision. Consider the situation where the held-out models have a higher nearness of a specific class than the marked dataset. This implies the held-in models have a lower normal nearness of a similar class, which will cause a bungle between the preparation and test information. Moreover, the

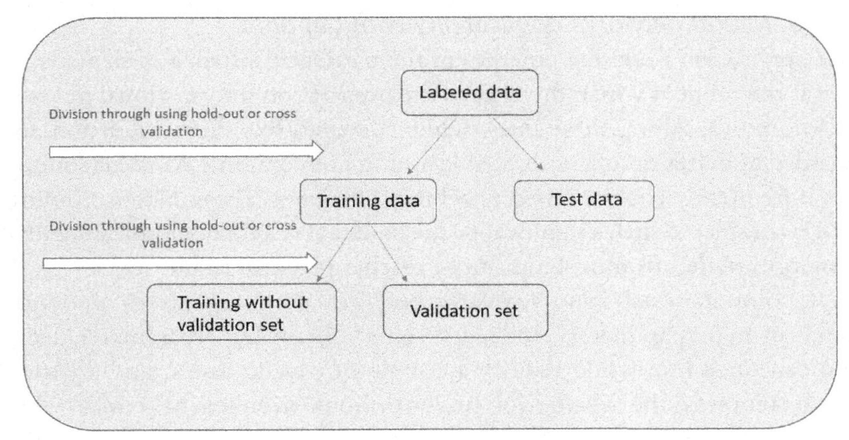

Fig. 2.13 Hierarchical segmentation into training validation and testing datasets.

class-wise recurrence of the held-in models will consistently be contrarily identified with that of the held-out models. This will prompt a reliable critical predisposition in the assessment. Regardless of these shortcomings, the hold-out technique has the upside of being basic and proficient, which settles on it a well known decision in enormous scale settings. From a deep learning point of view, this is a significant perception since large data sets are frequent.

Cross-validation

In the cross validation technique, the labeled information is separated into q equivalent fragments. One of the q sections is utilized for testing, and the remaining $(q-1)$ portions are utilized for training. This procedure is riterated q times by utilizing every one of the q fragments as the test set. The normal precision over the q distinctive test sets is accounted for. Note that this methodology can intently appraise the genuine precision when the estimation of q is huge. A unique case is one where q is picked to be equivalent to the quantity of marked information focuses and in this manner a solitary point is utilized for testing. Since this single point is left out from the preparation information, this methodology is alluded to as forget about one cross-approval. Albeit such a methodology can estimate the performance, it is typically too costly to even consider training the model countless times. Truth be told, cross validation is sparingly utilized in neural systems in view of productivity issues.

2.2.4.3 Neural networks applicability with big data

One application issue that emerges in the particular instance of neural systems is the point at which the sizes of the preparation informational indexes are enormous. Along these lines, while strategies like cross-validation are regarded as better decisions than hold-out in conventional AI, their soundness is frequently compromised to achieve efficiency. By and large, training time is considered such a significant issue in neural system demonstrating that numerous trade offs must be made to reach a practical result.

A computational issue regularly emerges with regards to network search of hyperparameters. Indeed, even a solitary hyperparameter decision can once in a while require a couple of days to assess, and a lattice search requires the testing of an enormous number of conceivable outcomes. Hence, a typical procedure is to run the training procedure of each setting for a fixed number of iterations. Various runs are executed over various decisions of hyperparameters in various runs. Those decisions of hyperparameters wherein great progress isn't made after a fixed number of iterations are ended. At last, just a few choices are permitted to raced to finish. One reason that such a methodology functions admirably is on the grounds that most by far of the advancement is frequently made in the early periods of the training.

2.2.4.4 Indicators of dataset suitability

The high generalization error in an ANN might be brought about by a few reasons. To start with, the information itself may be noisy, where case there is minimal one can do so as to improve performance. Second, neural systems are difficult to train, and the large error may be brought about by the poor convergence conduct of the calculation. The error may likewise be brought about by high bias, which is alluded to as under fitting. At long last, overfitting (i.e., high change) may cause an large prediction error. By and large, the error is a blend of more than one of these various variables. Actually, the expert can recognize overfitting in a particular training informational collection by inspecting the gap between the training and test accuracy. Overfitting is showed by an enormous gap among training and test accuracy. It is not phenomenal to have near 100% test accuracy on a little preparing set, notwithstanding when the test blunder is very low. The principal answer for this issue is to gather more information. With expanded preparing information, the training accuracy will lessen, though the test/approval performance is expected to become higher. In the occasion that more information isn't accessible, different strategies will be needed, like regularization so as to enhance the generalization performance.

Penalty-driven regularization

An error based regularization is the most widely recognized methodology for avoiding overfitting. So as to comprehend this point, let us revisit the case of the polynomial with degree d. For this situation, the forecast ŷ for a given estimation of x is given as in Eq. (2.9):

$$\hat{y} = \sum_{i=0}^{d} w_i x^i \tag{2.9}$$

It is conceivable to utilize a single layer NN with d inputs and a single bias neuron of weight w0 so as to display this expectation. The ith information is denoted as xi. This neural system utilizes linear activations, and the squared loss function for many training examples (x, y) from informational index D can be characterized as below:

$$L = \sum_{(x,\, y) \ni D} (y - \hat{y})^2 \tag{2.10}$$

A large value of d will increase overfitting. This problem can be compensated through the estimation of d. As it were, utilizing a model with fewer parameters forms a simpler model. For instance, diminishing d to 1 makes a direct model that has less degrees of freedom and will in general fit the information along these lines over various training data. As it were, oversimplification minimizes the performance of a neural system, with the goal that it can't modify adequately to the requirements of various kinds of informational collections. How might one hold a portion of this expressiveness without causing a lot of overfitting? Rather than lessening the quantity of parameters in a hard manner, one can utilize a soft penalty on the utilization of parameters. Besides, enormous (outright) estimations of the parameters are punished more than little qualities, since little qualities don't influence the expectation altogether. What sort of penalty would one be able to utilize? The most widely recognized decision is L2-regularization, which is additionally alluded to as Tikhonov regularization. In such a case, the extra penalty is characterized by the sum of squares of the estimations of the parameters. At that point, for the regularization parameter $\lambda > 0$, one can characterize the target work as given:

$$L = \sum_{(x,\, y) \ni D} (y - \hat{y})^2 + \lambda \sum_{i=0}^{d} w_i^2 \tag{2.11}$$

Expanding or diminishing the value of λ decreases the delicate quality of the penalty. One preferred position of this sort of parameterized penalty is that one can tune this parameter for ideal execution on a segment of the training data set that is not utilized for learning the parameters. This sort of methodology is called model validation. Utilizing this sort of methodology gives more noteworthy adaptability than fixing the economy of the model in advance. Consider the instance of polynomial regression examined previously. Limiting the quantity of parameters in advance seriously obliges the learned polynomial to a particular shape (e.g., a straight model), while a delicate penalty can regulate the state of the learned polynomial in an additional information driven way.

Generally, it has been practically seen that it is progressively alluring to utilize complex models (e.g., bigger neural systems) with regularization instead of basic models without regularization. The previous additionally gives more noteworthy adaptability by giving a tunable handle (i.e., regularization parameter), which can be picked in an driven data way. The estimation of the tunable handle is found out on a held-out bit of the informational index. Regarding the regularization influence on the updates in a NN, for some random load inside the NN, the updates are characterized by gradient descent (Eq. 2.12) (or its clustered form):

$$w_i \Leftarrow w_i - a\frac{\partial L}{\partial w_i} \tag{2.12}$$

where α denotes the learning rate. The use of $L2$-regularization is roughly equivalent to the use of decay imposition after each parameter update aw demonstrated in Eq. (2.13):

$$w_i \Leftarrow w_i - (1 - a\lambda) - a\frac{\partial L}{\partial w_i} \tag{2.13}$$

Note that the update above first inflates the weight with the decay factor $(1 - \alpha\lambda)$, and afterward utilizes the gradient based update. The decay of the weights can likewise be comprehended by a natural elucidation, on the off chance that we expect that the underlying estimations of the weightsare near 0. One can view weight decay as a sort of overlooking instrument, which carries the weights nearer to their underlying qualities. This guarantees just the rehashed updates significantly affect the volume of the weights. An overlooking system keeps a model from retaining the training dataset, in light of the fact that lone critical and rehashed updates will be reflected in the weights.

2.2.4.5 Convolutional autoencoder

The autoencoder recreates information by compressing them through a pressure stage. At times, the information isn't compressed even if the representations appear to be sparse. The bit before the most compressed layer of the design is called encoder, and the part after the compressed segment is called decoder. Fig. 2.14. demonstrates a visualization the encoder-decoder structure for the conventional case. The convolutional autoencoder has a comparative rule, which remakes images in the wake of going them through a compression stage. The primary distinction between a conventional autoencoder and a convolutional autoencoder is that the last is centered around utilizing spatial connections between points so as to produce features that have a visual meaning. The spatial convolution tasks in the middle layers accomplish exactly this objective. A delineation of the convolutional autoencoder appears in Fig. 2.15 (on the right side) compared with the conventional autoencoder in Fig. 2.15 (on the left side).

Note the 3-dimensional spatial state of the encoder and decoder in the second case. Be that as it may, it is conceivable to think about a few varieties to this fundamental construction. For instance, the codes in the center can

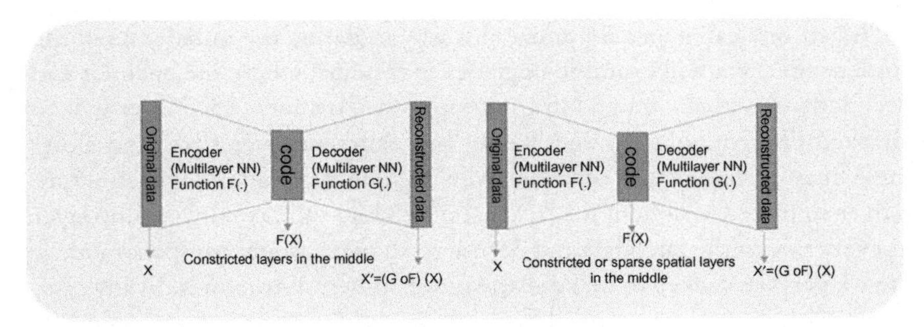

Fig. 2.14 Depiction of a typical autoencoder (left) and a convolutional one (right).

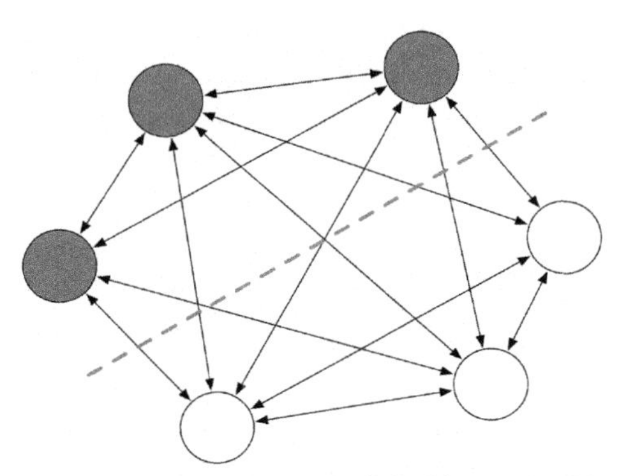

Fig. 2.15 A Boltzmann machine's structure with hidden units (denoted in blue color, black in monochrome version) and visible units (denoted in white color). The dashed line has been put, in order to symbolize the virtual separation (Chang & Bai, 2018).

either be spatial or they can be smoothed with the utilization of completely connected layers, contingent upon the current application. The completely associated layers would be important to make a multidimensional code that can be utilized with random applications (without stressing over spatial requirements among features.

Note that the layers are symmetrically positioned as far as how a corresponding layer in the decoder fixes the impact of a matching layer in the encoder. Nonetheless, there are numerous varieties to this situation. For instance, it is possible that the ReLU wil be set after the deconvolution. Moreover, in certain varieties (Makhzani & Frey, 2015), it is prescribed to utilize deeper encoders than the decoders with an asymmetric design. Be that as it may, with a stacked variety of the symmetric structure above, it is conceivable to train only the encoder with a classification output layer (and a supervised datast like ImageNet) and after that utilization its symmetric decoder (with transposed/inverted filters) to perform "deconvnet" visualization (Zeiler & Fergus, 2014).

Albeit one can generally utilize this way to define the initial state of the autoencoder, we will examine upgrades of this idea where the encoder and decoder are mutually trained in an unsupervised manner. Each layer will be checked like convolution and ReLU as a different layer here, and along these lines we have a sum of seven layers including the input. This structure is oversimplified in light of the fact that it utilizes a solitary convolution layer in every one of the encoders and decoders. In increasingly complex models, these layers are stacked to make all the more powerful structures. In any case, it is useful to delineate the relationship of the fundamental activities like unpooling and deconvolution to their encoding operations.

Another improvement is that the code is contained in a spatial layer, though one could likewise embed fully connected layers in the center. In spite of the fact that this model utilizes a spatial code, the utilization of fully connected layers in the center is progressively helpful for real life applications. Then again, the spatial layers in the center can be utilized for representation. Consider a circumstance where the encoder utilizes d2 square channels of size $F_1 \times F_1 \times d_1$ in the principal layer, where the principal layer is expected to be presented as a (spatially) square volume of size $L_1 \times L_1 \times d_1$. The (I, j, k)th passage of the pth channel in the primary layer will have weight $w_{ijk}^{(p,\,1)}$.

It isn't unexpected to utilize the exact degree of padding required in the convolution layer, so the element maps in the subsequent layer are likewise of size L_1. This degree of padding is F_1–1, which is alluded to as half-cushioning. Be that as it may, it is likewise conceivable to utilize no padding in the convolution layer, in the occasion that one uses full padding in the relating deconvolution layer. As a rule, the entirety of the paddings between the convolution and its relating deconvolution layer must aggregate to F_1–1 so as to keep up the spatial size of the layer in a convolution-deconvolution pair.

Here, it is critical to get that albeit each $W^{(p,1)} = \left[w_{ijk}{}^{(p,1)}\right]$ is a 3-dimensional tensor, one can make a 4-dimensional tensor by incorporating the index p in the tensor. The deconvolution task utilizes a transposition of this tensor, which looks like the methodology utilized in backpropagation. The associated deconvolution activity takes place from the 6th to the 7th layer (by checking the ReLU/pooling/unpooling layers in the center). Hence, the (deconvolution) tensor $U^{(s,1)} = \left[w_{ijk}{}^{(s,6)}\right]$ appears in connection to $W^{(p,1)}$. Layer 5 contains d_2 highlight maps, which were acquired from the convolution task in the primary layer (and unaltered by pooling/unpooling/ReLU activities). These d2 feature maps should be mapped into d1 layers, where d_1 is estimated equal to 3 for RGB color channels. Along these lines, the quantity of filters in the deconvolution layer is equivalent to the depth of the channels in the convolution layer and the opposite. This adjustment fit can be seen as a fiddle due to the transposition and spatial inversion of the 4-dimensional tensor made by the filters.

2.2.5 Bayesian networks

Bayesian systems, otherwise the so called belief systems or Bayes networks (BN), regularly utilize graphical structures to delineate the probabilistic connections among a lot of irregular factors (Friedman, Linial, Nachman, & Pe'er, 2000). In these BNs, every individual node denotes an arbitrary variable with explicit recommendations thinking about an uncertain area and the curves interfacing the nodes denote the probabilistic conditions among those random factors. The reliance between arbitrary variables is characterized by the condition likelihood related to every node.

Usually, there is just a single constraint on the arrows in a BN as one can't come back to a node basically by following directed arrows. BNs are fit for dealing with multivalued arbitrary variables in nature on the grounds that their random variables are made out of two dimensions, including a range of relational words and likelihood alloted to every one of the prepositions. With this plan, the design of BNs is a perfect model for coordinating earlier information and observations for different applications. BNs are, along these lines, fit for learning the causal-impact connections and anticipating future occasions in problem areas, even with missing information.

The idea of likelihood allocated to every one of the relational words in BNs requires further representation with the help of a Boltzmann machine in physics, forming a special learning process through Feed-Forward or Feed-Backward procedure. Boltzmann distribution, introduced by Ludwig Boltzmann, is the likelihood of particles in a framework over different potential states:

$$F(s) \propto e^{-\frac{E_s}{kT}} \tag{2.14}$$

where Es represents the state's s corresponding energy, k represents the Boltzmann's steady, and T denoteds the thermodynamic temperature. The Boltzmann factor is characterized as the proportion of two distributions featured by the difference of energies, as given below:

$$r = \frac{F(s_1)}{F(s_2)} = e^{\frac{Es2 - Es1}{kT}} \tag{2.15}$$

Through Eq. ((2.12)), the likelihood can be presented as the term of each normalized state:

$$P_{s_i} = \frac{e^{-(\mathrm{E}_{si}/\mathrm{kT})}}{\sum_j e^{-(\mathrm{E}_{si}/\mathrm{kT})}} \tag{2.16}$$

Boltzmann Machine is a stochastic Hopfield Network with a visible layer and a hidden layer (Fig. 2.15) in which just visible nodes identify with sources of info though hidden nodes are for the most part utilized as complementary to visible units in portraying the distribution of information (Ackley, Hinton, & Sejnowski, 1985). In this occasion, the model can be considered as comprising of two sections, comprising of the visible part and the hidden part. A state to State 1 can be set, paying little heed to the present state with the following likelihood:

$$p = \frac{1}{1 + e^{-(\Delta \mathrm{E}/\mathrm{kT})}} \tag{2.17}$$

where ΔE denotes the deference between energies when the state is on and off. In a perfect world, the higher the temperature T, the almost certain the state will change. The likelihood of a higher energy state changing to a lower state will be constantly bigger than that of the switch procedure, which is in accordance with the thermodynamic principle.

Restricted Boltzmann Machine (RBM), developed by Smolensky (1986), is an expanded version of Boltzmann Machine limited by one principle: there are no associations either between visible nodes or between hidden nodes. Then again, Deep Belief Networks (DBNs) is considered a probabilistic generative model comprising of numerous layers of stochastic hidden nodes being stacked, whose building squares are RBMs. Each RBM has two layers (Fig. 2.16).

The main layer comprises of visible nodes, like observable data variables and the subsequent layer comprises of concealed hubs, like latent variables. DBNs are not simply stacking RBMs together. Just the top layer has the bi-directional associations, though the base, while center layers don't. When considering an image handling issue, the visible layer can peruse out the picture at first and pass the pictures to hidden layers for learning and recovering the low-level features in the main hidden layer and the abnormal state includes in the second hidden layer in the stacked RBM, separately (Fig. 2.17).

Note that the bi-directional design in the RBM would permit the remaking of the weights forward and backward between the two hidden layers. Be that as it may, this is impossible between the visible layer and the first hidden layer due to the unidirectional design.

The energy of the joint setup of the visible and hidden nodes in a typical DBN is described through the Eq. (2.18) as follows:

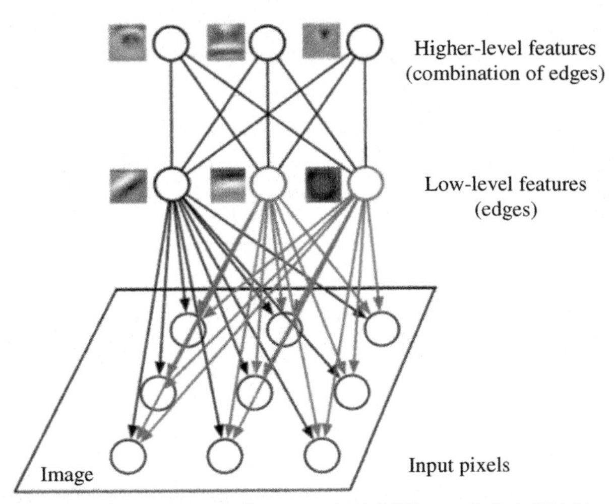

Fig. 2.16 Stacking up for image analysis in a DBN (Chang & Bai, 2018).

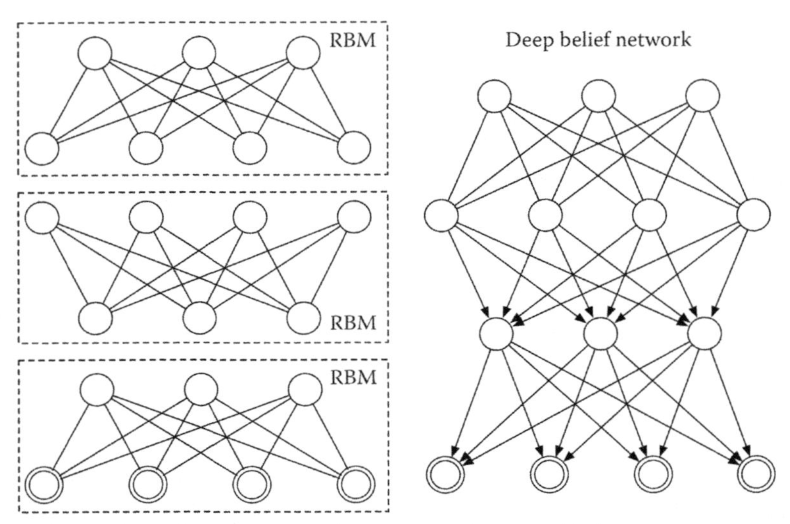

Fig. 2.17 A typical DBN structure with one visible layers of RBM |(seen at the bottom) and two hidden layers of RBM (seen in the middle and top) (Chang & Bai, 2018).

$$E(v, h) = -\sum_{i=1}^{V_N}\sum_{j=1}^{H_M} w_{i,j} v_i h_j - \sum_{i=1}^{V_N} a_i v_i - \sum_{j=1}^{H_M} b_j h_j \tag{2.18}$$

where model parameters are symbolized as $w_{i,j}$, a_i, while b_j. $w_{i,j}$ denote the symmetric weight between visible node vi and hidden node h_j. The factors a_i and b_j represent the bias terms for visible nodes and hidden nodes, respectively. The V_N and H_M are the numbers of visible nodes and hidden nodes, respectively. The optimal set of parameters ($w_{i,j}$, a_i, and b_j) can be estimated by maximizing the probability of visible nodes. The joint probability distribution of the visible-hidden node pair is symbolized as (Bishop, 2006a, 2006b) in Eq. (2.19):

$$P(v, h) = \frac{e^{-E(v, h)}}{\sum_{v'}\sum_{h'} e^{-E(v', h')}} \tag{2.19}$$

in which the denominator is composed for adding up all the probable pairs of visible-hidden units

$$P(v) = \sum_{h} P(v, h) = \frac{\sum_{h} e^{-E(v, h)}}{\sum_{v'}\sum_{h'} e^{-E(v', h')}} \tag{2.20}$$

The optimal parameter of $\{w_{i,j},\ a_i,\ b_i\}$ of an RBM with N training samples is estimated in Eq. (2.21):

$$\max_{w_{i,j},\, a_i,\, b_j} \left\{ \frac{1}{N} \sum_{k=1}^{N} \log P\left(v^k\right) \right\} \tag{2.21}$$

aiming enhance the model's probability for these given training samples (Hinton, 2010).

Given the training samples and the structure of a RBM, the ideal answer for Eq. (2.18) finishes the RBM. The DBN is stacked by layers of RMB. This sort of deep learning structure has three variations, be that as it may, as appeared in Fig. 2.17. Though a sigmoid belief net and a deep belief net have been modularized for various developments adaptably, a Deep Boltzmann Machine (DBM), as appeared in Fig. 2.18, is worked with a multilayer structure in which every unit of RBM captures complex, higher-order relationships between the activiation of hidden nodes includes in the layer below with a bi-directional connectivity.

Like DBN, DBMs have the capability of learning internal representations that become progressively perplexing, and have high level representations arising from an significant supply of unlabeled sensory data. Dissimilar to DBN, the approximate inference method in DBM can consolidate not just an underlying base up pass implanted in DBN, yet in addition top-down feedback, enabling DBMs to spread incertainties and limit biases (Salakhutdinov & Hinton, 2009).

For instance, the DBM in Fig. 2.19 comprises of three RBMs with a stacked multilayer structuree from which the ensemble learning ability might be connoted. The first RBM is 3 × 180 in size (visible nodes × hidden nodes) in which the RBM's image input by means of visible nodes is passed

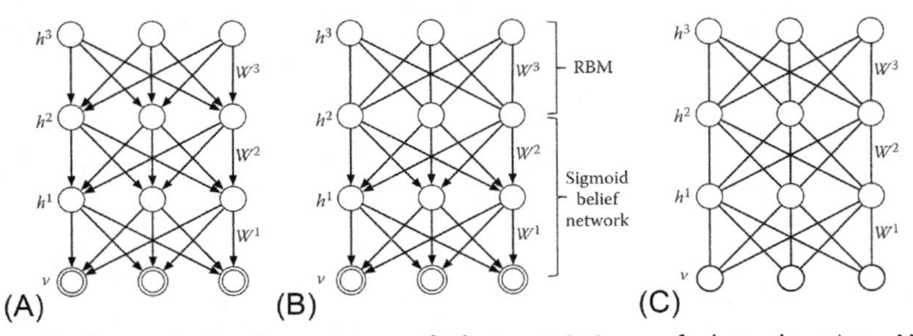

Fig. 2.18 Comparison of structures of three variations of deep learning ANN. (A) Sigmoid belief net, (B) deep belief net, and (C) Deep Boltzmann machine (Chang & Bai, 2018).

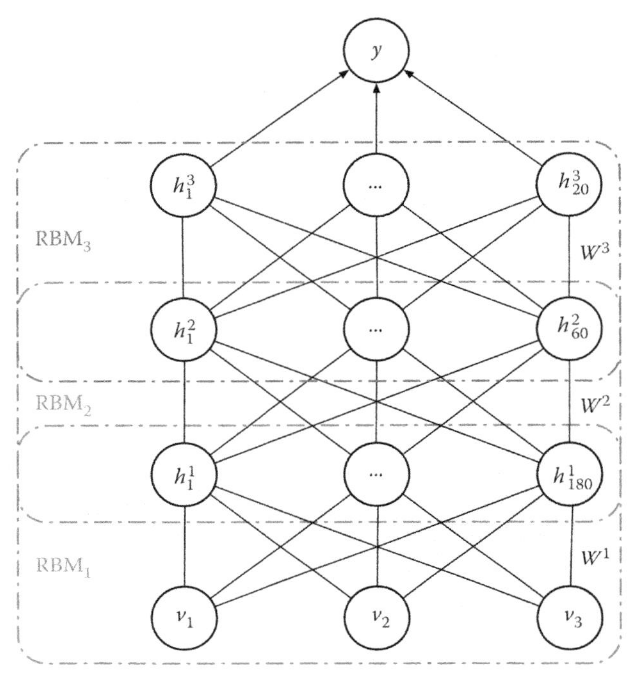

Fig. 2.19 The DBM structure for multilayer network construction, where the three inputs are denoted as v_1, v_2, v_3 and the unique output as y (Chang & Bai, 2018).

to the center level RBM unit (hidden nodes). The second RBM unit is 180×60 in size. After the center level RBM unit has taken in the low-level features, the RBM's output is passed to the upper level RBM unit for training high level features. The upper level RBM unit (i.e., third one from the base) is 60×20 in size. To improve the representation, the weight (w) is shown, while the inclinations (a and b) are neglected. After the layer-by-layer pretraining and last recovery in the DBM, every one of the parameters (w, a, and b) in the DBM are calibrated through backpropagation methods, shaping a mixture learning process.

2.2.6 Particle swarm optimization models

The Particle Swarm Optimization (PSO) model has been introduced by Kennedy and Eberhart (1995) and standardized by Kennedy et al. (2001). It is a biomimetic model inspired by the flock of birds which are searching or tracking a food target an the algorithm is adapted according to the distance from the food source. In PSO, a bird is denoted as a particle, and the flock of birds is the so called swarm. From a computational aspect, its area in respect to the available food is characterized by the fitness index. Every particle has one has to decide its speed, its direction, and separation from other particles, which comprise actions of inertia parameter and social and cognitive functions. Subsequently, the swarm (i.e., every one of the particles) has one fitness index characterized by a function, as the optimization procedure is to sort this function. Every one of the particles perform space scanning by following the most optical particle at present.

In a numerical setting, PSO is initiated as a grouping of arbitrary particles that advance naturally in a roaming space towards the ideal solution, while every particle is refreshed stepwise as it is bound by two limits. The first is known as personal best (pbest), which is characterized as an cognitive module, alluding to the ideal solution found by the particle itself. In similar manner, the other one is known as global best (gbest) and is characterized as a social part, alluding to the ideal solution found by the entire swarm. As a evolutionary algorithm, every particle in the PSO evolves naturally, which is for the most part reflected by the update of the location of each particle by means of a speed vector.

In this way, there are two viewpoints related with such a speed vector, specifically, the speed and heading. Basically, for every particle in the swarm, its current location in the search space is determined by its speed and location relative to the pbest and gbest. Because of the idea of global (gbest) and local (pbest) looking through search standards, particles in the swarm can be stochastically accelerated towards their best position at each iteration, in this way it is achieved that all particles to consistently scan for the optimal solution with the most appropriate found at each iteration in the solution space. Plainly the update of speed and position of every particle in the swarm is the most basic advance in the PSO estimation, as the enhancement procedure is identified with these updated steps. Numerically, the speed update of every particle in the swarm can be communicated as follows:

Numerically, denote $x_i = x_i^1, x_i^2, \ldots, x_1^D$, $i = 1,2,\ldots,N$ as the location of the ith particle (a swarm of N particles) in a D-dimension search space, and $v_i = (v_i^1, x_i^2 \ldots v_1^D)$ is the speed of of this particle. Providing $x_i = (x_i^1, x_i^2 \ldots x_1^D)$ represents the personal best position corresponding to the ith particle and $p_g = p_g^1, p_g^2 \ldots p_g^D$, denotes the global best position in the swarm, the procedure to fined the ideal configuration can be estimated as follows (Chen & Leou, 2012; Wang, Liu, Chen, & Wei, 2015):

$$v_i^d(t+1) = \omega \cdot v_i^d(\mathrm{t}) + c_i r_i(t) \cdot \left(p_i^d(t) - x_i^d(t)\right) + c_2 r_2(t) \cdot \left(p_g^d(t)\right) \quad (2.22)$$

$$x_i^d(t+1) = x_i^d(\mathrm{t}) + v_i^d(t+1)) \quad (2.23)$$

where d=1,2, ...,D and t is the cycle counter ω signifies the inertia factor controlling the effect of past speeds on the present speeds. c1 and c2 are two learning constraints that control the impacts of cognitive data to the local and global solutions, irrespectively. The factors r1 and r2 are two random scaling components which are located in the range of 0 and 1. A typical value range is [0, 1.2] for factor w and [0, 2] for factors c1 and c2. The three right terms in the update equation defining various segments in the PSO model (Kennedy et al., 2001). The principal term $\omega v_i^d\ (t)$ is the inertia component, denoting the essential solution. Regularly, a value in the range of 0.8 and 1.2 is given to the inertial coefficient w. Plainly a smaller value w will accelerate the convergence of the swarm to optima, though a bigger value will support the particles' investigation of the whole search space, along these lines making the optimization procedure to last longer.

The second term $r_1(t) \cdot (\mathrm{p}_i^d(t) - x_i^d(t))$ is generally known as the cognitive component, which denotes the particle's memory of the search history through space where the high individual fitness occurs. Usually, a value estimated near 2 is given to c1 to tweak the search steps in discovering its individual best solution $\mathrm{x}_i^d(\mathrm{t})$. The third term $\mathrm{c}_2 \cdot \mathrm{r}_2(t) \cdot (\mathrm{p}_i^d(t) - x_i^d(t))$ is known the social component, which is connected to control the particle to move to the best district that the swarm has found at present. Similarly, a approximate value equal to 2 is given to c2.

Generally, the PSO model requires just three stages towards an enhanced solution, including: assessment of the fitness of every molecule stepwise, update of individual and worldwide best fitness esteems and particle positions, and update of the speed and position of every particle towards the finish of each iteration. PSO can be utilized to upgrade the feature extraction algorithms created put together not just with respect to the statistics and decision science standards, for example, the linear discriminate analysis (Lin & Chen, 2009), yet additionally the AI and information data mining techniques, for example, SVM (Garšva & Danenas, 2014), and created advanced optimization approaches, for example, GA and PSO, for parameter optimisation in SVM, forming a hybrid feature extraction model. The SVM classifier is very performant for LULC mapping as it is depicted in Fig. 2.20 base on the image taken by the 2003 SPOT image in order to assess land use.

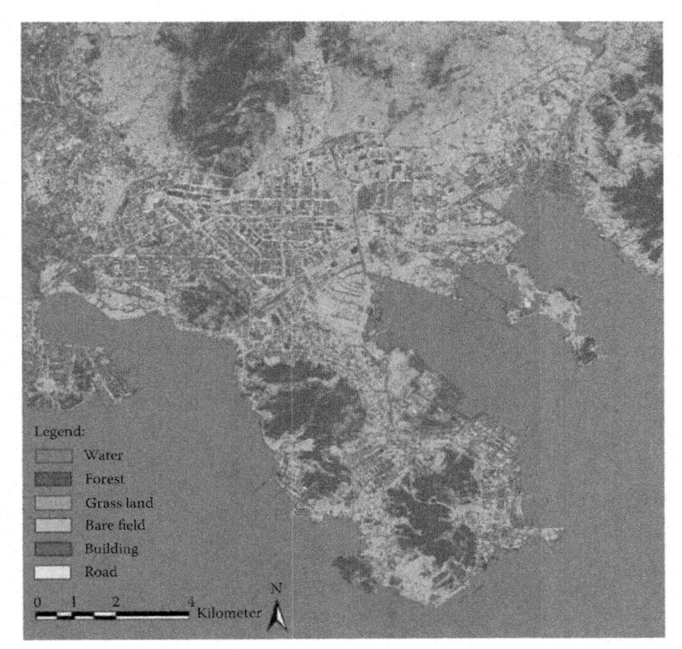

Fig. 2.20 Land Use and Land Cover (LULC) map a cultivated area situated in Dalian, China produced through employing the SVM classifier based on the 2007 SPOT 4 image (Chang et al., 2012).

2.2.7 Decision tree classifier

Decision tree classifiers are utilized as a well known classification technique in different pattern recognition issues, for example, image classification and character recognition (Safavian & Landgrebe, 1991). Decision tree classifiers perform more successfully, specifically for complex classification problems, due to their high adaptability and computationally effective features. Besides, decision tree classifiers exceed expectations over numerous typical supervised classification methods (Friedl & Brodley, 1997).

In particular, no distribution assumption is needed by decision tree classifiers regarding the input data. This particular feature gives to the Decision Tree Classifiers a higher adaptability to deal with different datasets, whether numeric or categorical, even with missing data. Also, decision tree classifiers are basically nonparametric. Also, decision trees are ideal for dealing with nonlinear relations among features and classes. At long last, the classification procedure through a tree-like structure is constantly natural and interpretable.

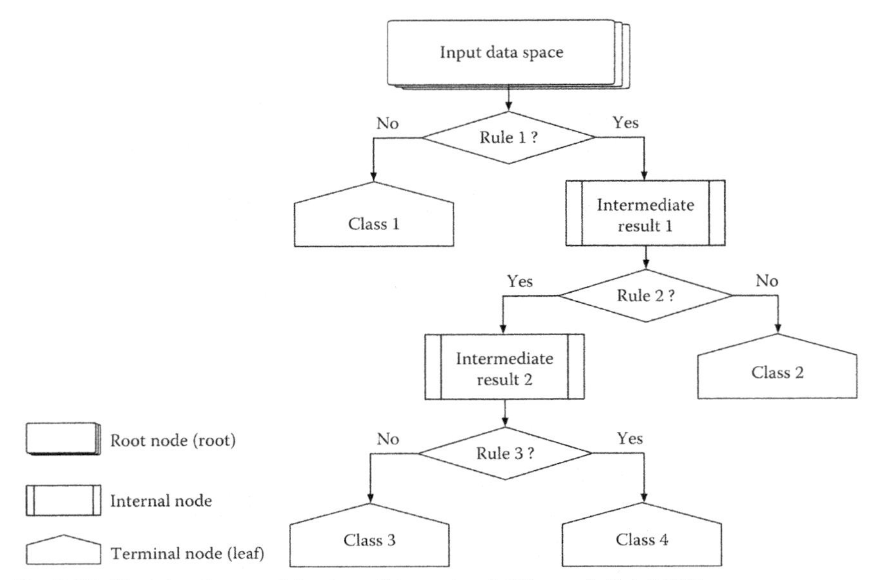

Fig. 2.21 Decision tree architecture (binary tree) (Chang & Bai, 2018).

Generally, a decision tree comprises of three basic segments including a root node, a few hidden nodes, and a lot of terminal nodes (known as leaves). An illustrative case of a decision tree structure is depicted in Fig. 2.21. As demonstrated, for each hidden and terminal node (known as child node), there should exist a parent node demonstrating the data source. In the interim, with respect to the root node and each hidden node (known as parent hub), at least two child nodes will be created from these parent nodes dependent on different decision rules.

In the event that each parent hub is part into two descendants, the decision tree is frequently known as a binary tree (e.g., Fig. 2.21, and the inherent decision rule can be communicated as a dyadic Boolean operator with the end goal that the data points focuses are split based on condition rules satisfaction. Among these three sorts of nodes, the root node includes the dataspace, while the other two sorts of nodes relate to divided subspaces. Instead of root and hidden nodes, the terminal hubs such as leaves, allude to the final decided outputs of the entire decision making procedure, which can't be additionally parceled; comparing class names will at that point be allotted. When constructing a decision tree, the most basic step is to part each internal node and the root node with different decision rules or learning models. Practically speaking, there are different learning models, which the most notable is the CART model, which is a binary recursive partitioning technique (Breiman, Friedman, Olshen, & Stone, 1984).

In the CART calculation, a splitting algorithm characterized as an determination function used to maximize the homogeneity of the training data as representing the descendant nodes. Regularly, an impurity function is characterized to assess the quality of split for every node, and the Gini diversity index is generally utilized as a well known measure for the impurity function. Numerically, the impurity estimation of the node t is generally characterized as follows:

$$i(t) = 1 - \sum_{j=1}^{K} P_j(t)^2 \tag{2.24}$$

where *Pj* (*t*) is the posterior probability corresponding to class *j* in node *t*. This probability represents the ratio formed by the training samples entering the node *t* labeled annotated as class *j* and all training samples in node *t*:

$$p_j(t) = \frac{N_j(t)}{N(t)}, \quad j = 1, 2, \ldots, K \tag{2.25}$$

In the case of binary node, the quality of the split *s* for the node *t* is estimated as follows:

$$\Delta i(\mathrm{s}, \mathrm{t}) = \mathrm{i}(\mathrm{t}) - \mathrm{P}_R i(t_R) - P_L i(t_L) \tag{2.26}$$

where PR and PL denote the proportion of the examples in node *t* that go to the correct descendant *tR* and the left descendant tL. Basically, the quality of the split s ought to be amplified to in the end accomplish the lowest impurity in each iteration towards the largest pupurity in the terminal nodes. Similar to an AI procedure, the halting model is likewise required by decision trees to derminate the split procedure. In the CART model, the stoping criterion is ordinarily formed as follows:

$$\max_s \Delta i(s, t) < \beta \tag{2.27}$$

where β symbolizes the predetermined threshold. The split procedure will proceed until it finds a stop criterion. The decision tree will stop developing, which suggests that the training procedure of the decision tree classifier completely terminated. Generally, a decision tree classifier-based procedure begins from the root node, and the unclassified data points in the root node are parceled into various various internal nodes following some splitting rules before they at last touch base at leaves where a class name will be assigned to every one of them.

Decision tree has been widely utilized in help with data mining and AI, aiming to obtain a target variable based on various input variables and decision rules (Liu, Zhang, & Lu, 2008). Due to their nonparametric and top-down structure, decision trees have been broadly applied in several remote sensing applications for feature extraction, for example, hyperspectral imagery classication (Chen & Wang, 2007), snow coverage extraction (Liu et al., 2008), invasive plant species identification (Ghulam, Porton, & Freeman, 2014) etc.

2.2.7.1 RF classifier

RF, or random decision forests, is a categories of decision trees made by creating a subset of training data through a bagging procedure (Breinman, 2001). All the more explicitly, RF comprises of a blend of decision trees where each tree is built utilizing an autonomously sampled random vector from the data set, while all trees in the forest keep up a consistent distribution (Pal, 2005; Tian, Zhang, Tian, & Sun, 2016). Preactically, around 2/3 of the first data will be randomly chosen to prepare these trees through the bagging procedure, while the rest of the data won't be utilized for training. Practically, that part of data is utilized for internal cross-validation so as to check the accuracy of of the trained trees (Belgiu & Drăguţ, 2016).

There is no compelling reason to perform cross-validation to get an accurate estimation of the set error since it has been accomplished during the time development of the RF model. As a rule, RF model attempts to develop various CART models by utilizing various examples and different initial variables, which creates RF, compensating for the handicap of overfitting related with traditional decision trees (Hastie, Tibshirani, & Friedman, 2009).

Generally, the RF-based classification needs two parameters to be employed. Moreover, the quantity of trees and the quantity of factors are arbitrarily picked at each split (Winham, Freimuth, & Biernacka, 2013). Every node in a tree will be split with a given number of arbitrarily chosen factors from the input feature space. In RF models, the Shannon entropy is routinely utilized as the splitting function to estimate the impurity of an attibute concerning the classes (Pal, 2005). Regarding prediction, each tree votes in favor of a class membership for data sample, and the class with a maximum vote will be selected as the winner class (Ni, Lin, & Zhang, 2017).

Unlikely to numerous other classification techniques that employ one classifier, multiple classifiers can be accommodated in RF model and a final prediction is rubustly made through joining every one of these decisions with an optimal solution. In conventional and group learning approaches utilize numerous learning models to acquire enhanced predictive performance over what would be achieved by single classifiers. Actually, the RF model is considered a group classifier of decision tree in which decision tree assumes the job of a meta model. This group learning nature renders RF numerous advantages, such as high performance, robustness against overfitting the training set, and integrated measures of variable level (Guo, Chehata, Mallet, & Boukir, 2011; Stumpf & Kerle, 2011).

Furthermore, no distribution assumption that is required for the input data set, henceforth it very well may be utilized to process different data sets. All things considered, following the same way as other measurable learning methods, the RF model is likewise seen to have a tendency to bias once the quantity of samples is distributed in an inconsistent manner between the classes of interest (Winham et al., 2013). In any case, due to its exceptional favorable circumstances, RF has been generally utilized for remote sensing classification in different case studies, including laser information data point cloud (Ni et al., 2017), LiDAR and multispectral urban scene classification (Guo et al., 2011), land cover classification and geospatial visualization (Fan, 2013; Tian et al., 2016), hyperspectral imagery characterization (Ham, Chen, Crawford, & Ghosh, 2005), and avalanche mapping (Chen, Li, Wang, Chen, & Liu, 2014).

2.2.8 GMM models

Distribution models are defined as models and limitations of clusters, as just items undoubtedly having a place with a similar dispersion will be assembled to shape the groups. The guideline of this methodology mimics the creation of artificial data sets, as items are arbitrarily collected dependent on a previously defined distribution. In practice, the most broadly utilized strategies are the so-called Gaussian Mixture Models (GMM) (Jian & Vemuri, 2011), which are created through employing the expectation-minimization algorithm (Dempster & Laird, 1977).

In fact, given a fitted GMM, the clustering algorithm will allocate every data point to the associated cluster producing the largest posterior probability. Each inquiry data can be allocated to more than one clusters, and subsequently this strategy can be viewed as a sort of fuzzy/soft clustering. Like most clustering strategies, the quantity of desired clusters ought to likewise be characterized before starting model fitting, and the model's complexity will unavoidably elevate as the amount of desired clusters goes higher. Since every data point will be associated with a score to each cluster, GMM clustering empowers the settlement of groups with various sizes and relationship structures. This adaptable property renders GMM clustering more fitting to use than fixed strategies like *k*-means clustering.

With their particular capacities in dimensionality decrease just as feature extraction, clustering techniques have been broadly utilized in the field of PC vision offering practical solutions including information mining, AI, image analysis, pattern recognition, information compression, etc. Generally, the determination of a proper clustering algorithm and its parameter settings depends to a great extent on the input data set just as the goal of the obtained results. Generally, the clustering method concerning a specific issue regularly should be chosen by experiment or from prior experience about the data along with the proposed utilization of the results.

Clustering strategies have for quite some time been generally utilized for feature learning and feature extraction to support several remote sensing applications; they incorporate aerial laser point cloud information (Tokunaga & Thuy Vu, 2007), SAR picture division by spectral clustering (Zhang, Jiao, Liu, Bo, & Gong, 2008), fluffy c-means clustering (Tian, Jiao, & Zhang, 2013), and numerous different practices, for example, coast board recognition (fuzzy clustering) (Modava & Akbarizadeh, 2017), geometrical structure recognition (density-distance-based clustering) (Wu, Chen, Dai, Chen, & Wang, 2017).

Recently, with the advances of high volume remote sensing data sets, for example, high resolution hyperspectral images, a large number of the current strategies cant manage these data sets because of the high dimensionality, which encourages the improvement of new clustering algorithms that are specialized in subspace clustering (Kriegel, Kröger, & Zimek, 2012). A typical clusterin algorithm is clustering QUEst (CLIQUE) (Agrawal, Gehrke, Gunopulos, & Raghavan, 2005). For acheiving hyperspectral imagery classification, Sun, Zhang, Du, Li, and Lai (2015) introduced an sparse subspace clustering technique to propel the band subset determination dependent on the assumption that band vectors can be tested from the coordinated low-dimensional subspaces and each band can be sparsely modeled as a linear or a linear mixture of other bands inside its subspace.

2.2.9 Principal component analysis (PCA)

PCA is a statistical strategy that is regularly used to change a group of interrelated factors into a set of linearly unrelated subsets dependent on a transformation that results in uncorrelated variables. Scientifically, it is characterized as a orthogonal linear transformation that makes a projection of the initial data set to another projection system with the end goal that the biggest variance has a projection of the first coordinate (it is the first principal component) while the second biggest variance has a projection of the second coordinate provided that it is vertical to the first component (Jolliffe, 2002). Basically, PCA plans to locate a linear transformation expressed as $z = W_k{}^T x$, where $x \in R^d$, and r<d, to enhance the variance of the data in the projected space (Prasad & Bruce, 2008). For a data matrix defined as $X = \{x_1, x_2, \ldots, x_i\}$, $x_i \in R^d$, $z \in R^r$ and $r < d$, the transformation can be characterized by a set of p-dimensional vectors of weights $W = \{w_1, w_2, \ldots\ldots, w_p\}$, $w_p \in R^k$ that matches every x_i vector of X to *a*

$$t_{k(i)} = W_{|(i)}{}^T x_i \tag{2.28}$$

For boosting the variance, an initial weight *W1* has to comply to the condition expressed as follows:

$$W_i = \arg\max_{|w|} = \left\{ \sum_i (x_i \cdot W)^2 \right\} \tag{2.29}$$

A further expansion of the previous condition is given below:

$$W_i = \arg\max_{\|w\|=1} \left\{ \|X \cdot W\|^2 \right\} = \arg\max_{\|w\|=1} \left\{ W^T X^T X W \right\} \tag{2.30}$$

A symmetric grid such as the $X^T X$, can be effectively analyzed by obtaining the biggest eigenvalue of the matrix, as W is the related eigenvector. Once W_1 is acquired, the primary principal component can be inferred by the projection of initial data matrix X onto the W_1 in the space that results from the transformation. The further segments can be procured along these lines following the subtraction of the recently obtained components.

Since the quantity of principal components is normally controlled of the most important eigenvalues, in relation to the global covariance matrix, the obtained components consistently are of lower dimensionality than the initial data set (Prasad & Bruce, 2008). These components frequently hold a large percentage of the variance of the initial data set. A group of six principal components obtained from a PCA analysis of Landsat TM multispectral

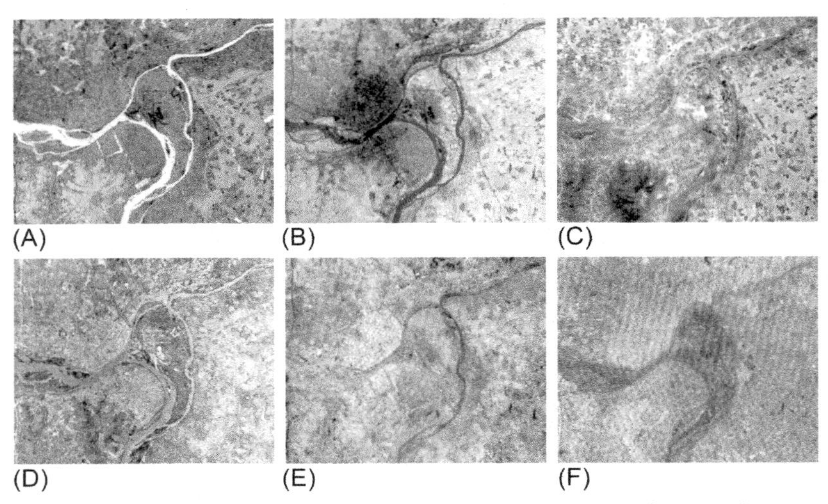

Fig. 2.22 Principal components produced by Landsat TM multispectral imagery. A total of 6 components are shown in (A)–(F) serially, with a corresponding variance of 68.5%, 27.2%, 3.3%, 0.6%, 0.3%, and 0.1%, respectively (Chang & Bai, 2018).

images as it is depicted in Fig. 2.22. Fig. 2.22A and B depict two principal components (Celik, 2009; Lian, 2012).

Moreover, there are various disadvantages and limitations associated with PCA, including the scaling effects (principal components are not robust enough to scale variations) (Prasad & Bruce, 2008). Recently, expanded PCA methods have been introduced offering solutions to many applications like the kernel PCA (Schölkopf, Smola, & Müller, 1997), the scale-invariant PCA (Han & Liu, 2014), and even more further developed methods such as independent component analysis (Wang & Chang, 2006) and projection pursuit (Chiang, Chang, & Ginsberg, 2001).

2.2.10 Maximum likelihood classifier (MLC)

Because of its probabilistic nature, MLC belongs to one of the most well-known classifications approaches in the field of remote sensing, in which the cluster of a pixel into one associating class will be just founded on its probability (or likelihood). Basically, MLC is considered a supervised classification approach derived from the Bayes hypothesis, as the categorization of each every cell to one explicit class related to the class samples (or training set) is resolved dependent on both variance and covariance of the class data vectors. Providing that the data vectors in each class test ought to follow a normal distribution, a class would possibly to be defined by the mean vector and the covariance matrix.

Provided these two properties for every cell value, the enrollment of the cells to the class can be defined based on their likelihood (probabilities). All the more explicitly, probability density functions will be constructed for each class dependent on the training data set, and afterward all unlabeled pixels will be arranged dependent on the relative likelihood of that pixel happening inside each class' probability density function. Scientifically, the probability that a pixel with feature vector defined as X is a member of a class k can be characterized as a posterior probability (Ahmad & Quegan, 2012):

$$P(k|\, X) = \frac{P(k)\mathrm{P}(X|\, k)}{p(X)} \tag{2.31}$$

where PXk()| represents the conditional probability of observation of X that belongs to class k (or alternatively to probability density function). The P(k) symbolizes the prior probability of class k, whose the values are estimated equal to each other because of the the low reference data availability. The P(X) denotes the X observation probability, which can be expressed as follows:

$$P(X) = \sum_{k=1}^{N} P(k)\mathrm{P}(X|\, k) \tag{2.32}$$

where N denotes the total classes number. Often, the P(X) is defined as the normalization constant aiming to assure that kNPkX=1()| equals to 1 (Ahmad & Quegan, 2012). A pixel x will be classified into the class k once it complies with the following condition:

$$x \in k \text{ if } P(k|X) > P(j|X) \text{ for all } j \neq k.$$

In mathematical terms, ML regularly makes the assumption that the distribution residing in a certain class of the data to be Gaussian distribution of many variables; Subsequently, the likelihood is defined as follows:

$$\mathrm{P}(\mathrm{k}|\, \mathrm{X}) = \frac{1}{(2\pi)^{\mathrm{N}/2}\left|\sum_k\right|^{1/2}} \exp\left(-\frac{1}{2}(X - u_k)^T\right) \tag{2.33}$$

where N defines the amount of data sets (e.g., bands derived from multispectral imagery) and X represents the entire data set of N bands. The u_k denotes the mean vector and Σk is the variance-covariance matrix corresponding to class k. The factor $|\Sigma k|$ is defined as the determinant of Σk.

Thanks to its roots, probability theory MLC has been employed in several remote sensing case studies for offering solutions mostly to classification problems. These applications include, rice crop recognition (Chen, Son, Chen, & Chang, 2011), land coverage alteration detection (Otukei & Blaschke, 2010) and water suitability mapping (Jay & Guillaume, 2014).

The accuracy of MLC has been extensively related and numerous other clasification methdos in the scientific literature, like decision trees, logistic regression, ANN, and SVMs. Other examples of further applications are presented in Frizzelle and Moody (2001), Hagner and Reese (2007), and Hogland, Billor, and Anderson (2013). Further examinations demonstrated that MLC might be inappropriate in some cases, classifiying categories based on spectral similarity (Kavzoglu & Reis, 2008). The PCA model overcomes these weakness, assisting the classification procedure. Moreover, many further expanded MLC methods have been proposed, like, hierarchical MLC (Ediriwickrema & Khorram, 1997) and calibrated MLC (Hagner & Reese, 2007).

2.3 Artificial neural networks applications in Biosystems engineering

The leading scientific journal published by the European Society of Agricultural Engineers (EurAgEng), Biosystems Engineering Journal, defines Biosystems Engineering term as the effective combination of the education and research in Physical Sciences and Engineering aiming to improve from a technical point of view, the biological systems function in terms of sustainability, food safety and quality, crop and land use management and the environmental protection.

According to the American Society of Biosystems Engineers (ASABE), Biosystems Engineering: involves the development of efficient and environmentally safe methods for food production as well as the use of renewable energy sources to meet the needs of an ever-increasing world population.

In conclusion, Biosystems Engineering incorporates engineering principles to sciences that are closely related to biology, agriculture and environmental sciences in such a way so as to guarantee the viable food and crop production that meet high national and qualitative standards taking into account the efficient utilization of natural resources.

More specifically, Biosystems Engineering has a broader scope than that of Agricultural Engineering field (including Agricultural Engineering as a central subset) because it is regarded as the application of a plethora of engineering sciences not only to agricultural systems but also to living organisms (design, analysis, technical support based on the engineering sciences). Biosystems Engineering cover the following scientific fields:

- Power and Machinery (MSC)

Advances in artificial intelligence and agriculture machinery engines, have enabled the utilization of biofuels by flexible adaptation of the engine operation to optimize the efficiency of the machine by fine-tuning the engine cycles and the emissions profile leading to cost reduction and minimal environmental footprint. Pantazi, Moshou, Kateris, Gravalos, and Xyradakis (2013) studied the vibrational profile of a four-stroke tractor engine fed by pure gasoline and gasoline mixtures with ethanol and methanol in different proportions (10%,20% and 30%). For the determination of optimal fuel blend, Machine Learning techniques including Neural Networks, Active Learning and original Novelty Detection algorithms have been utilized. An accuracy of 90% was achieved in detecting both type and mixture percentage of bioethanol.

- Post-harvest Technology

Photonic technologies like a reflectance sensing and chlorophyll fluorescence can produce signals that can automate the decision process needed in advanced quality sorting of apples in a nondestructive manner and in real time. Moshou et al. (2005) has proven the link between the chlorophyll fluorescence and the mealiness presence in apples, supposing mealiness and chlorophyll degradation are correlated. Their study has proven that through the employment advanced data mining architectures like Self Organizing Maps is more performant compared to other more conventional techniques like discriminant analysis and multilayer perceptrons demonstrating a high accuracy in the automated fruit sorting with the assistance of non-destructive methods.

- Livestock farming

To ensure the quality of animal and avian production, the volatiles and other emissions are necessary to be reduced to minimize health risk both to animals and to humans. Pan and Yang (2007) proposed an automated volatile sensing system consisted by an olfactory sensing developed specifically for registering and analyzing the odors from animal emissions in farms. The proposed tool called “Odor Expert” can be used both under lab and farm conditions by employing a typical backpropagation algorithm which achieved a performance of $r = 0.932$. Results indicated that the "Odor Expert" demonstrates a superior performance compared to a human panel. Therefore, is regarded as a useful tool for odor management practices in livestock and poultry farms.

The timely detection of health degradation in animals can result in better animal health management thus minimizing the impact of antibiotics and avoiding further complications. Moshou et al. (2001) proposed an ANN based technique concerning swine cough recognition. A neural network was trained in order to discriminate the swine cough from other ambient sound in the piggery area. The best results were obtained by a Self-Organizing map, which was able to recognize the coughs from other sounds by a percentage >95%.

- Food Security through early warning intelligent systems

Olive oil adulteration is a major concern and most of the times cause of economical profits loss and general threat for human health. Most of the times the olive oil adulteration with hazelnut oil is too difficult to be detected since their similar chemical composition. García-González, Mannina, D'Imperio, Segre, and Aparicio (2004) investigated this problem by proposing an artificial neural network based on hydrogen and carbon based nuclear magnetic resonance signatures (^{1}HNMR and ^{13}C NMR respectively). In terms of this study, A multilayer perceptron was employed to detect the type of blend of the two oils including pure olive oils, pure hazelnut oils, and mixtures of them.

- Crop Monitoring for Disease Detection and Stress Phenotyping

The growing conditions of crops should be monitored continuously to achieve incipient diseases, weeds and biotic and abiotic stress factors. Reduction of pesticides can be achieved by targeting the infected areas of the field with site-specific interventions (chemical or mechanical). This type of crop management results in much higher production with simultaneous reduction of inputs adjusted to the needs of the crop and to lower costs for maintaining the production at high level, while at the same time the environmental impact is minimized due to precise fertilization, phytosanitary interventions like spraying and targeted irrigation. DeChant et al. (2017) presented a phenotyping approach by recognizing stress symptoms in tomato leaves, with the help of a viral, fungal and bacterial and pest infection images by applying deep ANN architectures from the public domain including AlexNet and GoogLeNet. In the following Chapters are also presented specific application of crop monitoring for the detection of crop diseases and weeds.

All the sub disciplines of Biosystems Engineering rely on mathematical modeling and data acquisition to calibrate and use these models for prediction. Therefore, Data Mining is a indispensable tool for Biosystems Engineering applications since the knowledge discovery and model building enables the different applications that are needed for achieving practical engineering systems for the interpretation and control of biological commodities and environmental monitoring. A special domain integrating Biosystems applications and data mining in the form of sensor based interpretation and control of agricultural production systems is **Precision Agriculture**.

2.4 Contribution of artificial intelligence in precision agriculture (economical, practical etc.)

A variety of machine learning methods have been employed in the field of precision agriculture including Supervised and Unsupervised Learning for the assessment of plant health condition and invasive plant species recognition by interpreting spectral signatures and optical features. Advancements in sensor engineering, machine learning and geo-spatial analysis systems have enabled potential for precision agriculture enabling biotic and abiotic stress detection in crops at an early stage (Moshou, Deprez, & Ramon, 2004; Rumpf et al., 2010). The current techniques are capable of offering rapid, accurate and trustworthy decisions, in comparison to the conventional ones.

The precise invasive plant species recognition is considered a precondition for determine the type of suitable treatment. A number of machine learning methods have been applied to weed recognition; the majority of them rely on morphology features, weed leaf shape geometrical features for classification.

Detecting and forecasting crop health status are the principles for sustainable and efficient crop protection including approaches for the discrimination of plant biotic stress through machine learning (Moshou et al., 2004) reaching a classification range from 95% to 99%.

As regards the identification of insects through optical sensing and AI techniques the derivation of Vegetation Indices (Vis) from reflectance spectral data and the relationship to damage severity is more as more frequent approach. Prabhakar et al. (2011) investigated the detection of cotton infection infestation by leafhopper by utilizing specific wavebands emanating from linear intensity graphs associating reflectance and the level of infestation.

To sum up, a promising development of a variety of AI techniques in precision agriculture arises, contributing to the security and the sustainability in crop production. According to the current state of the art in sensing technologies and data analytics based on machine learning, the following projections and their respective contribution to precision agriculture are envisaged:

1. Customization of mainstream AI techniques to crop production. Attention should be paid to the fusion of information from the spatial, spectral and time series of crop parameters to detect trends related to crop status. The combination of advanced data analytics and the visualization mapping of the crop problems can lead to timely interventions that lead to reduction of treatment chemicals and crop yield loss avoidance. These projected outcomes guarantee a lower environmental impact and higher profit margins.
2. Technologies combining improvements in hardware of sensors will make them more compact and embeddable in field-deployed devices. These compact sensors will be equipped with board decision algorithms allowing the infield real time detection of crop anomalies and identification of the situation leading to event based interventions and management. The trend towards field automation by integrated sensors and software modules in filed deployable devices will make available big data for real time analysis of crop status and will allow event based decision algorithms that are available in autonomous machinery and the smart mobile devices to manage the crop status in real time. This will result in optimal crop production and quality leading to minimization of crop losses and higher competitiveness for crop producer.
3. The data fusion from different sources will enable the construction accurate and reliable models capable of assessing the potential of future states within a field—and fuse it with climate projections and soil data derived from soil maps to predict the yield potential and assess crop suitability concerning the examined fields. The effective combination of machine learning methods and sensor technologies can function as efficient tools for agricultural advisory boards and experts by assisting them into reaching conclusions, minimizing the risk of possible yield loss, adaptation of Best Management Practices (BMPs) to reduce quality variability and reducing inputs for sustainable crop production by ensuring higher economic benefits.

2.5 One class classifiers

In application where safety is a high priority, novelty detection is crucial for estimating the threshold of deviation from normality. Baseline system state

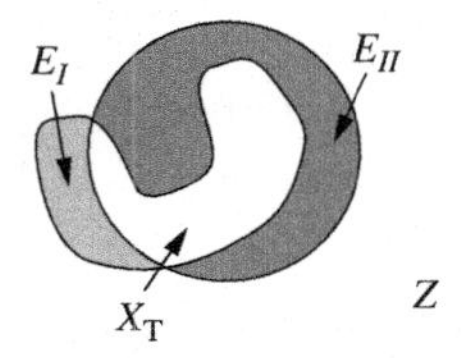

Fig. 2.23 Error areas defined in connection with the one-class classifier and the target domain (Tax & Duin, 2004).

may deviate due to wear, structural alterations, environmental changes are held responsible for altering the operational state of a system. The estimation of robust thresholds is important for credible novelty detection. For novelty detectors which utilize one class classifiers it is crucial to acquire all possible working stages that are attributed to normal operating conditions so that the one class classifier covers all possible normal operation states so that the robust thresholds signify faithful limits of these operations conditions leading to trustworthy outlier detection decisions.

One-class classification scheme can be described:

- Only baseline data are used as target without inclusions of outliers;

The boundary is set between two classes is defined only from target class data;

- The robustness of the boundary condition is associated with the criterion of maximizing the correct classification of as many target examples as possible while at the same time keeping the accepted outliers.

  Fig. 2.23, demonstrates a target dataset area X_T. E_I represents the error of the correclty rejected target examples while E_{II} symbolizes the incorrectly classified oulier objects as belonging to the baseline dataset. The area in the circle represents the area of definition of the one class classifier.

By assuming that the outlier data are distributed evenly, the consequences of shrinking the domain E_{II} to a minimal data description concerns the minimisation of false acceptance. This can be achieved by simultaneously minimisation of the volume of one class classifier and falsely accepted sector E_I.

2.6 Support vector machines (SVMs)

SVMs (Vapnik, 1998) are machine learning technique, mostly employed for pattern recognition, nonlinear function approximation. SVMs demonstrate important properties, when compared to other machine learning architectures including neural network learning and particularly algorithms for training architectures. In the last period, kernel methods, like SVM and RBN, became highly used image processing methods for data mining and recognition (Bovolo, Bruzzone, & Marconcini, 2008; Muñoz-Marí, Bovolo, Gómez-Chova, Bruzzone, & Camp-Valls, 2010). In the same fashion as RBN, SVM functions based on calculating a distance metric among data vectors to allocate data samples to clusters (Courant & Hilbert, 1954). With the help of the idea of structural risk minimization, this characteristic of a distance-calculated metric enhances the robustness of SVM by clearing noise in the training procedure, which induces a better performance in classification compared to other classifiers like decision trees, maximum likelihood, and neural network-based approaches (Huang, Davis, & Townshend, 2002; Bazi & Melgani, 2006). Contrary to the above referred classifiers, an SVM classifier is binary by definition. Provided that more than two classes are usually considered in remote sensing image processing, the way of building a multiple-class SVM classifier is a problem that needs solution. The main asset of SVMs is their capability of dealing with large volumes of data without encountering calibration bottlenecks that are usual in multilayer architectures that exhibit complex error surfaces with trapping regions. On the other hand, a further characteristic of SVM is the efficient learning with small amount of data, due to their simpler, more effective architecture and learning procedure. A classification mapping can be defined as follows:

$$f : \mathbf{x} \rightarrow \{\pm 1\} \tag{2.34}$$

A crucial step is to characterize and predict f due to the sensitivity and accuracy of separating overlapping classes, which is defined as *capacity* of the SVM (Tax & Duin, 2001a). Low value capacities cannot lead to trustworthy and accurate estimation of multi-parametric functions, while high value capacities will attempt to describe the data with superfluous paramaters leading to estimation error a phenomenon known as "overfitting".

Overfitting is a major problem in training of ANNs. To overcome this problem two main techniques are employed. The first one is known as "early stopping" and the second method is targeting on reducing the roughness of the error function, which corresponds to spurious noise contribution that might be due to local deviations of the data producing mechanism by penalizing such deviations through regularization of the error function. Similarly, in SVMs the overfitting is avoided by utilizing regularization functions. An advantage of SVMs is their ability of being calibrated with small datasets so they avoid the problem of overfitting (Vapnik, 1998). The less complex functions that are used for reaching decisions are the linear ones.

2.7 One class-support vector machines (SVMs)

2.7.1 Support vector data description (SVDD)

A specific type of One Class SVM classification model has been defined by Tax and Duin (2004) known as SVDD. In a set of targets $\{x_i\}$, $i=1, 2, 3, \ldots, N$ for training, the main objective of SVDD is to form a decision surface separating a new data into targets and outliers. A simplified boundary has the form of high dimensional sphere which is defined a couple of parameters, including the center (a) and radius (R). An optimal configuration that leaves out the non-relevant data would be a constricted hypersphere just tight enough in order enclose only the specific target data examples defined by the minimal radius R (Fig. 2.24).

The intention of SVDD is to form a hypersphere containing the training data (Fig. 2.35). The mathematic formulation of this type of requirement as an optimal configuration is given as follows:

$$\min R^2 + C\sum_{i=1}^{N}\xi_i$$

$$\text{s.t. } \|x_i - a\|^2 \le R^2 + \xi i, i = 1, 2, \ldots, \mathrm{N} \tag{2.35}$$

where C regularizes the loss term and ξ_i represents the slack variables, The parameter C is defined by the chosen upper threshold ν on falsely classified targets as outliers:

$$C = \frac{1}{N \cdot \nu} \tag{2.36}$$

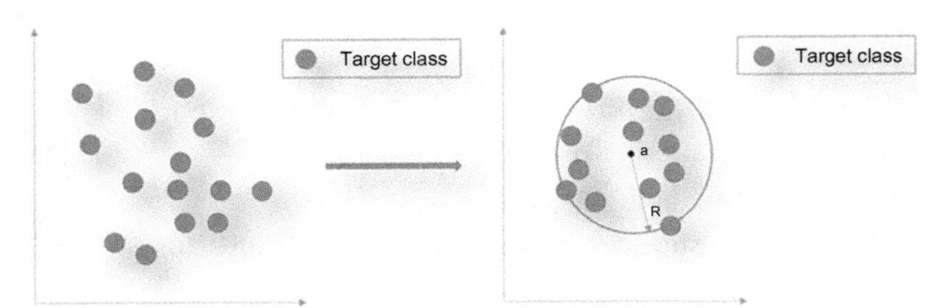

Fig. 2.24 Target data of one class (left) is enclosed by the circle (right) This circle represent a separating border between the target dataset and the rest of the input space.

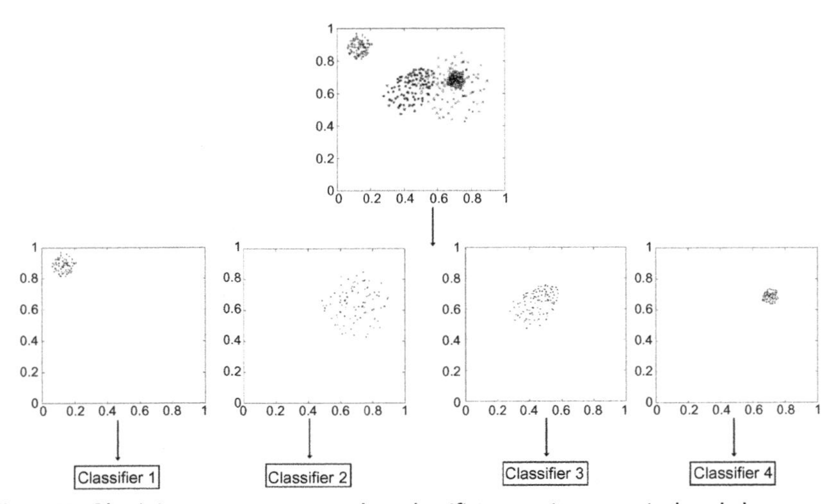

Fig. 2.25 Obtaining a separate one-class classifier mapping to an isolated class as part of classification problem with multiple classes (Sachs et al., 2006).

α is defined as follows:

$$a = \sum_{i=1}^{N} a_i x_i, \quad 0 \leq a_i \leq C \tag{2.37}$$

The α_i parameter may fall into three conditions:

$$a_i = 0 \Rightarrow \|x_i - a\| < R^2,$$
$$0 < a_i <^i C \Rightarrow \|x_i - a\|^2 = R^2,$$
$$a_i = C \Rightarrow \|x_i - a\|^2 < \mathrm{R}^2.$$

In order to classify a data sample ν, the proximity of ν to the center α is given as:

$$\|\nu - \alpha\|^2 \leq R^2 \Rightarrow \nu \text{ is target}$$
$$\|\nu - \alpha\|^2 > R \Rightarrow \nu \text{ is outlier} \tag{2.38}$$

2.7.2 One-class support-vector-learning for multi class problems

A working assumption would involve the full coverage of the calibration dataset, provided this assumption xi ∈ X the expectation error is null. Associated to the two class SVM methodology (Sachs, Thiel, & Schwenker, 2006) the limit error is expressed as follows:

$$E_{struct}(R, a) = R^2 \tag{2.39}$$

Therefore, the original optimal solution would be solved by minimizing the radius expressed as $\min R^2$. The one-class methodology is characterized insufficient for multiclass problems but it can be extended in order to tackle them as explained in the following. Let us assume that a dataset comprising k different classes belonging to the set $\Omega = \{1, \ldots, k\}$ (Fig. 2.25). For isolated class $t \in \Omega$, will construct a one-class mapping f_t, consisting of k center hyperspheres $\mathbf{a}_t$ and with radii R_t.

To decide the class of new data samples $\mathbf{z}$ a parallel calculation of membership functions takes place determining class probabilities:

$$f_t(z, a_t, R_t)^t = I\left(\|z - a_t\|_2^2 \le R_t^2\right) \tag{2.40}$$

During a class allocation stage of a new data sample z there are three possible outcomes:

1. In the event that one classifier is identified as responsible for indicating the class of the new data sample, the resulting class is assigned as the classification result for this data sample.
2. If multiple classifiers declare membership for the new data sample, a so conflict condition creates an allocation problem that has to be solved.
3. In the case that no classifier claims for a new data sample, this condition is called as an outlier situation.

The allocation of the latest two situations can be dealt with a couple of ways: The first one involves the rejection of the label allocation to the new pattern while the second one involves the assessment of the class label following a conflict resolution combined with an outlier estimation procedure (Fig. 2.3).

2.7.3 Nearest-center strategy—nearest-support-vector strategy

The nearest center strategy assigns a new data sample z, according to the distance of the center a_i to z. A conflict arises when this holds for multiple one class classifier and then the class remains due to multiple allocations meaning that the candidate set I(z) has more than one entries which is expressed mathematically as $I(z) := \{t\colon ft.(z) = 1\} \subset \Omega$ is of cardinality >1.

Accordingly the decision label estimation is expressed as:

$$f(z) = \mathrm{argmin} l \in C\|z - al\| \tag{2.41}$$

A graphical illustration of the functioning of this classifier is given in Fig. 2.26, indicating that it's enough to estimate the position of the classifier

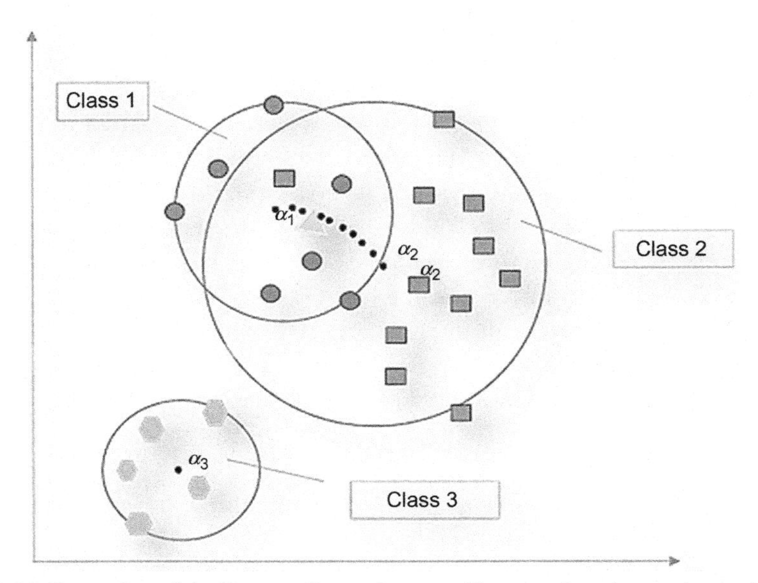

Fig. 2.26 Illustration of the Nearest-Center Strategy: The new data denoted as a triangle is allocated base on the proximity of the centers of the SVMs. In the scheme, two SVMs enclose the new data arriving at a conflict which has to be resolved by conflict resolution.

centers that belong to the candidate set I(z). For an outlier the set I(z) has no candidates, the argmin cover $C=\Omega$.

In the method of Nearest Support Vector, the class allocation of a new data z is decided by sorting the distances of $\mathbf{z}$ to the support vectors of all the available classifiers when a similar conflict can arise (Fig. 2.27). Hence, the class allocation is based on the classifier with the nearest support vector. Assuming that

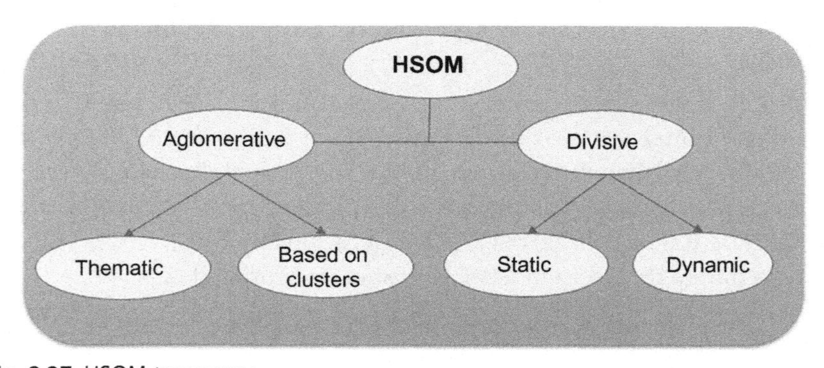

Fig. 2.27 HSOM taxonomy.

$$SV(t) = \{\hat{x}_1^t, \ldots, \hat{x}_{1t'}^t\} \quad (2.42)$$

denote the support vectors, while represents i_t nthe number of support vectors of classifier t.

$$f(z) = \text{argmin}_{l \in C} \|\hat{x}_i^l - z\| \quad (2.43)$$

The process of class allocation in this case involves the minimum distance from the set of distances between the data and the support vectors of all classifiers that have declared membership of the data. If a conflict situation arises, then C = I(z), and in the case that the data is an outlier then I(z) = Ω.

2.8 Active learning

ML community has been very innovative by selecting methods from different areas and combining them solve problems occurring in various areas (Jordan & Mitchell, 2015). If looked at from statistics angle, ML can be assigned to two views: Frequentist view and Bayesian view. Bayesian inference functions via probabilistic parametrisation and fixed data, while Frequentist view concerns fixed parameters working with stochastic data points. The Frequentist way, assumes not known parameters assumed as taking fixed not known values. It is not assuming random variables. Contrary to that, the Bayesian approach permits probabilities to be related to not known parameters. In ML research, we can recognize the paradigms that follow.

✓ Supervised learning (SL)
✓ Unsupervised learning (UL)
✓ Semi-supervised learning (SSL)
✓ Reinforcement learning (RL)
✓ Active learning (AL)

SL concerns the modeling of mapping function, $f: X \rightarrow Y$ formed from learning the relation between input data and their associated output (e.g., annotations of classes). The calibration set D = (X; Y) is used to learn the mapping (model) but carrying the minimum estimated loss. In case the output is of a category type,

SL is a characterization task; else it is a regression. Classification is supervised learning task where the annotations that we wish to anticipate take discrete values.

In UL, we are not given any output to use for training $\{D = x_1, \ldots, x_n\}$: UL includes clusters where we attempt to discover gatherings of information examples that are like one another. Loss criterion here is at times called objective criterion (e.g., K-Means calculation).

SSL, as its name infers, falls among UL and SL. It tends to be thought of as a class of directed discovering that likewise utilizes unlabeled information for training. SSL utilizes one of the accompanying presumptions.

- ✓ Smoothness assumption: directs which are close towards one another are bound to have a similar class value.
- ✓ Cluster assumption: the information will in general structure discrete groups and points in a similar cluster are bound to share a label (this is unique instance of the smoothness presumption)
- ✓ Manifold assumption: the information lies roughly on a complex of much lower dimensionality than the information space.

In RL, the yield is a reward given by the environment as indicated by its state because of certain gave information (activity) by the agent. Therefore, given a state s and an activity a, the domain response is a reward function R(s; a). In contrast to SL, where we get sources of info and yields from the environment and attempts to learn a model, RL specialist cooperates with the environment and attempts to learn a function f that maximizes cumulative rewards, where $f: A \times S \rightarrow R$ and A, S, R are the set of activities states and rewards individually.

AL is tied in with interfacing with an oracle to question the label of chosen information occasions. Rather than RL, the activities space is simply inquiries and the reward is the genuine name. Once acquired, the information occurrences and their labels are utilized to prepare SL calculations. We should bring up that AL can be utilized to pose inquiries of different structure than questioning certain information (Tong & Koller, 2000). Anyway in this exploration work, we consider AL that questions data labels for classification purposes. Then again, passive learning is a phrasing that will be utilized for any learner that gets latently the data provided (class name for our situation).

ML calculations can likewise be arranged by the manner in which information is handled. We can recognize two ideal models: online learning and offline learning. In online adapting, when the training stage is depleted, the model is not any more fit to take in further information from new occurrences, that is, it can't self-update later on. Contrarily, in online setting, information arrives in a type of streams and the model must foresee and adjust online, that is, it continues learning through time. For instance, a spam indicator gets online message and the errand is to foresee whether the messages are spam or not. The nature of the model's forecast is evaluated by a loss function that estimates the error between the anticipated mark and the genuine one. At that point, an adjustment of the model is performed if necessary.

Querying criteria

There exist three main approaches of AL (Settles, 2009) Membership Query Synthesis (MQS), Pool-based Selective Sampling (PSS) and Stream-based Selective Sampling (SSS). According to MQS, the learner generates new data samples from the feature space that will be queried. However, labeling such arbitrary instances may be impractical, especially if the oracle is a human annotator as we may end up querying instances (Baum & Lang, 1992) that are hard to explain. PSS is the most popular AL method, according to which the selection of instances is made by exhaustively searching in a large collection of unlabeled data gathered at once in a pool. Here, PSS evaluates and ranks the entire collection before selecting the best query. On the other hand, SSS scans through the data sequentially and makes query decisions individually. In the case of data streams, PSS is not appropriate, especially when memory or processing power is limited. It also assumes that the pool of data is stationary and uses the whole dataset. This will delay the adaptation and waste the resources. SSS, instead, adapts the classifier in real-time leading to fast adaptation. While there have been many AL studies on the offline variant, only few ones have investigated the online setting. Most of the AL sampling criteria have been first introduced for offline setting, then adapted to work online. Authors in (Lewis & Gale, 1994) introduce one of the most general frameworks for measuring informativeness, label uncertainty sampling criterion, where the queried instances are those which the model is most uncertain about their label. It has been since then used in many successful o_ine AL algorithms (Settles & Craven, 2008).

Another popular AL sampling criterion framework is the query-by-committee (Seung, Opper, & Sompolinsky, 1992a, 1992b). Here, a committee of models trained on the same dataset are maintained. They represent different hypotheses. The data label about which they most disagree is queried. To use the query-by-committee framework, one must construct a committee of models and have some measure of disagreement. Query-by-committee has shown both theoretical and empirical efficiency in several offline AL studies (Dagan & Engelson, 1995). Density-based is another AL sampling criterion that differs from uncertainty and query-by-committee in that it uses unlabelled data for measuring the instance informativeness (Settles & Craven, 2008). Density-based criterion assumes that the data instances in dense regions are more important. Many studies have involved density-based criterion by combining it with other criteria like uncertainty sampling and Query-by-committee sampling (Nguyen & Smeulders, 2004).

Authors in (Baram, Yaniv, & Luz, 2004) point out that there is no individual AL method superior to the others and that their combination could produce better results. A reasonable approach to combine an ensemble of active learning algorithms might be to evaluate their individual performance and dynamically switch to the best performer so far (Baram et al., 2004). This idea of rewarding is rooted in RL (Kaelbling, Littman, & Moore, 1996). There exists a classical trade-off in RL called the exploration/exploitation trade-off, which can be explained as follows. If we have already found a way to act in the domain that gives a reasonable reward, then is it better to continue exploiting the explored domain or should we try to explore a new domain in the hope that it may improve the reward (Tong, 2001). This concept dates back to the study of bandit problems in 1930s which are the most basic examples of sequential decision problems with an exploration-exploitation trade-off. Because of the strong theoretical foundation of bandit problems, they are exploited by many authors to formulate AL (Baram et al., 2004). However, none of them considers SSS active learning.

Online active learning

Online AL methods for data streams in presence of drift have been dealt with using batch-based learning, where data is divided into batches (Widyantoro & Yen, 2005). These methods assume that the data is stationary within each batch and then PSS AL strategies are applied. Authors in (Lindstrom, Delany, & Mac Namee, 2010) use a sliding window approach, where the oldest instances are discarded. Label uncertainty is then, used to label the most informative instances within each new batch. In (Ni et al., 2017) an online approach is compared against a batch based approach, finding that both have similar accuracy, but the batch-based one requires more resources. Moreover, the batch-based approach requires to specify the size of the batch. Some studies consider fixed size; Others use variable one. However, in both cases, more memory than the one with the online approach is needed. Another issue is that, in general, batch-based approaches cannot learn from the most recent examples until a new batch is full. This leads to more delay when responding to changes. This delay has a negative effect leading to late recovery. All these reasons make online learning more natural and suitable for AL.

Few SSS AL studies have been proposed (Attenberg & Provost, 2011). Methods i Online AL approaches that address the three data stream challenges: infinite length, concept drift and concept evolution are the rarest. Authors in (Masud, Gao, Khan, Han, & Thuraisingham, 2010) also deal with concept evolution and concept drift. They apply a hybrid batch-incremental learning approach, where the data is divided into fixed-size chunks and an offline classifier is trained from each chunk. An ensemble of M classifiers is maintained to classify the unlabeled data. To detect the novel classes, a clustering technique is used in order to isolate odd data instances. If the isolated samples are enough and sufficiently close to each other (coherent), they get queried. Otherwise, the algorithm considers them as noise. The algorithm also uses the uncertainty sampling within the current chunk to query the label of instances for which it is most uncertain. This algorithm does not address sampling bias and it needs to store the data instances in different batches.

Sampling bias

Sampling bias problem is, in general, associated with AL, where the sampled training set does not reflect on the underlying data distribution. Basically, AL seeks to query samples that, if labeled, significantly improve the learning. AL becomes increasingly confident about its sampling assessment. That confidence could lead, however, to negligence of valuable samples that reflect on the true data distribution. It, therefore, creates a bias towards a certain set of instances, which could become harmful. Sampling bias problem is more severe in online setting as the underlying classifier used by AL has to adapt. On the other hand, the adaptation can depend on the queried data. Hence, if drift occurs for samples, which the model is cofident about their labels, they will not be queried and the model will not be adapted.

Methods in (Chu, Zinkevich, Li, Thomas, & Tseng, 2011) takes sampling bias problem into account. In Žliobaitė, Bifet, Pfahringer, and Holmes (2013), SSS is adopted using randomization to avoid bias estimation of the class conditional distribution that may result from querying. This randomization is combined with label un- certainty to deal with concept drift. However, it results in wasting resources by randomly picking data to cover the whole input space. Moreover, randomization has no interaction when drift occurs and it naively keeps querying randomly. In (Chu et al., 2011) sampling bias is studied using importance weighting principle to reweight labeled data and remove the bias. Importance weighting principle has been theoretically proven to be effective (Beygelzimer, Dasgupta, & Langford, 2008). However, the method in (Chu et al., 2011) is restricted to binary classification. Both methods assume that the number of classes is fixed and known in advance, namely, there is no concept evolution. Furthermore, they ignore the effect of data distribution and do not benefit of abundant unlabeled data samples.

2.8.1 One class classification classifiers

One-class classifiers demonstrate the following features:

- The only available information concerns the target class;
- The determination of the threshold that defines the two classes is constrained on the estimation of data originating completely from the target class;
- A critical problem concerns the threshold definition around the target class having the objective to enclose the maximum of the target objects and simultaneously to exclude outliers.

A number of One Class classifiers have employed in the presented research:

- Support Vector Machines
- Autoencoder Neural Network
- Mixtures of Gaussians (MOG) and
- Self-Organizing Maps (SOM).

2.8.2 SVM based one-class classifier

One Class Support Vector Machines construct a model based on the target data. In the operational phase they allocate a new data by using the deviation from target data as a membership metric (Schölkopf, Platt, Shawe-Taylor, Smola, & Williamson, 2001). The value of the spread parameter determines the sensitivity of the classifier according to the kernel $K(\mathbf{x},\mathbf{z}) = \exp\{-\|\mathbf{x}-\mathbf{z}\|^2/\sigma^2\}$ which is more sensitive with a small spread in order to classify a data with nonlinear decision borders while a large spread works better for linearly separable data due to lower sensitivity.

2.8.3 Auto-encoder based one-class classifier

Auto-encoders (Japkowicz, Myers, & Gluck, 1995) are ANNs that are trained so they can reproduce input data as a mirror image in their output layer. In this classifier, only one hidden layer is used with less hidden neurons with sigmoidal transfer functions.

The main assumption in auto-encoder networks concerns the full reconstruction of objects from a target dataset resulting at a smaller error compared to outliers. In the special case of only one hidden layer, the network will find a similar solution to Principal Component Analysis (PCA) algorithm (Bourlard & Kamp, 1988). The network tries to achieve small reconstruction error and therefore manages to achieve that by developing a compact mapping from the input data to the subspace of the hidden neurons. The dimension of this subspace depends on the number of number of hidden neurons.

Auto-encoder networks offer many degrees of freedom but they demonstrate similar behavior to multilayer perceptron due to their similarity (Tax & Duin, 2001b). This behavior is mainly associated to improper selection of training parameters and reduced flexibility due to predefinition of the Auto-encoder construction.

2.8.4 MOG based one-class classifier

The simplest way for generating a one-class classifier is to estimate the topological variation of the training data samples (Tarassenko, Hayton, Cerneaz, & Brady, 1995) and accordingly to define a border delimiting this area while the same class data are situated. The optimal border defines a minimum volume connected to a model of a probability density of samples that belong to the target distribution. In order to obtain a model with more adaptive and sensitive modeling behavior, the basic spherical distribution can be extended to a mixture of Gaussians (MOG) which provides enhanced flexibility for modeling more complex data distributions (Duda, Hart, & Stork, 2001).

2.8.5 Augmentation of one class classifiers

A One Class Classifier is not flexible enough to be used for solving classification problems with multiple classes due to its inherent structure which is directed to One Class Classification. The augmentation of classified data of the target class and outliers allows performing a multi-step procedure where already classified data can be considered as data with a known class and data that do not have any label from a point of view that they do not belong to the already known target or outlier dataset. An iterative scheme where new data augmented with target data, while outlier data are augmented with an initial data in every step can lead to a multiclass classifier, which can classify an arbitrary number of classes. Such a classification scheme corresponds to Active Learning since the acquisition and the classification procedure are performed in an exploratory iteration. The following iteration outlines the Active Learning procedure:

1. The first training session concerned the calibration of the initial classifier that corresponds to the one class of the target data. After training, the resultant one-class classifier is validated with unknown data. The performance indicator is the successful classification of the new data as outlier with simultaneous classification of the new data belonging to the target set distribution.
2. At this phase, the new baseline set is the product of augmentation of the target set combined with outlier values from the already classified outlier class. The outlier detection process is continued but with the adaptation, concerning the new baseline set which is the augmented baseline of the already known targets and outliers. In the event that an unknown sample is classified as belonging to the augmented class then the procedure of outlier detection is repeated in order to classify internally in the augmented class aiming to find the identity of the new data sample with respect to the initial target dataset and the already known outlier class.
3. Steps 2 and 3 iterated. More precisely, the discovery of the outliers and the step of augmenting those takes place for newly added data that possibly are member of the existent categories or are outliers to the augmented classes.

A significant feature of the proposed active learning scheme concerns that the iterations between steps 1 to 3 do not require supervision but are based on outlier discovery and augmentation steps that are executed in an automatic manner.

2.9 Hierarchical self-organizing artificial neural networks

There is a need in practical applications to extend the original Self organizing Map with extra layers that are represented either as an output layer for Supervised learning or are more oriented to Unsupervised hierarchical extension to different layers or a combination of specialized SOMs in a higher level fusion SOM that is used for integration of the information originating from different sources.

Below a taxonomy of unsupervised HSOMs is presented, while in the following sections the Supervised HSOM architectures CPANN, SKN, XYF and the SOM with embedded output.

2.9.1 Taxonomy of unsupervised HSOM

There are essentially two explanations behind utilizing a Hierarchical SOM (HSOM) rather than a standard SOM:

- A HSOM can require less computer time than a standard SOM to accomplish certain objectives;
- A HSOM can be more qualified to model a problem that has, by its own tendency, some kind of multiple leveled structure.

The decrease of computational effort can be accomplished in two different ways: by diminishing the dimensionality of the input data to every SOM, and by lessening the quantity of units in every SOM. Rather than having a SOM that uses all components of the input data vectors, we may have a few SOMs, each utilizing a subset of those components, and along these lines we limit the impact of the "curse of dimensionality" (Shalev-Shwartz, 2012). The distance functions utilized for training the various SOMs will be less complex, and subsequently quicker to process. This simple design choice will more than counterbalance for the extra functions that must be processed. Speed increase can likewise be accomplished by utilizing less units in every SOM. The better refinement between various units can be accomplished in upper level SOMs that will just need to manage a portion of the input data. This "divide and conquer" methodology will abstain from employing Spatial Clustering Using Hierarchical SOM distances and neighborhoods to units that are altogether different from the input data being trained in every moment.

The second purpose behind utilizing HSOMs is that, they are more qualified to manage issues that present a hierarchical/thematic structure. In these cases, HSOM can map the natural structure of the problem, by utilizing an alternate SOM for each hierarchy level or specific plane. This partition of the global clustering or classification problem into various levels may not just represent the true structure of phenomena yet it might likewise give a simpler translation of the results, by enabling the user to perceive what clustering was performed at each level. GIS science applications, have a solid thematic structure that can be communicated with a custom specialized SOM for each thematic component, and an upper level (progressively predominant) SOM, that fuses the data to create globally characteristic clusters. HSOMs are regularly utilized in application fields where an organized decomposition into smaller and layered problems is helpful. A few models include: remote sensing classification (Lee & Ersoy, 2005), image compression (Barbalho et al., 2001), ontology (Chifu & Letia, 2008), speech recognition (Kasabov & Peev, 1994) pattern classification and extraction using wellbeing information, species information (Vallejo, Cody, & Taylor, 2007), financial information (Tsao & Chou, 2008), climate information (Salas et al., 2007), music information (Law & Phon-Amnuaisuk, 2008) and electric power information (Carpinteiro & Alves da Silva, 2001).

In view of the study of the work made on the field, we propose the accompanying taxonomy to group the HSOM strategies (Fig. 2.23). This is a conceivable taxonomy for the HSOM dependent on their target and on the sort of structure utilized.

Hence, the primary partition groups HSOM techniques in two fundamental sorts: the agglomerative and divisive HSOMs (Fig. 2.27). This partition results from the kind of methodology embraced in each HSOM technique. In an agglomerative HSOM, we for the most part have a few SOMs in the primary layer (i.e., the layer straightforwardly associated with the first data examples), and after fuse the activations in a higher level SOM, while in the divsive HSOM, we will normally have a solitary SOM in the main layer, and afterward have a few SOMs in the subsequent layer.

In the agglomerative HSOM (Fig. 2.28A), the degree of data abstraction increments as we advance up the hierarchy. Consequently, the principal level on the HSOM is the more detailed depiction (or a depiction of a specific part of the information) and, as we rise in the structure, the primary target is to make clusters that will be progressively broad and give a less complex method for visualizing the information.

In the divisive HSOM (Fig. 2.28B), the principal level is typically less exact and utilizes small networks. The primary goal of this level is to make rough partitions, which will be progressively detailed and precise as we rise in the levels of HSOM. In the second taxonomic level, agglomerative HSOMs can be split into thematic and dependent on clusters while

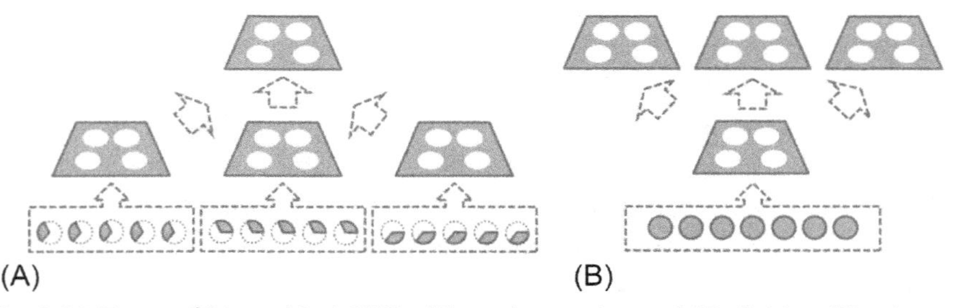

Fig. 2.28 Types of hierarchical SOMs: (A) agglomerative and (B) divisive. (Henriques, Lobo, Bação, & Johnsson, 2012).

divisive HSOMs can be split into static or dynamic. In the accompanying, we will introduce a depiction on every class.

2.9.1.1 Thematic agglomerative HSOM

The first class of agglomerative HSOMs is named Thematic. The name results from the way that the info space is viewed as a gathering of subspaces, every one shaping a theme. Fig. 2.29 presents a chart representing how HSOM strategies are commonly organized in this class.

In a thematic HSOM, the factors of the information examples are gathered by certain criteria, framing a variety of themes. For example, on account of census data, factors can be gathered into various themes, for example, financial, social, statistic or other. Every one of these subjects shapes a subspace that is then exhibited to a SOM, and its output activation will be utilized to train a last merging SOM. As officially expressed, the kind of output sent from the lower level SOM to the upper level can be different in various applications.

In Fig. 2.29, each theme is represented by a subset of the original variables. Accepting that every unique data pattern (with every one of its variables) would get represented to by a gray circle, a segment of that circle is utilized to represent the subset of every data example utilized in each theme. This structure exhibits a few advantages when performing multidimensional clustering. The first benefit is the decrease of calculation brought about by the partition of the info space into a few themes. This partition likewise permits the production of thematic clusters that, in essence, might intrigue the investigator. Consequently, since various clustering points of view are displayed in the lower level, these can be contrasted with the global clustering arrangement enabling the user to more readily comprehend and investigate the emerging patterns.

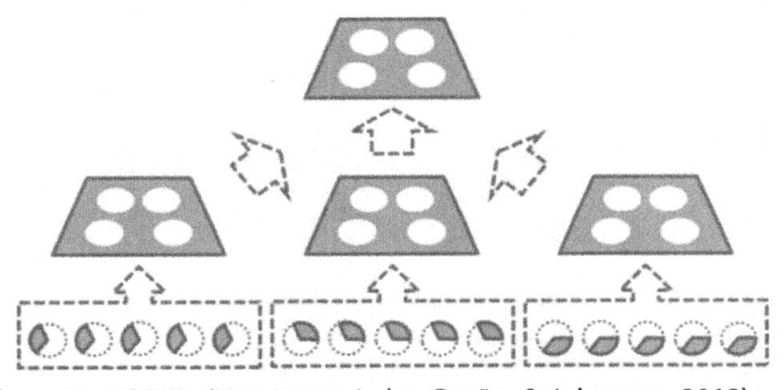

Fig. 2.29 Thematic HSOMs (Henriques, Lobo, Bação, & Johnsson, 2012).

2.9.1.2 Agglomerative HSOM based on clusters

This class is created by two levels, each utilizing a standard SOM (Fig. 2.30). The first level SOM is trained from the original input data, while its output is utilized in the second level SOM. The second level SOM is normally littler, permitting a coarser, however presumably simpler to utilize, definition of the clusters. In this design, if just the coordinates of the base level SOM are passed as contributions to the top level, every unit of the top level SOM is BMU for a few units from the first level. For this situation, the top level is essentially clustering together units of the first level, and the final outcome is like utilizing a smaller typical SOM. Be that as it may, this strategy has the benefit of showing two SOMs mapping the same data with various degrees of detail, without training the top level directly with the original data. Fig. 2.30 presents an illustration of this type of HSOM.

A HSOM dependent on clusters will be altogether unique in relation to a standard SOM if, rather than utilizing just the coordinates of the BMU, more data is passed as contribution to the top level. For instance, one may utilize both the coordinates and the quantization error of the input data as contributions to the top level. For this situation, the top level SOM will presumably cluster together examples that have high quantization error (for example patterns that are not represented enough) in the base level. In this manner the top level SOM could be utilized to recognize input data that, by

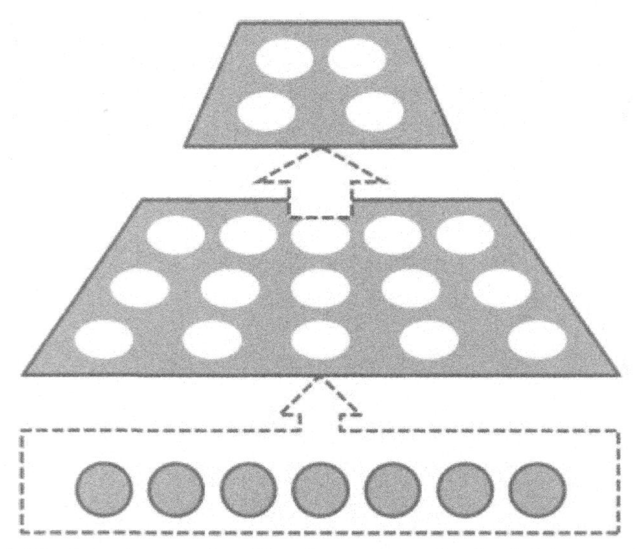

Fig. 2.30 HSOMs based on clusters (Henriques, Lobo, Bação, & Johnsson, 2012).

being distorted in the base level, require further consideration. The name proposed for this class (HSOM dependent on clusters) comes from the way that the base level SOM utilizes the full examples to form clusters, and the data about those clusters is the contribution to the top level SOM. Contingent upon what cluster information is passed on, the HSOM dependent on clusters might be comparable or altogether different from the standard SOM.

2.9.1.3 Static divisive SOM

In this class, the HSOM has a static structure, characterized by the user. The quantity of levels also, the associations between SOMs are predefined as indicated by the objective. Fig. 2.31 presents two instances of HSOM structures conceivable in this class.

In the main case (Fig. 2.31A) the base level SOM makes a rough partition of the dataset and, in a subsequent level, a SOM is made for every unit of the base level SOM. Every one of these second level SOMs get as information just the information examples presented by its origin unit in the base level that functions as a gating unit.

In the subsequent case (Fig. 2.31B), each top level SOM gets information from a few base level units. This permits various degrees of detail for various regions of the base level SOM. The main benefits of Static Divisive SOMs over enormous standard SOMs are the decrease of computational load because of the modest number of first level units (and just a portion of the top level units will be utilized for each situation), and the likelihood of having varying detail levels for various areas of the SOM. For instance, if we need to train a 100×100 unit SOM, we may utilize a base level SOM with 10×10 units, and a progression of 10×10 unit SOMs to build

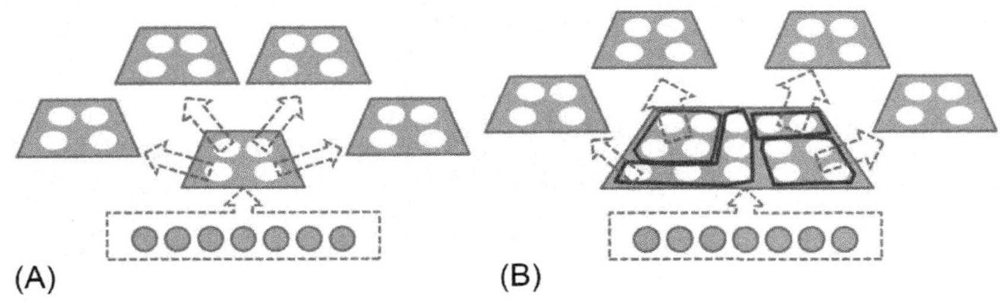

Fig. 2.31 Static HSOMs: (A) structure in which every unit will begin another SOM and (B) structure in which a gathering of units will start another SOM (Henriques, Lobo, Bação, & Johnsson, 2012).

a mosaic in the subsequent level. While each training sample will require the calculation of 10.000 distances in the first case, it will require just 100+100=200 distances in the second.

2.9.1.4 *Dynamic divisive HSOM*

At long last, the class of dynamic divisive HSOMs is described by the structure's self-adjustment to information. These techniques, otherwise called Growing HSOM (Dittenbach, Merkl, & Rauber, 2002), permit the development of the structure during the learning stage. Two kinds of development are permitted: horizontal and vertical development. The first concerns the expansion in the quantity of units of every SOM, while the second concerns the increment of the quantity of layers in the HSOM (Fig. 2.32). A depiction of this sort of HSOM appears in Fig. 2.32. The size of each level SOM and the quantity of levels is decided dynamically during the learning stage and depends on certain criteria, for example, the quantization error.

2.9.2 HSOM implementation and applications

One of the principal works identified with HSOM was proposed by Luttrell (1989). In his work, hierarchical vector quantization is proposed as a particular instance of multistage vector quantization. This work focuses on the

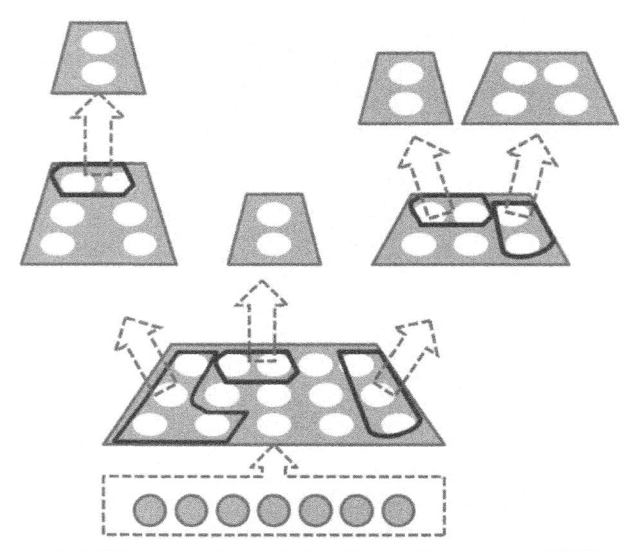

Fig. 2.32 Dynamic HSOMs (Henriques, Lobo, Bação, & Johnsson, 2012).

distinction in the data dimensionality among standard and multi-leveled vector quantization and demonstrates that distortion in a multistage encoder is limited by utilizing SOM. Lampinen and Oja (1992) examines the HSOM as a clustering instrument. The structure proposed depends on picking, for each data vector, the matching unit from the first level to train the subsequent level map. The first level produces numerous small scale clusters, while the second creates fewer more extensive and increasingly understandable clusters.

HSOM has demonstrated to be very important for training temporal data, regularly utilizing various time scales at various hierarchical levels. A model is crafted by Carpinteiro (1999), where the creators use HSOM to perform sequence classification in music and electric power load datasets. Another model is (Guimarães & Urfer, 2000) where HSOM is utilized to process rest apnea information.

Another class of HSOM is proposed in Dittenbach et al. (2002) with the Growing Hierarchical Self-Organizing Map (GHSOM). This neural system model is made out of a few SOMs, every one of which is permitted to develop vertically and horizontally. During the training procedure, until a given triggering condition is met, every SOM is permitted to develop in size (horizontal development) and the quantity of layers is permitted to develop (vertical development) to shape a layered architecture with the end goal that relations between input data examples are additionally detailed at the next levels of the structure. One of the problems of GHSOM is the meaning of the two conditions used to control the two kinds of development. A few authors proposed a few variations to this strategy to all the more likely characterize these criteria.

One model is the Enrich-GHSOM. Its primary contrast is the likelihood to drive the development of the hierarchy along some predefined routes. This model arranges information into a predefined ordered structure. Another case of a GHSOM variation is the RoFlex-HSOM expansion (Salas et al., 2007). This technique is fit to non-stationary time-dependent environments by fusing strength and adaptability in the gradual learning algorithm. RoFlex-HSOM demonstrates plasticity when finding the structure of the information, and step by step forgets (however not disastrously) past learned examples. Additionally, Pampalk, Widmer, and Chan (2004) proposed a Tension and Mapping Ratio expansion (TMR) to the GHSOM. Two new indices are presented, the mapping proportion (MR) and the strain (T) that will control the development of the GHSOM. MR estimates the proportion of data patterns that show signs of improvement spoken to by a virtual unit, set between two existent units. T estimates how comparative are the distances between every one of the units.

Another case of HSOM is proposed in Smolensky (1986) with the Hierarchical Overlapped SOM (HOSOM). The procedure begins by utilizing only one SOM. In the wake of finishing the unsupervised learning, every unit is annotated. At that point, a directed learning strategy is utilized (LVQ2) and units are consolidated or expelled, in light of the quantity of mapped patterns. After this, another LVQ2 is presented and, in light of the order quality, extra layers can be appended. The procedure is then rehashed for every one of these layers.

A comparative structure is introduced in Endo, Ueno, and Tanabe (2002), which proposes a cooperative learning scheme for the hierarchical SOM. In the base layer, some BMUs are chosen, and for every one of these BMUs a SOM in the subsequent layer is made. Information examples utilized in this second level SOM are emanating from the first BMU.

Ichiki, Hagiwara, and Nakagawa (1991) propose a hierarchical SOM that is ideal for managing semantic maps. In this proposition, each input pattern is created by two components: the attribute and the symbol, $X_i = X_{ai}\ X_{si}$. The attribute part X_{ai} is formed by the factors characterizing the input pattern, while the symbol part X_{si} is a binary vector. The base level SOM is trained utilizing the two pieces of the examples, while the second level SOM just uses the symbol set and data from the base level. HSOM has additionally been utilized for phoneme classification (Kasabov & Peev, 1994). The authors utilize sound sign traits in a first level SOM to characterize the phonemes into pause, vocalized phoneme, non vocalized phoneme, and fricative fragment. After phonemes are classified, a component frequency-scale vector is utilized to prepare the associated second level SOM.

An alternate methodology called tree structured topological feature map (TSTFM) is introduced in (Koikkalainen & Oja, 1990). This methodology utilizes a hierarchical structure to look for the BMU, in this way diminishing calculation times. While the reason for this methodology is mainly to decrease calculation times, its tree looking through procedure is in actuality a progression of static divisive HSOMs.

Miikkulainen (1990) suggests a hierarchical feature map to perceive an input story (content) as a case of a specific text by classifying it in three levels: scripts, tracks and role bindings. At the base level, a standard SOM is utilized for a coarse script classification. The second level SOMs gets just the input patterns similar to its contents (scripts), and various tracks are recognized at this level. At long last, in the third level a role recognition takes place.

At the lowest level, a standard SOM is used for a gross classification of the scripts. The second level SOMs receives only the input patterns relative to its scripts, and different tracks are classified at this level. Finally, in the third level a role classification is made.

2.9.3 Counter-propagation artificial neural networks (CP-ANNs)

CP-ANNs are considered as neural network algorithms which are capable of both supervised but and unsupervised feature learning (Zupan, Novič, & Gasteiger, 1995). They are comprised of a Kohonen and an output layer. The vector that models the classes defines a matrix C, with I rows and G columns. The I symbol shows the number of samples while G symbol indicates the classes (Fig. 2.33). The membership degree of an ith sample to the gth class is defined with a binary expression and is indicated by individual entries c_{ig} of C. In the event that consecutive training is executed, the rth weights of the neurons belonging to the output layer (y_r) are optimized in

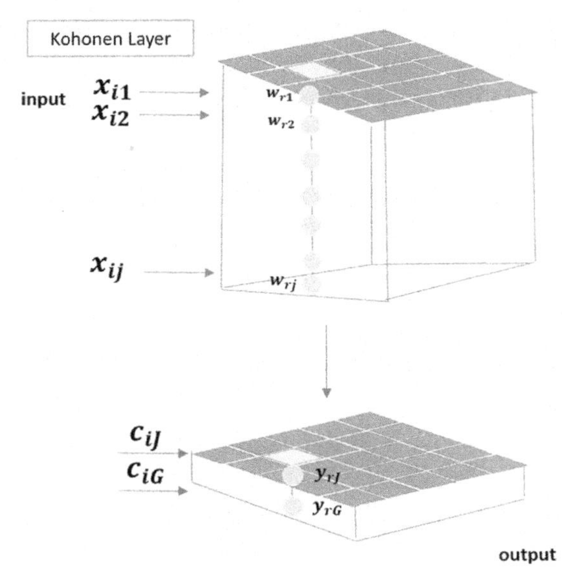

Fig. 2.33 CP-ANN depiction concerning a example dataset consisting of J variables and G classes. The symbols in the figure are given as follows: x_{ij} denotes the the jth variable for the ith sample, w_{rj} denotes the jth Kohonen weight for the rth neuron, c_{ig} denotes the degree of membership of the ith sample to the gth class taking binary value, and y_{rg} represents the gth output weight for the rth neuron.

a supervised manner (Ballabio & Vasighi, 2012). For each sample i, the update Δy_r is calculated based on the following equation:

$$\Delta y_r = \eta\left(1 - \frac{d_n}{d_{\max} + 1}\right)\left(c_i - y_r^{old}\right) \tag{2.44}$$

where d_{ri} symbolizes the grid based distance between neuron r and the winning neuron in the Kohonen layer; c_i represents the ith row of the matrix C denoting classes and a binary G-dimensional vector indicates class membership for the ith sample. At the final phase of training, each neuron, that belongs the Kohonen layer can be assigned a class label derived from output weights and correspondingly a new sample that correspond to that neuron will be assigned to the same class automatically.

2.9.4 Supervised Kohonen networks (SKNs)

The Supervised Kohonen Networks (SKNs) models (Melssen, Wehrens, & Buydens, 2006) are supervised neural networks demonstrating similarities with the CP-ANNs and XY-Fs, inspired by SOMs offering solutions for several classification problems. What differentiates the SKNs architecture is that the combined layer formed by augmenting the Kohonen and output layers which updated by using the training algorithm of SOMs. Each data sample (x_i) and the associated class vector (c_i) are intercorrelated and formulate a training input for the network. A normalization procedure must be applied xi and ci so as to guarantee an accurate prediction capability. Hence, a normalization factor concerning c_i is introduced so as control the class vector impact on the derivation of the model (Ballabio & Vasighi, 2012). (Fig. 2.34).

2.9.5 XY-fused networks (XY-F)

XY-fused Networks (XY-Fs) (Melssen et al., 2006) are viewed as supervised neural systems equipped for creating classifiers that are derived from SOMs. In these systems, the best matching neuron is dictated by evaluating the Euclidean distances between (a) input data vector (x_i) and the weights of the Kohonen layer, (b) the class membership vector (c_i) and the weights of the output layer. At that point, these two Euclidean distances are consolidated to make a fused metric, which is used for the best matching neuron selection (Ballabio & Vasighi, 2012) (Fig. 2.35).

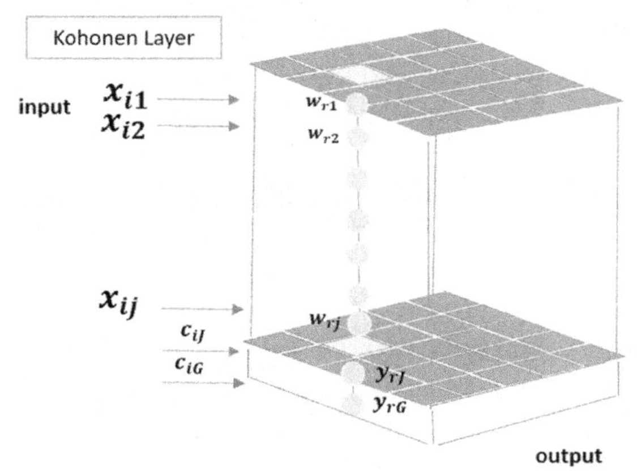

Fig. 2.34 Illustration of SKN for a model dataset that consists of J variables and G classes where x_{ij} is the value of the jth variable for the i-th sample, w_{rj} denotes the j-th Kohonen weight for the r-th unit, c_{ig} is the membership of the i-th data sample belonging to the gth class denoted with a binary code, and yrg represents the g-th output weight for the r-th unit.

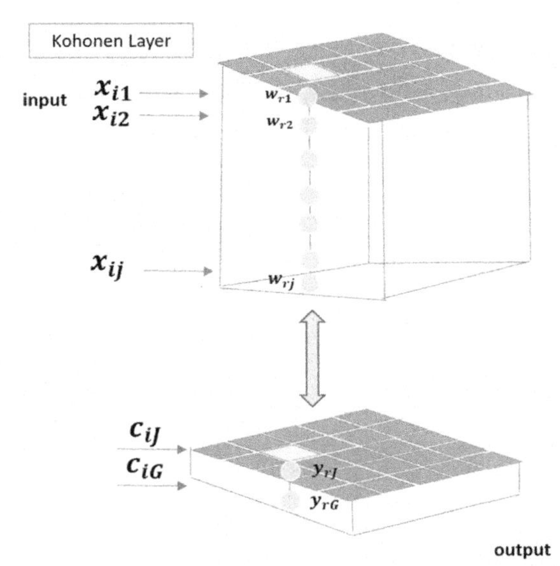

Fig. 2.35 Structure of XYF for a conventional dataset established by J variables and G classes. Symbolism in the figure alludes to symbols utilized in the text: x_{ij} means the estimation of the j_{th} variable for the i-th sample, w_{rj} means the estimation of the j-th Kohonen weight for the r-th neuron, c_{ig} means the degree of member of the $_{I}$-th test to the g_{th} class depicted as a binary code, and y_{rg} means the estimation of the g-th output weight for the r-th neuron.

2.9.6 Supervised self-organizing map with embedded output

When SOM is enhanced with output weights $\mathbf{w}_s^{(out)}$, it has the capability of supervised learning of the output mapping $\mathbf{y}=f(\mathbf{x})$ when encountering both training input and output data. This linking is illustrated in Fig. 2.36.

A training regime appropriate for training an extended map in a in a self-organizing mode (when compared to supervised) is to connect the self-organization phase to a randomly chosen input vector part, that will dominate the training process of Self -Organization (Moshou et al., 2004). The output component can be scaled by using a very small weight, much less than 1. By this inspiration, the input component influences maximally the self-organizing process driven by the selection of the best matching unit. In the operational environment the map is examined by a new data sample which is unknown, with an empty output part of the SOM. The SOM will select the best matching unit and recall the linked output that has been stored during the learning phase. The learning algorithms for the augmented weights is provided in Eq. (2.45).

$$\begin{bmatrix} \mathbf{\Delta w}_\mathbf{s}^{(in)} \\ \mathbf{\Delta w}_\mathbf{s}^{(out)} \end{bmatrix} = \varepsilon h \left(\begin{bmatrix} \mathbf{x} \\ a^*\mathbf{y} \end{bmatrix} - \begin{bmatrix} \mathbf{w}_\mathbf{s}^{(in)} \\ \mathbf{w}_\mathbf{s}^{(out)} \end{bmatrix} \right) \tag{2.45}$$

The tuning parameters ε and h denotes the learning increment and neighborhood kernel respectively. During the retrieval procedure, a new

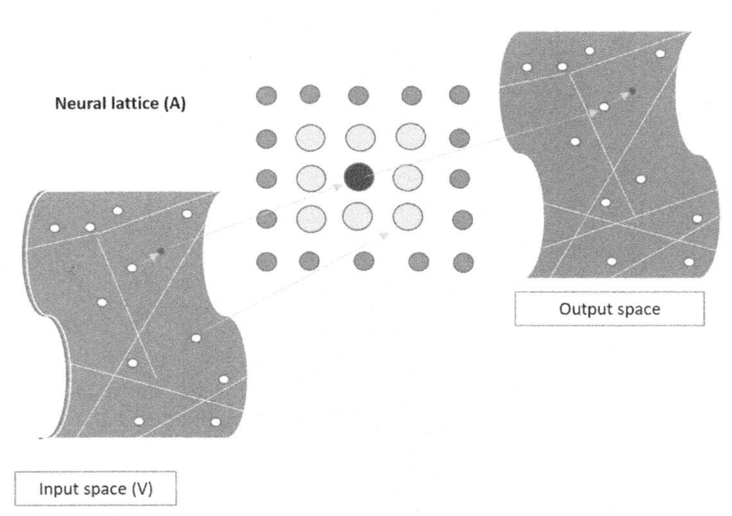

Fig. 2.36 Input-output mapping concept showing the training phase of a Self-Organizing Map.

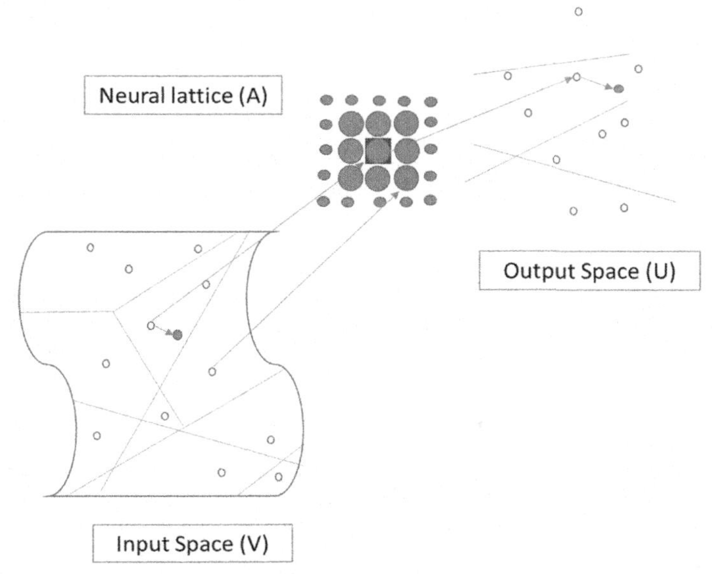

Fig. 2.37 The prediction of a learned output value of a function from the associated output space is retrieved from the competitive selection of the best matching neuron in the input.

input is introduced to the input map. The neuron in the input map that becomes selected based on the smallest Euclidean distance to the input vector is the winning neuron and through the mapping that has been learned during the training phase the associated output is retrieved. The whole retrieval process is illustrated in Fig. 2.37.

The output module can be employed for function approximation and time series prediction which is defined as a mapping associating high-dimensional vector spaces. The value prediction of a function output is driven by the competitive selection of unit (i) which associates the output component $\mathbf{w}_s^{(out)}$ of the neuron. Similar to simple classification, the prediction of associated output data, SSOM may be characterized as an extended version of SOM labelling method enhanced with a set of arbitrary metadata of high dimension.

References

Abe, S. (2005). *Support vector machines for pattern classification* (Vol. 2, p. 44). London: Springer.

Ackley, D. H., Hinton, G. E., & Sejnowski, T. J. (1985). A learning algorithm for Boltzmann machine. *Cognitive Science*, *9*, 147–169.

Agrawal, R., Gehrke, J., Gunopulos, D., & Raghavan, P. (2005). Automatic subspace clustering of high dimensional data. *Data Mining and Knowledge Discovery*, *11*, 5–33.
Ahmad, A., & Quegan, S. (2012). Analysis of maximum likelihood classification on multispectral data. *Applied Mathematical Sciences*, *6*, 6425–6436.
Albelwi, S., & Mahmood, A. (2017). A framework for designing the architectures of deep convolutional neural networks. *Entropy*, *19*, 242. https://doi.org/10.3390/e19060242.
Attenberg, J., & Provost, F. (2011). Online active inference and learning. In *Proceedings of the 17th ACM SIGKDD international conference on Knowledge discovery and data mining* (pp. 186–194): ACM.
Bai, Z., Huang, G. B., Wang, D. W., Wang, H., & Westover, M. B. (2014). Sparse extreme learning machine for classification. *IEEE Transactions on Cybernetics*, *44*, 1858–1870.
Bai, Q., & Zhang, M. (2006). Coordinating agent interactions under open environments. In J. Fulcher (Ed.), *Advances in Applied Artificial Intelligence* (pp. 52–67). Hershey, PA: Idea Group.
Ballabio, D., & Vasighi, M. (2012). A MATLAB toolbox for self organizing maps and supervised neural network learning strategies. *Chemometrics and Intelligent Laboratory Systems*, *118*, 24–32.
Baram, Y., Yaniv, R. E., & Luz, K. (2004). Online choice of active learning algorithms. *Journal of Machine Learning Research*, *5*(Mar), 255–291.
Barbalho, J. M., et al. (2001). Hierarchical SOM applied to image compression. *International Joint Conference on Neural Networks, IJCNN'01. 2001. Washington, DC.*
Baum, E. B., & Lang, K. (1992, November). Query learning can work poorly when a human oracle is used. *International joint conference on neural networks (Vol. 8, p. 8).*
Bazi, Y., & Melgani, F. (2006). Toward an optimal SVM classification system for hyperspectral remote sensing images. *IEEE Transactions on Geoscience and Remote Sensing*, *44*, 3374–3385.
Belgiu, M., & Drăguţ, L. (2016). Random forest in remote sensing: A review of applications and future directions. *ISPRS Journal of Photogrammetry and Remote Sensing*, *114*, 24–31.
Bergenti, F., Giezes, M. -P., & Zambonelli, F. (2004). *Methodologies and Software Engineering for Agent Systems: The Agent-Oriented Software Engineering Handbook*. Berlin: Springer-Verlag.
Beygelzimer, A., Dasgupta, S., & Langford, J. (2008). Importance weighted active learning. arXiv preprint arXiv:0812.4952.
Bishop, C. M. (2006a). *Pattern Recognition and Machine Learning*. Berlin: Springer.
Bishop, C. (2006b). *Pattern Recognition and Machine Learning, Technometrics*. New York: Springer-Verlag.
Botros, N. M., & Abdul-Aziz, M. (1994). Hardware implementation of an ANN using field programmable gate arrays (FPGAs). *IEEE Transactions on Industrial Electronics*, *41*(6), 665–667.
Bovolo, F., Bruzzone, L., & Marconcini, M. (2008). A novel approach to unsupervised change detection based on a semisupervised SVM and a similarity measure. *IEEE Transactions on Geoscience and Remote Sensing*, *46*(7), 2070–2082.
Boukerche, A., Juc'a, K. R. L., Sobral, J. B., & Notare, M. S. M. A. (2004). An artificial immune based intrusion detection model for computer and telecommunication systems. *Parallel Computing*, *30*(5–6), 629–646.
Bourlard, H., & Kamp, Y. (1988). Auto-association by multilayer perceptrons and singular value decomposition. *Biological Cybernetics*, *59*(4–5), 291–294.
Breiman, L., Friedman, J. H., Olshen, R. A., & Stone, C. J. (1984). *Classification and Regression Trees*. Wadsworth & Brooks/Cole Advanced Books & Software.
Breinman, L. (2001). Random forests. *Machine Learning*, *45*, 5–32.

Brooks, R. A. (1991). Intelligence without representation. *Artificial Intelligence, 47*(1–3), 139–159.

Butcher, J. B., Verstraeten, D., Schrauwen, B., Day, C. R., & Haycock, P. W. (2013). Reservoir computing and extreme learning machines for non-linear time-series data analysis. *Neural Networks, 38*, 76–89.

Carpinteiro, O. A. S. (1999). A hierarchical self-organizing map model for sequence recognition. *Neural Processing Letters, 9*(3), 209–220.

Carpinteiro, O. A. S., & Alves da Silva, A. P. (2001). A hierarchical self-organizing map model in short-term load forecasting. *Journal of Intelligent and Robotic Systems*, 105–113.

Celik, T. (2009). Unsupervised change detection in satellite images using principal component analysis and k-means clustering. *IEEE Geoscience and Remote Sensing Letters, 6*, 772–776.

Chang, N. B., & Bai, K. (2018). *Multisensor Data Fusion and Machine Learning for Environmental Remote Sensing*. CRC Press.

Chang, N. B., Bai, K. X., & Chen, C. F. (2015). Smart information reconstruction via time-space-spectrum continuum for cloud removal in satellite images. *IEEE Journal of Selected Topics in Applied Earth Observations, 99*, 1–19.

Chang, N. B., Han, M., Yao, W., & Chen, L. C. (2012). Remote sensing assessment of coastal land reclamation impact in Dalian, China, using high resolution SPOT images and support vector machine. In *Environmental remote sensing and systems analysis*. Boca Raton, FL: CRC Press.

Chen, H. Y., & Leou, J. J. (2012). Multispectral and multiresolution image fusion using particle swarm optimization. *Multimedia Tools and Applications, 60*, 495–518.

Chen, W., Li, X., Wang, Y., Chen, G., & Liu, S. (2014). Forested landslide detection using LiDAR data and the random forest algorithm: A case study of the Three Gorges, China. *Remote Sensing of Environment, 152*, 291–301.

Chen, C. -F., Son, N. -T., Chen, C. -R., & Chang, L. -Y. (2011). Wavelet filtering of time-series moderate resolution imaging spectroradiometer data for rice crop mapping using support vector machines and maximum likelihood classifier. *Journal of Applied Remote Sensing, 5*, 53525.

Chen, J., & Wang, R. (2007). A pairwise decision tree framework for hyperspectral classification. *International Journal of Remote Sensing, 28*, 2821–2830.

Chiang, S. -S., Chang, C. -I., & Ginsberg, I. W. (2001). Unsupervised target detection in hyperspectral images using projection pursuit. *IEEE Transactions on Geoscience and Remote Sensing, 39*, 1380–1391.

Chifu, E. S., & Letia, I. A. (2008). Text-based ontology enrichment using hierarchical self-organizing maps. In *Nature inspired Reasoning for the Semantic Web (NatuReS), Karlsruhe, Germany*.

Chu, W., Zinkevich, M., Li, L., Thomas, A., & Tseng, B. (2011). Unbiased online active learning in data streams. In *Proceedings of the 17th ACM SIGKDD international conference on Knowledge discovery and data mining (pp. 195–203)*: ACM.

Courant, R., & Hilbert, D. (1954). Methods of mathematical physics: Vol. I. *Physics Today*, 7, 17.

Dagan, I., & Engelson, S. P. (1995). Committee-based sampling for training probabilistic classifiers. In *Machine Learning Proceedings 1995* (pp. 150–157): Morgan Kaufmann.

DeChant, C., Wiesner-Hanks, T., Chen, S., Stewart, E. L., Yosinski, J., Gore, M. A., et al. (2017). Automated identification of northern leaf blight-infected maize plants from field imagery using deep learning. *Phytopathology, 107*(11), 1426–1432.

Dempster, A., & Laird, N. (1977). Maximum likelihood from incomplete data via the EM algorithm. *Journal of the Royal Statistical Society, Series B, 39*, 1–38.

Deng, W., Zheng, Q., & Chen, L. (2009). Regularized extreme learning machine. In *IEEE symposium on computational intelligence and data mining* (pp. 389–395).

Dittenbach, M., Merkl, D., & Rauber, A. (2002). Organizing and exploring high-dimensional data with the growing hierarchical self-organizing map. In *Proceedings of the 1st International Conference on Fuzzy Systems and Knowledge Discovery (FSKD 2002) Orchid Country Club*.

Duda, R. O., Hart, P. E., & Stork, D. G. (2001). *Pattern classification* (2nd ed). New York: John Wiley & Sons.

Ediriwickrema, J., & Khorram, S. (1997). Hierarchical maximum-likelihood classification for improved accuracies. *IEEE Transactions on Geoscience and Remote Sensing, 35*, 810–816.

Endo, M., Ueno, M., & Tanabe, T. (2002). A clustering method using hierarchical self-organizing maps. *The Journal of VLSI Signal Processing, 32*(1), 105–118.

Fan, H. (2013). Land-cover mapping in the Nujiang Grand Canyon: Integrating spectral, textural, and topographic data in a random forest classifier. *International Journal of Remote Sensing, 34*, 7545–7567.

Friedl, M. A., & Brodley, C. E. (1997). Decision tree classification of land cover from remotely sensed data. *Remote Sensing of Environment, 61*, 399–409.

Friedman, N., Linial, M., Nachman, I., & Pe'er, D. (2000). Using Bayesian networks to analyze expression data. *Journal of Computational Biology*, 7, 601–620.

Frizzelle, B. G., & Moody, A. (2001). Mapping continuous distributions of land cover: A comparison of maximum-likelihood estimation and artificial neural networks. *Photogrammetric Engineering & Remote Sensing, 67*, 693–705.

García-González, D. L., Mannina, L., D'Imperio, M., Segre, A. L., & Aparicio, R. (2004). Using 1H and 13C NMR techniques and artificial neural networks to detect the adulteration of olive oil with hazelnut oil. *European Food Research and Technology, 219*(5), 545–548.

Garšva, G. and Danenas, P., 2014. Particle swarm optimization for linear support vector machines based classifier selection. Nonlinear Analysis: Modelling and Control, 19, 26–42.

Ghulam, A., Porton, I., & Freeman, K. (2014). Detecting subcanopy invasive plant species in tropical rainforest by integrating optical and microwave (InSAR/PolInSAR) remote sensing data, and a decision tree algorithm. *ISPRS Journal of Photogrammetry and Remote Sensing, 88*, 174–192.

Grossberg, S. (2013). Recurrent neural networks. *Scholarpedia, 8*, 1888.

Guimarães, G., & Urfer, W. (2000). *Self-Organizing Maps and its Applications in Sleep Apnea Research and Molecular Genetics*. University of Dortmund-Statistics Department.

Guo, L., Chehata, N., Mallet, C., & Boukir, S. (2011). Relevance of airborne lidar and multispectral image data for urban scene classification using Random forests. *ISPRS Journal of Photogrammetry and Remote Sensing, 66*, 56–66.

Hagner, O., & Reese, H. (2007). A method for calibrated maximum likelihood classification of forest types. *Remote Sensing of Environment, 110*, 438–444.

Ham, J., Chen, Y., Crawford, M. M., & Ghosh, J. (2005). Investigation of the random forest framework for classification of hyperspectral data. *IEEE Transactions on Geoscience and Remote Sensing, 43*, 492–501.

Han, F., & Liu, H. (2014). High dimensional semiparametric scale-invariant principal component analysis. *IEEE Transactions on Pattern Analysis and Machine Intelligence, 36*, 2016–2032.

Hastie, T., Tibshirani, R., & Friedman, J. (2009). The elements of statistical learning, Methods. In *Springer Series in Statistics*. New York: Springer.

Henriques, R., Lobo, V., Bação, F., & Johnsson, M. (2012). Spatial clustering using hierarchical SOM. *Applications of Self-Organizing Maps*, 231–250.

Hinton, G. E. (2010). A practical guide to training restricted Boltzmann machines. *Momentum, 9*, 926.

Hinton, G. E., & Salakhutdinov, R. R. (2006). Reducing the dimensionality of data with neural networks. *Science, 313*, 504–507.

Hogland, J., Billor, N., & Anderson, N. (2013). Comparison of standard maximum likelihood classification and polytomous logistic regression used in remote sensing. *European Journal of Remote Sensing, 46*, 623–640.

Horata, P., Chiewchanwattana, S., & Sunat, K. (2013). Robust extreme learning machine. *Neurocomputing, 102*, 31–44.

Huang, C., Davis, L. S., & Townshend, J. R. G. (2002). An assessment of support vector machines for land cover classification. *International Journal of Remote Sensing, 23*, 725–749.

Huang, G. B., Zhu, Q. Y., & Siew, C. K. (2006a). Real-time learning capability of neural networks. *IEEE Transactions on Neural Networks, 17*, 863–878.

Huang, G. B., Zhu, Q. Y., & Siew, C. K. (2006b). Extreme learning machine: Theory and applications. *Neurocomputing, 70*, 489–501.

Huynh, H. T., & Won, Y. (2008). Weighted least squares scheme for reducing effects of outliers in regression based on extreme learning machine. *International Journal of Digital Content Technology and its Applications, 2*, 40–46.

Ichiki, H., Hagiwara, M., & Nakagawa, M. (1991). Self-organizing multilayer semantic maps. In *International joint conference on neural networks, IJCNN-91. Seattle.*

Jain, A. K., Mao, J., & Mohiuddin, K. M. (1996). Artificial neural networks: A tutorial. *Computer, 3*, 31–44.

Japkowicz, N., Myers, C., & Gluck, M. A. (1995). A novelty detection approach to classification. In *Proceedings of the fourteenth joint conference on artificial intelligence* (pp. 518–523).

Jay, S., & Guillaume, M. (2014). A novel maximum likelihood based method for mapping depth and water quality from hyperspectral remote-sensing data. *Remote Sensing of Environment, 147*, 121–132.

Jian, B., & Vemuri, B. C. (2011). Robust point set registration using Gaussian mixture models. *IEEE Transactions on Pattern Analysis and Machine Intelligence, 33*, 1633–1645.

Jolliffe, I. T. (2002). Principal component analysis. In *Springer Series in Statistics*. New York: Springer-Verlag.

Jordan, M. I., & Mitchell, T. M. (2015). Machine learning: Trends, perspectives, and prospects. *Science, 349*(6245), 255–260.

Kaelbling, L. P., Littman, M. L., & Moore, A. W. (1996). Reinforcement learning: A survey. *Journal of Artificial Intelligence Research, 4*, 237–285.

Kasabov, N., & Peev, E. (1994). Phoneme recognition with hierarchical self organised neural networks and fuzzy systems—a case study. In *Proceedings of ICANN'94, International Conference on Artificial Neural Networks, Springer.*

Kavzoglu, T., & Reis, S. (2008). Performance analysis of maximum likelihood and artificial neural network classifiers for training sets with mixed pixels. *GIScience & Remote Sensing, 45*, 330–342.

Kennedy, J., & Eberhart, R. (1995). Particle swarm optimization. In *vol. IV. Proceedings of the IEEE international conference on neural networks, Piscataway, NJ* (pp. 1942–1948): IEEE Press.

Kennedy, J., Eberhart, R. C., & Shi, Y. (2001). *Swarm Intelligence*. San Francisco, CA: Morgan Kaufmann Publishers.

Koikkalainen, P., & Oja, E. (1990). Self-organizing hierarchical feature maps. In *International Joint Conference on Neural Networks*. Washington, DC: IJCNN.

Kriegel, H. -P., Kröger, P., & Zimek, A. (2012). Subspace clustering. WIREs. *Data Mining and Knowledge Discovery, 2*, 351–364.

Krizhevsky, A., & Hinton, G. (2010). Convolutional deep belief networks on cifar-10. Unpublished Manuscript, 1–9.

Lampinen, J., & Oja, E. (1992). Clustering properties of hierarchical self-organizing maps. *Journal of Mathematical Imaging and Vision*, 261–272.

Law, E., & Phon-Amnuaisuk, S. (2008). Towards music fitness evaluation with the hierarchical SOM. In *Applications of evolutionary computing* (pp. 443–452).
Lee, J., & Ersoy, O. K. (2005). Classification of remote sensing data by multistage selforganizing maps with rejection schemes. In *Proceedings of 2nd international conference on recent advances in space technologies, RAST 2005. Istanbul, Turkey*.
Lewis, D. D., & Gale, W. A. (1994). A sequential algorithm for training text classifiers. In *SIGIR'94* (pp. 3–12). London: Springer.
Lian, H. (2012). On feature selection with principal component analysis for one-class SVM. *Pattern Recognition Letters, 33*, 1027–1031.
Lin, S. W., & Chen, S. C. (2009). PSOLDA: A particle swarm optimization approach for enhancing classification accuracy rate of linear discriminant analysis. *Applied Soft Computing, 9*, 1008–1015.
Lindstrom, P., Delany, S. J., & Mac Namee, B. (2010). Handling concept drift in a text data stream constrained by high labelling cost. In *Twenty-third international FLAIRS conference*, May.
Lipton, Z. C., Berkowitz, J., and Elkan, C., 2015. A critical review of recurrent neural networks for sequence learning. arXiv preprint arXiv:1506.00019.
Liu, Y., Zhang, D., & Lu, G. (2008). Region-based image retrieval with high-level semantics using decision tree learning. *Pattern Recognition, 41*, 2554–2570.
Luttrell, S. P. (1989). Hierarchical vector quantisation. *Communications, Speech and Vision, IEE Proceedings I, 136*(6), 405–413.
Makhzani, A., & Frey, B. J. (2015). Winner-take-all autoencoders. In *Advances in neural information processing systems* (pp. 2791–2799).
Masud, M. M., Gao, J., Khan, L., Han, J., & Thuraisingham, B. (2010). Classification and novel class detection in data streams with active mining. In *Pacific-Asia conference on knowledge discovery and data mining* (pp. 311–324). Berlin, Heidelberg: Springer.
McCulloch, W. S., & Pitts, W. (1943). A logical calculus of the ideas immanent in nervous activity. *Bulletin Mathematical Physics, 5*, 115–117.
Melssen, W., Wehrens, R., & Buydens, L. (2006). Supervised Kohonen networks for classification problems. *Chemometrics and Intelligent Laboratory Systems, 83*, 99–113.
Miikkulainen, R. (1990). Script recognition with hierarchical feature maps. *Connection Science, 2*(1), 83–101.
Modava, M., & Akbarizadeh, G. (2017). Coastline extraction from SAR images using spatial fuzzy clustering and the active contour method. *International Journal of Remote Sensing, 38*, 355–370.
Moshou, D., Bravo, C., Oberti, R., West, J., Bodria, L., McCartney, A., et al. (2005). Plant disease detection based on data fusion of hyper-spectral and multi-spectral fluorescence imaging using Kohonen maps. *Real-Time Imaging, 11*(2), 75–83.
Moshou, D., Chedad, A., Van Hirtum, A., De Baerdemaeker, J., Berckmans, D., & Ramon, H. (2001). Neural recognition system for swine cough. *Mathematics and Computers in Simulation, 56*(4–5), 475–487.
Moshou, D., Deprez, K., & Ramon, H. (2004). Prediction of spreading processes using a supervised Self-Organizing Map. *Mathematics and Computers in Simulation, 65*(1–2), 77–85.
Muñoz-Marí, J., Bovolo, F., Gómez-Chova, L., Bruzzone, L., & Camp-Valls, G. (2010). Semisupervised one-class support vector machines for classification of remote sensing data. *IEEE Transactions on Geoscience and Remote Sensing, 48*(8), 3188–3197.
Nguyen, H. T., & Smeulders, A. (2004). Active learning using pre-clustering. In *Proceedings of the twenty-first international conference on Machine learning* (p.79): ACM.
Ni, H., Lin, X., & Zhang, J. (2017). Classification of ALS point cloud with improved point cloud segmentation and random forests. *Remote Sensing, 9*, 288.
Otukei, J. R., & Blaschke, T. (2010). Land cover change assessment using decision trees, support vector machines and maximum likelihood classification algorithms. *International Journal of Applied Earth Observation and Geoinformation, 12*, S27–S31.

Pal, M. (2005). Random forest classifier for remote sensing classification. *International Journal of Remote Sensing, 26*, 217–222.

Pampalk, E., Widmer, G., & Chan, A. (2004). A new approach to hierarchical clustering and structuring of data with self-organizing maps. *Intelligent Data Analysis, 8*(2), 131–149.

Pan, L., & Yang, S. X. (2007). Analysing livestock farm odour using an adaptive neuro-fuzzy approach. *Biosystems Engineering, 97*(3), 387–393.

Pantazi, X. E., Moshou, D., Kateris, D., Gravalos, I., & Xyradakis, P. (2013). Automatic identification of gasoline–biofuel blend type in an internal combustion four-stroke engine based on unsupervised novelty detection and active learning. *Procedia Technology, 8*, 229–237.

Pfeifer, R., & Bongard, J. (2007). *How the body shapes the way we think: A new view of artificial intelligence*. Cambridge, MA: MIT Press.

Prabhakar, M., Prasad, Y. G., Thirupathi, M., Sreedevi, G., Dharajothi, B., & Venkateswarlu, B. (2011). Use of ground based hyperspectral remote sensing for detection of stress in cotton caused by leafhopper (Hemiptera: Cicadellidae). *Computers and Electronics in Agriculture, 79*(2), 189–198.

Prasad, S., & Bruce, L. M. (2008). Limitations of principal components analysis for hyperspectral target recognition. *IEEE Geoscience and Remote Sensing Letters, 5*, 625–629.

Rumpf, T., Mahlein, A. K., Steiner, U., Oerke, E. C., Dehne, H. W., & Plümer, L. (2010). Early detection and classification of plant diseases with support vector machines based on hyperspectral reflectance. *Computers and Electronics in Agriculture, 74*(1), 91–99.

Russell, S., & Norvig, P. (2001). *Artificial intelligence: A modern approach* (2nd ed.). Englewood Cliffs, NJ: Prentice Hall.

Sachs, A., Thiel, C., & Schwenker, F. (2006). One-class support-vector machines for the classification of bioacoustic time series. International Journal of Artificial Intelligence and. *Machine Learning, 6*(4), 29–34.

Safavian, S. R., & Landgrebe, D. (1991). A survey of decision tree classifier methodology. *IEEE Transactions on Systems, Man, and Cybernetics: Systems, 21*, 660–674.

Salakhutdinov, R., & Hinton, G. (2009). Semantic hashing. *International Journal of Approximate Reasoning, 50*(7), 969–978.

Salas, R., et al. (2007). A robust and flexible model of hierarchical self-organizing maps for non-stationary environments. *Neurocomputing, 70*(16–18), 2744–2757.

Schölkopf, B., Platt, J. C., Shawe-Taylor, J., Smola, A. J., & Williamson, R. C. (2001). Estimating the support of a high-dimensional distribution. *Neural Computation, 13*(7), 1443–1471.

Schölkopf, B., Smola, A., & Müller, K. R. (1997). Kernel principal component analysis. In *International conference on artificial neural networks, Lausanne, Switzerland* (pp. 583–588).

Schwenker, F., Kestler, H. A., & Palm, G. (2001). Three learning phases for radial-basis-function networks. *Neural Networks, 14*, 439–458.

Settles, B. (2009). *Active learning literature survey*. University of Wisconsin-Madison, Department of Computer Sciences.

Settles, B., & Craven, M. (2008). An analysis of active learning strategies for sequence labeling tasks. In *Proceedings of the conference on empirical methods in natural language processing* (pp. 1070–1079): Association for Computational Linguistics.

Seung, H. S., Opper, M., & Sompolinsky, H. (1992a, July). Query by committee. In *Proceedings of the fifth annual workshop on computational learning theory* (pp. 287–294): ACM.

Seung, H. S., Opper, M., & Sompolinsky, H. (1992b, July). Query by committee. In *Proceedings of the fifth annual workshop on computational learning theory* (pp. 287–294): ACM.

Shalev-Shwartz, S. (2012). Online learning and online convex optimization. *Foundations and Trends® in Machine Learning, 4*(2), 107–194.

Smolensky, P., 1986. Information Processing in Dynamical Systems: Foundations of Harmony Theory. Technical report, DTIC Document.

Stumpf, A., & Kerle, N. (2011). Object-oriented mapping of landslides using random forests. *Remote Sensing of Environment, 115*, 2564–2577.

Sun, W., Zhang, L., Du, B., Li, W., & Lai, M. Y. (2015). Band selection using improved sparse subspace clustering for hyperspectral imagery classification. *IEEE Journal of Selected Topics in Applied Earth Observations and Remote Sensing, 8*, 2784–2797.

Tang, J., Deng, C., & Huang, G. B. (2016). Extreme learning machine for multilayer perceptron. *IEEE Transactions on Neural Networks and Learning Systems, 27*, 809–821.

Tarassenko, L., Hayton, P., Cerneaz, N., & Brady, M. (1995). Novelty detection for the identification of masses in mammograms. *Proceedings fourth IEE international conference on artificial neural networks, Cambridge* (pp. 442–447).

Tax, D. M., & Duin, R. P. (2001a). Uniform object generation for optimizing one-class classifiers. *Journal of Machine Learning Research, 2*(Dec), 155–173.

Tax, D. M., & Duin, R. P. (2001b). Combining one-class classifiers. In *International Workshop on Multiple Classifier Systems* (pp. 299–308). Berlin, Heidelberg: Springer.

Tax, D., & Duin, R. (2004). Support vector data description. *Machine Learning, 54*(1), 45–66.

Tian, X., Jiao, L., & Zhang, X. (2013). A clustering algorithm with optimized multiscale spatial texture information: Application to SAR image segmentation. *International Journal of Remote Sensing, 34*, 1111–1126.

Tian, S., Zhang, X., Tian, J., & Sun, Q. (2016). Random forest classification of wetland land covers from multi-sensor data in the arid region of Xinjiang, China. *Remote Sensing, 8*, 954.

Tokunaga, M., & Thuy Vu, T. (2007). Clustering method to extract buildings from airborne laser data. In *IEEE 2007 international geoscience and remote sensing symposium, Barcelona, Spain* (pp. 2018–2021).

Simon Tong. Active learning: Theory and applications. PhD thesis, Citeseer, 2001.

Tong, D., & Koller, D. (2000). Active learning for parameter estimation in Bayesian networks. In *NIPS*.

Tsao, C. Y., & Chou, C. H. (2008). Discovering intraday price patterns by using hierarchical self-organizing maps. In *JCIS-2008 Proceedings, Advances in Intelligent Systems Research*. Shenzhen: China Atlantis Press.

Vallejo, E., Cody, M., & Taylor, C. (2007). Unsupervised acoustic classification of bird species using hierarchical self-organizing maps. *Progress in Artificial Life*, 212–221.

Vapnik, V. (1998). *Statistical learning theory*. New York: John Wiley and Sons, Inc.

Von Neumann, J. (1958). *The computer and the brain*. New Haven, CT: Yale University Press.

Wang, C., Liu, Y., Chen, Y., & Wei, Y. (2015). Self-adapting hybrid strategy particle swarm optimization algorithm. *Soft Computing, 20*, 4933–4963.

Wang, H., & Raj, B. (2017). *On the origin of deep learning*. arXiv:1702.07800v4. https://arxiv.org/pdf/1702.07800.pdf Accessed on July 2017.

Wang, J., & Chang, C. -I. (2006). Applications of independent component analysis in endmember extraction and abundance quantification for hyperspectral imagery. *IEEE Transactions on Geoscience and Remote Sensing, 44*, 2601–2616.

Wehrens, R., & Buydens, L. M. C. (2007). Self- and super-organizing maps in R: The Kohonen package. *Journal of Statistical Software, 21*, 1–19.

Widyantoro, D. H., & Yen, J. (2005). Relevant data expansion for learning concept drift from sparsely labeled data. *IEEE Transactions on Knowledge and Data Engineering, 17*(3), 401–412.

Wiener, N. (1948). *Cybernetics*. New York, NY: Wiley.

Winham, S. J., Freimuth, R. R., & Biernacka, J. M. (2013). A weighted random forests approach to improve predictive performance. *Statistical Analysis and Data Mining, 6*, 496–505.

Wu, J., Chen, Y., Dai, D., Chen, S., & Wang, X. (2017). Clustering-based geometrical structure retrieval of man-made target in SAR images. *IEEE Geoscience and Remote Sensing Letters, 14*, 279–283.

Yu, X., Yang, J., & Zhang, J. (2012). A transductive support vector machine algorithm based on spectral clustering. *AASRI Procedia, 1*, 384–388.
Zeiler, M. D., & Fergus, R. (2014). Visualizing and understanding convolutional networks. In *European conference on computer vision* (pp. 818–833). Cham: Springer.
Zhang, X., Jiao, L., Liu, F., Bo, L., & Gong, M. (2008). Spectral clustering ensemble applied to SAR image segmentation. *IEEE Transactions on Geoscience and Remote Sensing, 46*, 2126–2136.
Zhao, Z. P., Li, P., & Xu, X. Z. (2013). Forecasting model of coal mine water inrush based on extreme learning machine. *Applied Mathematics & Information Sciences*, 7, 1243–1250.
Žliobaitė, I., Bifet, A., Pfahringer, B., & Holmes, G. (2013). Active learning with drifting streaming data. *IEEE Transactions on Neural Networks and Learning Systems, 25*(1), 27–39.
Zupan, J., Novič, M., & Gasteiger, J. (1995). Neural networks with counter-propagation learning strategy used for modelling. *Chemometrics and Intelligent Laboratory Systems, 27*(2), 175–187.

Further reading

Bouneffouf, D., Laroche, R., Urvoy, T., Féraud, R., & Allesiardo, R. (2014). Contextual bandit for active learning: Active Thompson sampling. In *International Conference on Neural Information Processing* (pp. 405–412). Cham: Springer.
Friedman, N., Geiger, D., & Goldszmidt, M. (1997). Bayesian network classifiers. *Machine Learning, 29*, 131–163.
Ganti, R., & Gray, A. (2013). *Building bridges: Viewing active learning from the multi-armed bandit lens*. arXiv preprint arXiv:1309.6830, 2013.
McCallumzy, A. K., & Nigamy, K. (1998). Employing EM and pool-based active learning for text classification. In *Proceedings of international conference on machine learning (ICML)*, (pp. 359–367).
Penny, W. D., & Roberts, S. J. (1999). EEG-based communication via dynamic neural network models. In *Vol. 5. IJCNN'99. International joint conference on neural networks. Proceedings (Cat. No. 99CH36339)* (pp. 3586–3590): IEEE.
Read, J., Bifet, A., Holmes, G., & Pfahringer, B. (2012). Scalable and efficient multi-label classification for evolving data streams. *Machine Learning, 88*(1–2), 243–272.
Sculley, D. (2007). Online active learning methods for fast label-efficient spam filtering. In *Vol. 7. CEAS* (p. 143).
Settles, B. (2010). *Active learning literature survey*. Madison: University of Wisconsin.
Suganthan, P. N. (1999). Hierarchical overlapped SOM's for pattern classification. *IEEE Transactions on Neural Networks, 10*(1), 193–196.
Wu, X., Kumar, V., Ross Quinlan, J., Ghosh, J., Yang, Q., Motoda, H., et al. (2008). Top 10 algorithms in data mining. *Knowledge and Information Systems, 14*, 1–37.
Zheng, C., et al. (2007). Hierarchical SOMs: Segmentation of cell-migration images. In *Advances in neural networks-ISNN 2007* (pp. 938–946).

CHAPTER 3

Utilization of multisensors and data fusion in precision agriculture

Contents

3.1 The necessity of multisensors network utilization for agriculture monitoring and control

Multisensory networks in agriculture are mostly designed aiming to serve different needs regarding event detection and timely intervention for site-specific operations like monitoring and control. These type of networks are often deployed in order to gather enough information so as to give actionable insights that can enable timely intervention in the case of preventing situations leading to yield loss or quality degradation. The processing affects the relevance of the data to the final result. So from this point of view, information fusion can be defined as the discipline that maximizes the information content of the combined data in order to reach improved inferences

https://doi.org/10.1016/B978-0-12-814391-9.00003-0

compared to handling the gathered data in isolation. To sum up, data fusion in precision agriculture comprises of the efficient and proper combination of numerous and heterogeneous data from different sensors aiming to provide actionable insights, which lead to decisions for most cost effective and sustainable crop management adaptation.

The necessity to offer a customized solution with the help of multisensory networks depends heavily on the application objectives. Inevitably, this affects the type of processing that needs to be applied as well as the delivery and assessment. The effective combination of multisensory networking and several information fusion algorithms, are capable of reducing the amount of data traffic leading to bottlenecks avoidance due to extracting features or decisions deriving from the raw data, filter noisy measurements for improving predictions and deriving the context of crop status and gaining insight on the type intervention that is needed per case.

A variety of heterogeneous sensors produces responses that are constrained by the nature of each sensor resulting in different information "windows" from the observed phenomenon. Sensor fission occurs in the event that heterogeneous sensors windows "overlap" timely or spatially. In this occasion, both sensors' character and their relative observations can describe sufficiently the phenomenon. The information content of sensors is filtered in such a way that the signal characteristics are altered or becoming noisy, data fusion seeks to retrieve the crucial aspects of the phenomenon that allows the phenomenon prediction with high accuracy.

3.1.1 Proximal sensing

In contrast with remote sensing techniques, in proximal sensing the sensor's distance from the observed object is less than a defined threshold which is usually in the range of few meters (Viscarra Rossel, Adamchuk, Sudduth, McKenzie, & Lobsey, 2011). Similarly for soil and plant sensing, proximal sensing can be realized as a handheld apparatus, or vehicle mounted (in this case the sensing is performed while a vehicle that carries the sensor, scans a field). In spite of the fact that the handheld version is not difficult to utilize, it is laborious and ineffective from a data collection point of view. contrasted with those gathered by the on-line versions, which can end up collecting 1000 readings/ha (Mouazen & Kuang, 2016). This is fundamental to investigate the inside field spatial variability at high resolution to ensure effective site specific administration of both soil and crop utilizing variable rate approaches.

3.1.2 Proximal crop sensors

Recently, proximal crop sensing has become highly appreciated as a way to assessing crop health status in natural, forestry and agricultural environments. In situ strategies incorporate leaf-level to canopy level estimations from different devices including portable sensors, static equipment (e.g., tripods), platforms and vehicles that carry the sensors in the field area. Furthermore, laboratory proximal sensors can assess the crop health status for the efficient agrarian administration.

A few proximal crop sensors are being utilized for serving research or business operations, through adapting optical, laser scanners and ultrasound sensing technologies. The most ordinarily utilized sensing techniques in proximal crop sensing are based on optical sensing principles. Data obtained utilizing ordinary cameras are 2D projections of this present reality in a single frequency (monochromatic) or multiple frequencies (applied to multifrequency cameras). RGB information usually concerns the red, green and blue wavebands, mimicking the human vision.

Information at other, non-perceptive frequencies captured by sensors, for example, a Color Infra-Red (CIR) camera, which gets multispectral information at certain VIS-NIR frequencies. The standard method to visualize these wavebands is utilizing separate dark white pictures or false-shading pictures.

Thermography is a method that acquires 2D warm radiation pictures, covering the range from 3 to 100 μm. Contingent upon its temperature and emissivity, the item under examination produces a specific warm radiation. The crop's temperature can be assessed by the warm radiation caught by a thermal camera, provided that emissivity is already known. This temperature is highly related to the plant's stomatal conductance. At the point when the leaf stomata shut during daytime, basic metabolic activities of the plant such as the circulation of CO_2 and H_2O within the plant, provoke an inner nursery impact in the leaf area. A worldwide increment in leaf temperature can be considered normal in the occasion that warm radiation is high (for example, due to sunlight or halogen illumination, and so on).

TIR radiation sensing can help with estimating the biotic and abiotic crop stresses occurrence though identifying temperature alterations (Alchanatis et al., 2010). Crops under stress tend to have a higher temperature contrasted with unstressed crops. For instance, the utilization of thermal cameras for water administration relies upon the way that the crop canopy surface temperature is a relation of transpiration rate, and that, itself, is an element of atmospheric evaporative capacity and crop accessible soil water availability (Khanal, Fulton, & Shearer, 2017).

A common reaction of plants is the closure of their stomata when are subjected to water stress, subsequently this condition increases their temperature. Thermal imaging employs a special lens that concentrates the infrared radiation (IR) radiated by plant canopy or different plants areas. Nevertheless, plants have the tendency to shut their stomata when they are experiencing water stress and by this, their temperature increases. Other ecological factors, for example, environmental temperature, daylight, precipitation, or wind speed can influence canopy or leaf temperature, thus limiting its functional activities (Mahlein, 2016). At long last, by documenting the forced excitation of the plant's photosynthesis apparatus and the sensing of the reactions, knowledge about its metabolic condition can be acquired. This is generally accomplished through fluorescence. Fluorescence is the emanation of light after electromagnetic radiation was consumed at a lower wavelength. The chlorophyll complex belongs to a part of the plant taking part to the fluorescence procedure. Illuminating the chloroplasts with blue or actinic light (420 nm) will bring about some re-discharge of the retained light by the chlorophyll.

The proportion of re-transmitted light to illumination, is variable and relies upon the capacity of the plant to use the reaped light. The photosynthetic process in supposed in dark-adapted leaves is equal to zero. The plant's respiration is done though mitochondrial operation. At the point when the light-harvesting complexes (HCL) of the chloroplasts are all of a sudden be hit with an extremely solid blue light shaft, they become totally saturated. The energized chlorophyll can't discharge its excitation energy immediately into the photons of lower energy, often appearing as red outflows. This fluorescence plainly relies upon the chlorophyll capacity and is considered the main indicator of the plant's potential to assimilate actinic light.

Also, consolidating an actinic light source with brief saturating blue pulses, it is conceivable to assess the plant's potential of photograph absorption, non-photochemical extinguishing (comprising a proportion of the thermal dispersal of the excited energy) and other physiological plant parameters. There are two kinds of fluorescence that can be discovered. The first one is the blue-green fluorescence which is obtained at the range of 400–600 nm, while the second is the chlorophyll fluorescence between 650 and 800 nm. Chlorophyll fluorescence is legitimately identified with the rate of photosynthesis as is a significant crop stress indicators (Zwiggelaar, 1998).

During the most recent 20 years, laser-induced fluorescence has been utilized for vegetative examinations, for example, crop status monitoring and crop metabolic activities (Belasque Jr, Gasparoto, & Marcassa, 2008). Crop sensors using reflectance principles are applied widely for canopy level assessment, while fluorescence is received for leaf scale estimation. Obviously, varieties of the crops spectral signatures demonstrate that reflectance sensors can recognize biotic and abiotic stresses for operational and effective crop management (Lee et al., 2010). Along these lines, the discourse beneath is centered around optical strategies for discovery of possible abiotic and biotic stresses.

3.1.2.1 Proximal sensors for crop biotic stresses

The current section concerns the weed detection and crop infections.

Weed detection

Weed detection has demonstrated to be industrially suitable though implying proximal sensing with Weedseeker® (Trimble Ag). The Weedseeker® (Hanks & Beck, 1998) utilizes a functioning light sensor producing red (R) and NIR radiation, and finds alterations in reflectance from exposed ground versus weeds between crop rows lines (Sui, Thomasson, Hanks, & Wooten, 2008). However, the Weedseeker® is not capable of discriminating weeds from main crops or weed species from another. Weedseeker® is best used to direct pre-plant spot uses of herbicides.

The way of thinking of utilizing Weedseeker® is opposite to traditional herbicide use, which is based on uniform application with glyphosate into crops that have developed resistance. This regular administration technique is risky, due to expanding resistance of weeds. Weedseeker® innovation has effectively demonstrated that herbicide application rates can be decreased by 90% in respect to uniform application rates, along these lines lessening herbicide expenses and environmental impact (https://www.agrioptics.co.nz/portfolio/weedseeker/).

A subsequent way to deal with local weed detection is computer imaging, which depends on shape recognition. Weeds and plants have various shapes, and dependent on this distinction, weed detection is conceivably realistic though image processing techniques. Weed detection incorporates two stages: (a) recognize weeds that have attacked a field and separate them from the crops and (b) identify the weed species, if different weed species have been set up in the field, so as to have the option to utilize the correct herbicide for application.

While it may be anything but difficult to detect weeds, the recognition of their species isn't a simple task because of a wide range of weed classes. Image processing has the efficiency of separating among weeds and crop plants and is an undeniably used technique, as its results become increasingly exact. In particular, computerized imaging by utilization of optical sensors (different kinds of cameras) gives the capacity to online detection location with enhanced performance (Pantazi, Moshou, Alexandridis, Whetton, & Mouazen, 2016) and, thus, result in precise weed treatment (Tillett, Hague, Grundy, & Dedousis, 2008). Digital images facilitate the formation of customized algorithms for the detection of specific weed classes recognition (Bakhshipour, Jafari, Nassiri, & Zare, 2017).

Crop recognition is a complementary task with respect to weed identification. In specific circumstances, the assignment is less complicated as crops have a uniform outlook contrasted with the weeds and morphology is less variable. Such a methodology is utilized for site explicit herbicide treatment (Lee, Slaughter, & Giles, 1999) just as mechanical weeding systems, following that the on-line weeding capacity is running progressively and the crops must be protected (Van der Weide et al., 2008).

Such a methodology is utilized for site explicit herbicide application (Lee et al., 1999) just as mechanical weeding approach, following that the on-line weeding process is running perpetually and the crops must be avoided by keeping a safe distance from them (Van der Weide et al., 2008). The detection of crop depends on fusing shading and textural features, prepared through classifiers, for instance 2D wavelets (Tillett et al., 2008). In the investigation of Lottes, Hörferlin, Sander, and Stachniss (2017) image processing was utilized to apply a two-strategy approach, to be specific, keypoint-based arrangement and Object Based Image Analysis examination (OBIA), to effectively separate between sugar beets and different weeds.

Following a variation, crop spatial location can be defined and enlisted when seeding or transplanting. Be that as it may, to precisely accomplish that, constant RTK-GPS locating with high exactness is required (Sun et al., 2009). Precision at the range of 2–3 cm have been accomplished in deciding the crop position, empowering this technique to be utilized for site explicit herbicide application. Be that as it may, one of the most broadly examined subjects is the weed detection in acquired pictures and the right identification of the species. To accomplish this, the morphological or spectral features of the weeds are utilized as training sets for AI classifiers. Sometimes, the location of the plants relative to the rows is utilized as an additional feature for picking the class (crop or weed) (Shrestha, Steward, &

Birrell, 2004). In the event that position of the rows isn't available, shading and morphological attributes supplement the feature set.

The reflectance of plant leaves has been utilized in a few published works so as to discriminate crops from weeds. In research facility conditions it has been demonstrated that crops and weeds can be recognized and, additionally, weed species classification can be accomplished, by utilizing plant spectral reflectance as an indicator (Borregaard, Nielsen, Nørgaard, & Have, 2000). Under field conditions, spectral reflectance is integrated because of many disturbing factors (corrupted from soil, stems and leaf shadows), adversely influencing the weed detection procedure, bringing about basic affirmation of territories contaminated by weeds, inside which species recognition is not feasible (Goel, Prasher, Patel, Smith, & DiTommaso, 2002). Proof exists that spectral features enable the classification for explicit scenarios of crop/weed sets. Different wavelengths can be used for each crop/weed pair to accomplish classification.

Data attained during different crop seasons are regarded as necessary for investigating a fixed pattern of spectral classifiers to accomplish recognition recognition for field plants (Zwiggelaar, 1998). In another report, canopy reflectance procured by a VIS-NIR spectrometer was employed by Shirzadifar, Bajwa, Mireei, Howatt, and Nowatzki (2018) to recognize diverse weed species.

The VIS-NIR bands of the spectrum was assessed for its efficiency to separate different weed species. Critical wavebands in NIR area (indicated at 1078, 1435, 1490, and 1615 nm) gave the best outcomes to effective characterization, alongside with the red and red-edge in the VIS region (indicated at 640, 676, 730 nm).

The recognition of weed species utilizing spectral imaging (Pantazi et al., 2017b) and their subsequent class mapping (Pantazi et al., 2017a), has a critical impact on sustainable farming because of weeds difficulty to be separated and isolated from crops. AI methods supplied with data attained from sensory systems can find effectively the location of weed patches, keeping away from environmental impact and adverse responses. Data mining techniques implemented for weed species classification, empower advance in weed handling techniques and automated weeding including both mechanical and chemical, which limits the herbicide necessity (Pantazi et al., 2016).

Plant disease detection in field conditions

Any infection that causes enough plant stress to corrupt the reflectance features of the crop leaf is a possible case for proximal detection.

Taubenhaus, Ezekiel, and Neblette (1929) as early as the late 1920s, utilized a standard portable camera furnished with panchromatic film (highly contrasting) to take pictures of cotton fields swarmed with cotton root rot. The high contrast photos were utilized to find root spots of Phymatotrichum in various sizes and shapes in the field. All the more as of late, Pantazi, Moshou, and Tamouridou (2019) utilized an automated method for crop disease classification on different leaf pictures comparing to various crop species, utilizing Local Binary Patterns (LBPs) for feature extraction and One Class SVMs. The proposed system utilizes a committed One Class Classifier for each plant health status, including induct, downy mildew, powdery mildew and black rot. The algorithm was calibrated on vine leaves and then it was tested in a variety of crops accomplishing an extremely high prediction performance when evaluated in different crops.

The feasibility of detection of plant disease under field conditions by proximal sensors has shown high evidence. Relevant data is transmitted by light or in more general terms by electromagnetic radiation. The standard emerges from healthy crops and how they behave with respect to VIS and IR light in other spectral bands (the way they retain, reflect, produce, transmit and fluoresce) in comparison with contaminated crops. Plants have various reactions to light, which are alluded to as optical characteristics. Certain optical attributes are noticeable by the naked human eye, which can detect just VIS part of the electromagnetic range, or then again, transducers that can make visible different pieces of the range not obvious by the human eye.

Disease detection uses the range of spectral bands from crop canopies (or isolated leaves), which is the outcome of different physical procedures interfacing with the incoming light inside the plant and progressively specific plant issues. The reflected light has a unique signature as a spectral reflectance bend in the VIS range between 400–700 nm, the NIR area (700–1200 nm) and the SWIR (1200–2400 nm). The interaction incoming light with plants changes with wavelength. Healthy leaves demonstrate the corresponding responses to emitted light:

- Diminished reflectance at VIS wavelengths due to high absorbance of photoactive pigments.
- Increased reflectance in NIR due to different scattering occurring that happens at the air-cell interface inside the leaf tissue.
- Decreased reflectance in the SWIR, coming about because of water absorption, and carbon based substances (Jacquemoud & Ustin, 2001).

– Healthy leaves contain high water concentrations (emissivity range is somewhere in the range of 0.97 and 0.99), accordingly they carry the phenology of "dark bodies", emanating radiation at TIR band ≈ 10 μm proportionally to their temperature.

Alterations of plant leaf reflectance are related with leaf structural changes. Diseases reflect optical attributes of leaves at a few wavelengths, henceforth, disease detection techniques can be performed dependent on light reflection for various wavelengths or by fusing wavelengths through chemometric models. Intact crops are commonly green since green wavelength (ca. 550 nm) is reflected by photoactive pigments, as opposed to the blue and red wavelengths that are ingested during photosynthesis.

Plants with a contamination have local lesions, which are due to cell breakdown, and relatively raised reflection in the VIS range, all the more explicitly at chlorophyll absorbance wavelengths. In detail, the reflectance proportion changes at the waveband at roughly 670 nm, influence the red edge (characterized as the intense change between low reflectance at VIS wavelengths and the high NIR reflectance, happening at around 730 nm) moving it towards higher recurrence or lower wavelengths in the range. Despite what might be expected, the bringing down of the biomass related with senescence is connected with diminished development and thus leaf area, which thusly brings about less canopy reflection in the NIR wavelengths.

The most significant estimating systems for the location of foliar diseases are demonstrated quickly in Fig. 3.1. Fluorescence is the most fitting system during the initial phase of disease since it records the health status regarding photosynthetic efficiency. After the early metabolic changes, the disease radially spreads around the infected spots. By this stage, the underlying corruption of the infected area loses pigmentation, the photosynthetic system gets degraded and the cell walls breakdown. At this point, disease symptoms become obvious and the light reflection analysis is capable of identifying any possible infested area at this stage.

Pathogen propagules might be recognized in different regions of the spectrum, contingent upon the effect of the pathogen: (1) the VIS and red-edge chlorophyll corruption occurring between 650 and 720 nm); (2) the VIS-NIR senescence is detectable at 680–800 nm because of the brown color and the decreased leaf thickness, and (3) SWIR (1400–1600 nm and 1900–2100 nm) because of dryness. The disease steadily assumes responsibility for the whole plant, which will demonstrate a global stress that prompts a

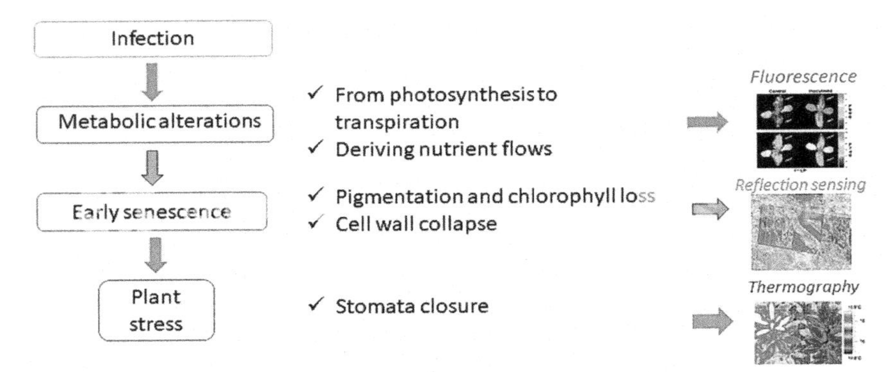

Fig. 3.1 Acquisition methods that offering solutions at various infection phases of a foliar infection.

general shut down of the stomata so as to diminish water loss. This transmission alteration can be seen with the help of through thermography. The overall leaf temperature changes quickly, in any case, and is intensely reliant on surrounding temperature, lighting and wind. In this manner, thermography may give poor outcomes utilizing proximal detecting stages in light of changing natural variables. The strategies give increasingly solid outcomes if illnesses are fully developed and infections are high. Certain alterations in the spectral features of plants profoundly offer the possibility to utilize optical signals to distinguish the occurrences of diseases in crops.

Disease detection using light reflection

Provided that the initial impacts of different infections are different shift, affecting the plant's chlorophyll, water content, and the leaf temperature, various wavebands can be utilized to recognize various diseases (Dudka, Langton, Shuler, Kurle, & Grau, 1999). For instance, changes in reflectance of violet-blue and NIR wavebands (380–450 nm and 750–1200 nm) were utilized to identify early contaminations of cucumber leaves by the fungus called *Colletotrichum orbiculare* (Sasaki, Okamoto, Imou, & Torii, 1998). Contaminated plants characterized by reflectance alterations in chlorophyll absorption wavelengths (470 nm and 670 nm) and, with lower noteworthiness, in the NIR spectral range. Afterward, plants were confirmed as contaminated when obvious symptoms showed up.

Early finding of indications in Nicotiana debneyi plants at different phases of tomato mosaic tobamovirus disease was made utilizing leaf spectral reflectance estimations (Polischuk et al., 1997). These spectral reflectance estimations could recognize a decrease of their chlorophyll content inside 10 days after innoculation, albeit discrete differences between healthy and disease plant. A few techniques for disease evaluation were assessed for downy mildew in quinoa, distinguishing reflectance estimations in the red (640–660 nm) and NIR (790–810 nm) to furnish the highest correlation with yield loss (Danielsen & Munk, 2004). Plant defoliation is the pathogen's fundamental effect, which is featured in the previously mentioned wavelengths. Tall fescue reflectance in the 810 nm band demonstrated the highest correlation ($R^2 = 19$–63%) with Rhizoctonia blight and grey leaf spot visual severity assessment (Green, Burpee, & Stevenson, 1998).

Malthus and Madeira (1993) utilized field spectra to investigate the impact of the necrotrophic parasite *Botrytis fabae* on field bean leaves (*Vicia faba*). Diseased plants were described by decreased photosynthetic activity because of diminished stomatal conductance. Notwithstanding, before visual side effects were observed, no alterations in unearthly reflectance were apparent. The most significant changes in the unearthly reflectance related with the disease were the decrease of the reaction in the VIS locale and the decrease of the NIR reflectance shoulder to 800 nm. Both these reflectance changes can be credited to the breakdown of the leaf cell structure with the spread of the fungus.

Muhammed and Larsolle (2003) examined the spectra of wheat and presumed that the parasite *Drechslera tritici-repentis* primarily influenced the spectral signature by flattening of the green reflectance crest together with a general decrease of reflectance in the NIR area and bringing down the shoulder of the NIR reflectance level together with a general increment in the VIS region (550–750 nm). Devadas, Lamb, Simpfendorfer, and Backhouse (2009) attempted to distinguish various types of wheat rust utilizing narrowband indices. To begin with, they employed the Anthocyanin Reflectance Index (ARI) to isolate intact, yellow rust and intermingled stem rust/leaf rust classes. They also employed the altered chlorophyll absorption and reflectance file (TCARI) to discriminate leaf from stem rust classes.

Their mix could give methods for wheat rust species separation. The Photochemical Reflectance File (PRI) was demonstrated an exceptionally vigorous spectral ratio list for yellow rust contamination evaluation (Huang et al., 2007). The adequacy of the PRI has been clarified by the way that it is profoundly correlated with biomass and foliar N concentration, and contrarily associated with occurrence of rust. The disease occurrence can likewise be assessed in a indirect way. Utilizing canopy reflectance at 810 nm, models had the option to justify 12–15% deviations in alfa alfa yield and LAI parameter more than that anticipated with a model dependent on rate defoliation set up through visual evaluation as stand alone variable (Guan & Nutter Jr, 2002). Defoliation was brought about by pandemics of foliar illness and the absence of fungicide spraying.

Like parasites and bacterial disease, insects assault can likewise be recognized with reflectance sensors. During insects assault on crops, the plants pass on exceptionally quick and senescence occurs, clarifying the radical spectral changes that can be observed. Yang and Cheng (2001) noticed that the blue, red and NIR spectral areas were exceptionally subject to the pervasion of brown plant hoppers in rice crop, while green bands were totally untouched. At the point when plant hoppers eat the rice leaves, the inquiry here is whether the deliberate spectral signature is absolutely bug related or a blend of soil and senescent leaves whose reflectance raises with infestation. For wheat crop infected by *Schizaphis graminum Rondani* and *Diuraphis noxia Mordvilko*, noticeable outcomes have been achieved (Riedell & Blackmer, 1999)

Early detection of grapevine leafroll sickness can be distinguished even at a pre-symptomatic stage (Naidu, Perry, Pierce, & Mekuria, 2009). This was for the most part because of the negative impact of the virus infection on the plant physiology bringing about changes in metabolic activities and pigmentation.

In outline, there are a few factors that control the adequacy of disease recognition methods that depend on light reflection, for example, changes in pigments concentrations other than chlorophyll, chlorophyll degradation and the structure change of the plant tissue, all have distinctive randomly appearing effects on the spectral features of the contaminated leaf regions. Because of the time between the disease and the breakdown of the inner leaf structure, correlated features that use NIR reflectance are relied upon to be less effective at early period of contamination. The phase of the disease and the spread of its relevant symptoms is along these lines a vital factor for identifying diseases and assessing them dependent on light reflection systems.

The viability of the identification of infections relies upon the algorithms utilized for data processing. Hyperspectral imaging strategies can be utilized for each crop disease framework to improve and naturally identify diseases. This should be possible through simple ANN models. Moshou, Bravo, West, McCartney, and Ramon (2004) and Bravo, Moshou, West, McCartney, and Ramon (2003) utilized image processing algorithms to separate among background and wheat canopy by using their reflectance at 675 nm and 750 nm and after that to recognize intact leaf tissue and yellow rust disease injuries in wheat crop by labeling mixes of spectral wavebands (West et al., 2003). Higher accuracy was accomplished thought applying multisensor fusion of spectral and fluorescence features from chosen wavebands of a spectrograph (Moshou et al., 2005).

These information were fused with lesion indices originating because of fluorescence imaging of the same plants. The development of a ground-based continuous proximal sensing system that can be propelled by tractors or robotic platforms, has been presented by Moshou et al. (2011). This platform makes feasible the recognition of crop infections in arable crops in an automated way naturally at an initial stage of infection by utilizing a multipoint spectral camera and a multispectral camera. The recognition between concurrent yellow rust contamination and N stress has been demonstrated by utilizing hyperspectral imaging and ANNs by Moshou et al. (2006). Enhanced precision for recognizing disease from N stress has been performed by ANN based models like hierarchical SOMs (Pantazi et al., 2017).

Smartphone apps for proximal sensing of biotic stresses

Farmers, confronted the basic requirement for early detection of biotic stresses incurred by weeds, bugs and infections, have to a great extent picked to depend on yield exploring approaches, with the special case (referenced already) of weed identification utilizing Weedseeker. Crop exploring is upgraded by the presentation of a wide selection of free and low cost, simple to utilize and user friendly applications for with smartphones (Logan et al., 2018). Smartphones are ideally fit to help crop scouts as they stroll through fields, since they can take georeferenced high resolution photos, which can be contrasted with photographic libraries of known weed, insects or disease stresses. There are various degrees of refinement for smartphones versatile applications to identify biotic stresses.

The most basic concern applications that allows the client to query to photographic libraries for explicit weeds, bugs or diseases. This methodology necessitates that the ability of crop scout in looking at field photographs of the unknown stress with library based photographs. Instances of these incorporate IOS or Android operating system applications, for example, ID Weeds (Meng & Bradley, 2016), which enables clients to look photographs for a particular weed or enter phenological attributes of a weed so as to let the application distinguish it. The University of Nebraska in the USA built up an application, which permits crop scouts to enter the quantity of plants infested by aphids to decide if the quantity of aphids surpasses a limit of 250 aphids for each plant (Ohnesorg, Hunt, & Wright, 2011).

South Dakota State University in the USA built up an application to help crop scouts distinguish diseases in soybeans (Osborne & Deneke, 2010). This application incorporates photographs of basic crop diseases and a calendar demonstrating the most likely time period regarding the occurrence of the infection. More advanced applications like these have additionally been produced for use in developing countries (Patel & Patel, 2016). Progressively complex applications ordinarily include the harvest scout to take photographs of the weed, insects or diseased plants (Pongnumkul, Chaovalit, & Surasvadi, 2015). These photographs are sent to a local specialist who at that point analyzes the assess the severity and disease type. Shortcomings of this methodology incorporate the requirement for high quality photographs, uniform lighting conditions and nonappearance o interference from glare or shadows. What's more, delays can appear as the local expert will be unable to assess the photographs.

Proximal sensors to measure crop abiotic stresses

The primary causes of abiotic stress are the absence of water, extremely high and low temperatures and insufficiency of crop nutrition (e.g., N, P and K). Plant reactions to abiotic stress are frequently fundamentally the same as the reactions to biotic stress, for example discoloration of leaves such a yellowing and browning, loss of leaves, delayed growth, and diminished crop yield. The principle focal point of this segment is on photography and imaging and non-imaging (point) spectroscopy procedures for the estimation of abiotic stresses.

3.1.2.2 Basic principles of visible, near infrared and mid infrared spectroscopy

In the course of recent years, diffuse reflectance spectroscopy has attracted more interest in soil examination. This is because of the event actualities that spectral reflectance techniques are quick, financially savvy, non-destructive, and give spectra that are highly indicative for the dirt soil type and texture, thus enabling the examination of many soil properties (Kuang et al., 2012). Moreover, the two VIS-NIR (400–2500 nm) and MIR (2500–25,000 nm) diffuse reflectance spectroscopy are techniques with low environmental impact, as they evade the utilization of unsafe extractants in the examination. They require insignificant or no sample processing (Soriano-Disla, Janik, Viscarra Rossel, Macdonald, & McLaughlin, 2014), and are effectively versatile for field soil conditions, particularly the VIS-NIR spectroscopy (Kuang et al., 2012).

Since soils are a blend of mineral and Organic Matter (OM) with a physical structure made out of aggregates visible totals of particles, which may contain water or air (Ben-Dor et al., 2009), they are an exceedingly light absorbing and dispersing medium. Once a soil medium is exposed to a light source, some portion of the light is absorbed, and part is diffusely reflected out of the soil. The final appearance of soil spectra either in the VIS-NIR or MIR is reflection of both the light diffusion and absorption that varies because of the sample physical and chemical, separately.

3.1.2.3 X-ray fluorescence (XRF) spectroscopy

XRF spectrometry is one of the most effective proximal soil detecting strategies that can surpass the limitations of traditional systems (Soodan, Pakade, Nagpal, & Katnoria, 2014). The energy and intensity of the fluorescence radiation in a given sample are utilized to perceive and decide their concentrations (Kaniu, Angeyo, Mwala, & Mangala, 2012). At the point when a solitary molecule is energized by an outside energy source, it discharges X-Ray photons of a specific wavelength. Components present in a given sample might be distinguished and quantitated by tallying the quantity of photons of every energy produced (Gates, 2006). Fluorescence is a type of iridescence happening when the produced light has a more extended wavelength (lower energy) than the absorbed radiation (Weindorf, Bakr, & Zhu, 2014).

Natural distinguishing proof in a given example depends on the connection between emission wavelength and the atomic number, while the elementary concentration can be assessed from the line intensities (Bosco, 2013). XRF has numerous advantages as it is quick, non-destructive, and compact, which is fundamental for in situ estimation. Estimation performance increments with element consideration. XRF has additionally a few impediments as it is volatile to matrix effects and moderately costly to run; in addition its investigation is slower contrasted with VIS-NIR and MIR spectroscopy.

With contemporary technical advancements, portable XRF (PXRF) spectrometry has now turned out to be accessible, for on location, fast and financially savvy estimation of soil contaminants (Horta et al., 2015). It needs a negligible example processing, being quick and non-destructive, multi-components can be estimated in one run, and it can give moderately low recognition limits (Parsons et al., 2013). PXRF was accounted to investigate a huge scope of elements from F to U, in spite of the fact that it is more qualified for the examination of highly homogeneous materials.

It has been frequently utilized for elemental investigation including contaminants in soils, despite the fact that its utilization for physicochemical soil properties and macronutrients (K, P, Mg, Ca, and S) is as yet constrained. A generally modest number of concentrates revealed the utilization of XRF for measuring soil properties for rural use. For instance, Weindorf, Herrero, Castañeda, Bakr, and Swanhart (2013) utilized PXRF spectroscopy, as an intermediary for direct $CaSO_4$ measurement with extraordinary achievement ($R^2 = 0.91$). Swanhart et al. (2013) utilized PXRF to assess salinity of soils attained from coastal areas in Louisiana, USA utilizing chlorine (Cl) as an intermediary, detailing strong connections between PXRF Cl and saturated paste electrical conductivity ($R^2 = 0.83$) and between PXRF Cl, S, K, Ca and saturated paste electrical conductivity ($R^2 = 0.866$).

Sharma, Weindorf, Man, Aldabaa, and Chakraborty (2014) assessed the possibility of utilizing PXRF for pH determination utilizing elemental data as a proxy for soil pH. A MLR with auxiliary input of CC, sand substance and OM gave the best predictive model ($R^2 = 0.825$, RMSE$= 0.54$), while a MLR with pure PXRF data came about in deteriorated results ($R^2 = 0.772$, RMSE $= 0.685$). Sharma, Weindorf, Wang, and Chakraborty (2015) researched the capability of XRF for soil CEC forecast by utilizing 450 soil tests from California and Nebraska, USA. Numerous linear models with and without auxiliary data has a similar accuracy, with auxiliary input model yielding a better R^2 (0.92 versus 0.90) and marginally lower RMSE (2.23 versus 2.49), contrasted with pure elemental data models. Independent validation datasets were considered credible for both pure elemental models ($R^2 = 0.90$) and auxiliary input models ($R^2 = 0.95$).

Zhu, Weindorf, and Zhang (2011) researched the practicability of predicting soil clay, silt and sand capacities from PXRF data using 584 soil samples acquired from two different locations in Louisiana and north New Mexico, (USA), reaching the conclusion that PXRF may be a promising tool for fast assessment of soil texture fractions with high performance where R^2 and RMSE are ranging from 0.682 to 0.975 and from 2.68% to 6.26%, respectively. Generally, PXRF instruments are found in the market with ready calibrations for various soil properties. But, like within the VIS-NIR or MIR spectroscopic analysis, mentioned before, XRF spectra are often employed to develop customer-built calibration, using statistics or machine learning tools.

XRF spectroscopic analysis yielded inferior accuracy for the estimation of the low-Z components (i.e., K, P, Ca, Mg) (Kaniu, Angeyo, Mwala, & Mwangi, 2012), which might be attributed to the complicated nature of the soil structure that puts more challenges on XRF analysis like the high background signal levels (Weindorf et al., 2014). As a result, the advanced chemometrics and data mining techniques that may lower the spectral noise and handle spectral interferences and also the advanced soil matrix effects are required (Kaniu, Angeyo, Mwala, & Mwangi, 2012). However, the chemometrics and data mining strategies haven't drawn enough attention in XRF spectroscopic analysis, wherever additional analysis is required to indicate the particular potential of this novel technology that employs such high performance data analysis tools.

3.1.2.4 Gamma ray spectroscopy

Gamma beam is a generally new soil detecting strategy that estimates gamma radiation transmitted from the normal decay of radioactive isotopes that are available in all soils (Viscarra Rossel, Taylor, & McBratney, 2007). Gamma-beam spectrometers measure the distribution of the power of gamma (y) radiation versus the energy of every photon. Gamma beam sensors might be either dynamic, utilizing a radioactive source (e.g., 137Cs) or passive sensors. The last version is increasingly appealing, and they are financially accessible and vary as indicated by the detector type that comprises of various scintillation crystals, including bismuth germanate (BGO), sodium iodide (NaI), and caesium iodide (CsI) (Kuang et al., 2012).

The radiometer dependent on the CsI detector is a powerful option in contrast to both NaI and BGO ones. The BGO is characterized by low peak resolution, while the CsI is of high thickness contrasted with NaI and shows better discovery sensitivity, particularly for stones of smaller size (Van Egmond, Loonstra, & Limburg, 2010). Data about mineralogy, weathering, and chemical synthesis of soil can be acquired. Around 95% of the deliberate gamma beams are produced from the upper 0.5 m of soil (Taylor, Smettem, Pracilio, & Verboom, 2002). Gamma-beam spectrometers ordinarily measure from 256 to 512 channels that involve a energy range extending between 0 and 3 MeV.

The energy groups at 1.4, 1.7, and 2.55 MeV denote the diagnostic energy peaks corresponding to individual elements such as total potassium (K), uranium (U) and thorium (Th), individually, which are attributed to the physical decay of these components. The radiation that doesn't come from the earth's surface is considered as background and is filtered out during data processing. The primary causes of background radiation are climatic radon (222Rn), cosmic radiation and instrumental.

The working rule of VIS-NIR reflectance comes from the vibrations of elemental bonds by exciting mineral blends in this way making reverberation vibration of various modes, such as twisting or extending. Vibrations bring about light retention, in various levels, having a specific energy quantum related towards the difference between two energy levels. Due to the fact that the energy quantum is really connected with frequency, the subsequent absorbtion features of the spectral bend can be utilized for scientific purposes (Stenberg, Viscarra Rossel, Mouazen, & Wetterlind, 2010). In the VIS range (400–780 nm), absorption groups identified with soil shading are because of electron excitations. This is a significant element, as shading can be in a roundabout way connected with key soil properties, for example, Organic Matter (OM), Fe and MC.

Inside the NIR go (780–2500 nm), overtones of OH, combinations and blends of C–H + C–H, C–H + C–C, OH+ minerals, and N-H of basic vibrations (e.g., C–H, N–H, O–H, C–O, Si–O) that happen in the MIR spectral range (2500–25,000 nm or wave number of 4000–400 cm^{-1}) (Soriano-Disla et al., 2014) are the major spectral features for the discovery and evaluation of key soil properties having direct spectral reactions, e.g., MC, OM, clay minerals, and total nitrogen (TN) (Mouazen, Kuang, De Baerdemaeker, & Ramon, 2010). These properties can be anticipated with VIS-NIR spectroscopy with apparent exactness, contrasted with properties with indirect spectral reactions (P, K, Mg, Ca, Na, CEC, pH) that are possibly estimable through covariation with properties having direct spectral reactions in the VIS-NIR unearthly range (Stenberg et al., 2010).

The NIR overtones and mixes are of broad bands that overlap and are 10–100 times more fragile, contrasted with MIR central vibrations-based spectral features, which are increasingly settled, allowing the structure of a sample to be better clarified (Stenberg et al., 2010). The higher energy of NIR radiation and the generally low absorptivity of water position NIR spectroscopy as a superior strategy when wet soils and sediments tests are contrasted with MIR spectroscopy (McClure, 2003). Be that as it may, MIR is better suited for soil examination because of its sensitivity to both the organic and inorganic constituents (Bellon-Maurel & McBratney, 2011), on account of basic vibrations of atoms. Along these lines, MIR can give substantial and significant data about the compound and physical properties of the samples (Viscarra Rossel & Behrens, 2010).

Notwithstanding the central vibrations in the MIR, VIS-NIR spectroscopy is generally adjusted to decide essential soil synthesis, especially OM, clay minerals, texture, nutrients, like heavy metals and hydrocarbons contaminants (Okparanma, Coulon, & Mouazen, 2014; Douglas, Nawar, Alamar, Coulon, & Mouazen, 2019). This is especially valid for in situ and on-line applications and is ascribed mostly to robustness, versatility and the moderate impact of MC on VIS-NIR contrasted with MIR spectroscopy. Be that as it may, one of the significant restrictions related with VIS-NIR is that the spectra are nonspecific, because of the overlapping absorptions among overtones and blends came about because of their broad bands that influence the estimation precision of the models determined (Malley, Martin, & Ben-Dor, 2004). The disperse impacts brought about by soil structure or explicit constituents, for example, quartz increment the trademark absence of specificity (Stenberg et al., 2010).

This requires advanced spectral processing and data mining approaches, to obtain usable information from spectra, limit noise and scattering effect and expand the estimation ability of the produced algorithms. MIR spectroscopy give basic data identified with both organic and some inorganic mixes, and has been recognized as a superior recognition procedure than VIS-NIR spectroscopy. In addition, organic functional groups have specific assimilation bands in MIR extend, so spectra can be utilized to recognize various atoms and describe their structure dependent on the various blends of their functional groups (Soriano-Disla et al., 2014). They can be utilized likewise to evaluate the individual related parts due the tight band features and absorption potential.

All the more critically, adjustment data is considerably more nonexclusive than that in the VIS-NIRS, and can be all the more promptly transferable from instrument to instrument (Soriano-Disla et al., 2014). Be that as it may, MIR spectra are unfavorably influenced by water presence and this impact is more critical than that on the VIS-NIR spectra (Kuang et al., 2012), which requires sample preparation to acquire the ideal spectra for amplified model prediction accuracy (Soriano-Disla et al., 2014). In addition, MIR performs inadequately in anticipating compound properties that depend on the soil solution as opposed to those in the soil matrix or solid constituents, for example, extractable K and P (Minasny et al., 2011). The principal vibrations in the MIR range lead to clear absorption crests than in the VIS-NIR extend (Fig. 3.2).

MIR spectroscopy is ordinarily utilized even industrially to quantify soil properties, fundamentally under research center conditions, for example in Australia (Minasny et al., 2011). Notwithstanding, current MIR installation are fragile and this fact regularly confines the utilization of MIR techniques for field applications (Ji et al., 2016). With the accessibility of portable

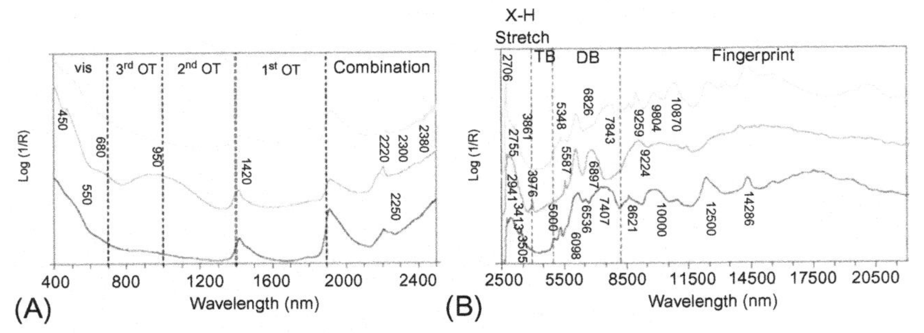

Fig. 3.2 Soil diffuse reflectance spectra for: (A) Visible and near infrared (VIS–NIR) range (400–2500 nm, 25,000–4000 cm^{-1}), demonstrating an approximation of the, and the first, second, and third overtone (OT) vibrations; (B) The mid infrared (MIR) range (2500–15,000 nm, 4000–680 cm^{-1}), demonstrating an approximation of the characteristic signature, double-bond (DB), triple-bond (TB), and X–H stretch regions (Stenberg et al., 2010).

systems as of late, an expanding number of field studies is being accounted for. Among the few published works, Dhawale et al. (2015) explored the practicality of a portable prototype MIR (Wilks Enterprise, Inc., East Norwalk, Connecticut, USA) to anticipate chosen soil properties of samples as two moisture conditions gathered from an test field, located in Canada. The creators revealed R2 and RMSEP estimations of 0.74–0.91% and 7.82–11.98% for sand and CC, and 0.82% and 0.76% for OM, individually.

Correspondingly, Ji et al. (2016) utilized a portable prototype model MIR spectrometer (Spectrum Technologies, Inc., Aurora, Illinois, USA) to gather soil spectra at 120 selected sampling points from two horticultural fields (natural and mineral soils) both situated at McGill University, Canada. The creators revealed a higher precision for OM, BD, CEC, Ca, and Mg (R^2 = 0.60–0.86 and RPD = 1.55–2.62), contrasted with soil pH, Fe, Cu, P, nitrate-N, K or Na (R^2 <0.56 and RPD <1.48). More investigations can be found in the literature that show the capacity of MIR to give increasingly stable and exact estimation of key soil properties when contrasted with VIS-NIR. Subsequently, the ongoing utilization of MIR spectroscopy for business services, e.g., SoilCares in The Netherlands.

Distinctive modeling strategies for soil spectra are mentioned in the literature, such as straight and nonlinear techniques. The linear regression techniques are generally utilized, including yet not restricted to SMLR (Shi et al., 2012), PCR (Mouazen et al., 2010), and PLSR (Nawar, Buddenbaum, Hill, Kozak, & Mouazen, 2016). PLSR is by a wide margin the most utilized linear regression technique to estimate soil parameters (Viscarra Rossel, Walvoort, McBratney, Janik, & Skjemstad, 2006). In the early occasions of spectroscopy research, reports proposed that a better estimation accuracy can be accomplished by choosing the most relevant spectral factors rather than utilizing the full-range. PLSR joined with feature selection strategies, for example, iterative stepwise elimination PLS (ISE-PLS) (Kawamura et al., 2017), irrelevant variable elimination PLS (UVE-PLS), and a genetic algorithm (GA-PLS) (Vohland & Emmerling, 2011) have been utilized in soil spectroscopy.

As opposed to the direct strategy recorded above, information mining methods, for example, AI, e.g., ANN (Mouazen et al., 2010), SVM (Vohland, Besold, Hill, & Fründ, 2011), RF (Nawar & Mouazen, 2017) multivariate adaptive regression (MARS) (Nawar et al., 2016), and boosted regression trees (BT) (Brown, Shepherd, Walsh, Dewayne Mays, & Reinsch, 2006) can deal with both linearity and nonlinearity in the informational index. Most of studies contrasting linear and nonlinear techniques revealed the last to give preferred outcomes over the linear strategies.

In soil application, the estimation of γ-beam spectrometry lies chiefly in the way that diverse rock types contain various concentration of radioisotopes of K, U, and Th, as do the soil profiles, to which they weather (Dickson & Scott, 1997). A few components can influence the γ-beam estimations, for example, soil MC, air temperature, precipitation, thick vegetation, background radiation, and nonradioactive overburden (Kuang et al., 2012). For the most part, high soil MC with expanded BD can diminish the radiation flux, particularly in K and Th rot arrangement (Grasty, 1997). This can be clarified by the way that water causes radiation attenuation, and an elevation in soil MC of 1% because of precipitation will diminish the measured nuclide concentration by a similar quantity (Cook, Corner, Groves, & Grealish, 1996). Also, a negative linear relationship exists between absolute MC and estimated nuclide concentration (De Groot, van der Graaf, de Meijer, & Maučec, 2009).

This impact is clarified by changes in the mass attenuation coefficient of water in the soil. Hence, the most proper time to get the peak gamma radiation from soil is in summer when temperature is high and soil is dry. Thick vegetation or dynamic biomass (plant cover and roots) does lessen the gamma radiation, however also functions like a source of gamma beam on its own (Kuang et al., 2012). Consequently, translating gamma beam spectra and discovering correlations with soil data are a challenge (Dierke & Werban, 2013), and heavy amount of pre-processing is necessary (Viscarra Rossel et al., 2007). Be that as it may, γ-beam is a sensitive strategy to map soil with high resolution when a compact γ-beam sensor mounted on a vehicle (on-line sensor stage) is utilized (Kuang et al., 2012).

A few examinations have investigated the utilization of ground gamma beam for estimating soil texture (Heggemann et al., 2017), accessible plant K (Dierke & Werban, 2013), OC (Mahmood, Hoogmoed, & van Henten, 2013), and pH (Huang et al., 2014). Recent work likewise demonstrates the utilization of an automaton conceived γ-beam sensor for soil estimation (Medusa Radiometrics B.V., Netherlands (http://the.medusa.institute/show/GW/Drone-borne+gamma-ray#). Out of these reports, texture was the frequent and and effectively estimated soil property.

Heggemann et al. (2017) built up a site-independent estimation model for topsoil texture on γ-beam spectra combined with linear and SVM regression techniques. The creators announced that SVM beat the linear models and brought about estimations of averaged vR^2 of 0.96 (sand), 0.93 (sediment), and 0.78 (clay), and corresponding averaged absolute prediction errors from 2% to 4%. Though the linear models came about in R^2 of 0.73 for sand, 0.61 for silt, and 0.18 for clay and arrived at averaged absolute estimation errors of 9–5%. They reasoned that γ-spectrometry-based texture is feasible, offering credible texture estimation at low expenses and at a less time consuming way.

Accessible plant K is a primary macronutrient that can conceivably be estimated with γ-beam spectroscopy. Viscarra Rossel et al. (2007) utilizing bugging PLSR acquired reasonable prediction K (RPD = 1.63). In any case, Dierke and Werban (2013) revealed that γ-K was not relevant for indirect mapping of accessible K at their field site, since is was representative of a minimal fraction of the whole K and consequently was not quantifiable in connection to it, and the fixation of total K in the Aeolian loess insoluble remain in another field location was the fundamental deciding element for gamma K in this field. Hence, they inferred that it is conceivable to utilize gamma K for estimation of accessible K in soils, if the last is a significant percentage of the total K.

Different analysts assessed the capability of γ-beam for the estimation of OC and pH utilizing ground-based γ-beam spectrometry. Wong and Harper (1999) prepared out a γ- beam survey of soils at the central and eastern wheat belt in Western Australia, finding a solid positive connection between K gamma and OC ($R^2 = 0.89$) for a high scope of OC (0–3.5%). Mahmood et al. (2013) assessed the range of a γ-beam spectrometer to predict a few soil properties including OC and pH utilizing energy windows and full-range examination strategies in two fields with a similar sandy topsoil structure with various administrations (ordinary and organic) at Flevoland territory, Netherlands. They detailed a model for good estimation of OC and pH with R^2 equal to 0.42–0.65 and RMSEP equal to 0.54–0.59 mg/g, and R^2 = 0.28–0.65 and RMSEP of 0.01–0.04, separately. They presumed that γ-beam spectroscopy can assume a job in displaying and mapping the examined soil properties for the best 0–15 cm layer superior to the 15–30 cm soil layer. In another investigation, Viscarra Rossel et al. (2007) oppressed γ-beam spectral information to bagging PLSR for fruitful expectation of soil pH (R^2 = 0.63–0.75, RMSEP = 0.41–0.43, RPD = 1.67–1.90) in various fields in New South Wales, Australia.

There have been further developed research to fuse γ-beam spectral information together with other soil sensing data outputs for various applications in agribusiness. Rampant and Abuzar (2004) utilized γ-beam counts combined with EMI and an Digital Elevation Model (DEM) to distinguish yield MZs. A decision tree classifier was utilized to classify soil types and yield information arising from the different combinations of both geophysical and territory properties. The creators found that soil types are efficiently assessed, with under 2% of the zone being misclassified. They could likewise assess yield zones effectively, by misclassifying only 5% of the region, despite the fact that the yield predictions for an individual year were in every case more lower than for those concerning soil types classification and yield zones assessment. Castrignanò, Wong, Stelluti, De Benedetto, and Sollitto (2012) integrated multi-sensor information of a gamma beam, EMI estimated with EM38 and EM31, and RTK GPS, in a geostatistics approach for depicting locations of homogeneous soil. Both EM31 and EM38 maps were like gamma-U, Th and TC maps, assuming that they possessed similar soil properties, however were to some extent, different in relation to the γ-K maps.

Dennerley, Huang, Nielson, Sefton, and Triantafilis (2018) utilized soil parameters (clay, sand, silt, CEC, ESP and pH) and subordinate information including EMI (DUALEM-421S) and gamma-beam spectrometry information combined with Fuzzy k-Means (FKM) grouping to distinguish MZs over a sugarcane field close to Helens Hill in the Herbert River Valley, Queensland, Australia. They found the higher correlation among's TC and Th (0.97) and TC and K (0.96) as for the γ- beam information, demonstrating likenesses between both γ-beam and DUALEM-421S information as far as mean, skewness and kurtosis over the field and the adjustment destinations. They presumed that the methodology was fruitful in recognizing soil MZs that can be utilized to for VRAs of gypsum and lime.

It tends to be inferred that the γ-beam is a valuable device for the forecast and mapping of soil surface, and available K specifically, and that further research is expected to uncover its potential for other soil properties. Plainly the future holds guarantee for the multi-sensor fusion approach, where γ-beam can assume a significant job. The best blend of sensor technology and information data mining instruments will rely upon the application required, which will require application rules to be created sooner rather than later.

3.1.3 Remote sensing

Remote detecting is characterized as the innovation empowering the estimation of soil or yield attributes, by UAVs, plane flights or satellites from distances running from several meters, to numerous kilometers from the objective. Remote detecting sensing in farming depend on the collaboration of electromagnetic radiation with soil or plant material. Ordinarily, remote sensing includes the estimation of reflected or discharged radiation, as opposed to transmitted or retained radiation. Remote sensing alludes to non-contact estimations of radiation reflected or produced from horticultural fields (Mulla, 2013). Notwithstanding reflectance, transmittance and absorption, plant leaves can emanate energy by fluorescence (Apostol et al., 2003) or thermal discharge (Cohen, Alchanatis, Meron, Saranga, & Tsipris, 2005).

The utilization of remote sensing resources is accounted predominantly for crop classification, which incorporates crop development demonstrated as Normalized Difference Vegetation Index (NDVI), Leaf Area Index (LAI), biomass, crop thickness, yield potential estimation, recognition and mapping of biotic, caused by insects, weed and ailments and abiotic stresses like water pressure, N, P, K and OM and salinity (Mulla, 2013). The utilization of remote sensing for gathering information on soil is confined to the main 2–3 mm of the soil, and the crop cover presence prevents from the actuation of high resolution pictures or spectral information on soil to be effectively gathered. Besides, spatial goals influences the territory of the littlest pixel that can be distinguished (Mulla, 2013).

As spatial detail increases, the region of the tiniest pixel diminishes, and the homogeneity of soil or crop attributes inside that pixel increments. Poor spatial resolution creates enormous pixels with expanded heterogeneity in soil or plant qualities. Return frequency is significant for evaluation of transient examples in soil or plant attributes utilizing satellites. The accessibility of remote sensing pictures from satellite and flying platforms is regularly seriously constrained by cloudiness (Moran, Inoue, & Barnes, 1997).

Remote sensing offers the likelihood to gain data about a zone by watching it remotely. A few imaging sensors are utilized to gain information while on board different platforms, for example, satellites, planes or UAVs (Bregaglio et al., 2015). UAVs acquires information from 2 m to 100 m distance (contingent upon regional flight guidelines), though satellite sensing attains information from a several hundreds kilometers over the crop (Adao et al., 2017). The imaging sensors cover VIS, IR, thermal and microwave spectral range. They are intended to procure single band pictures, like thermal or microwave multispectral pictures with a few bands in the VIS-NIR, or hyperspectral pictures with hundreds of bands in the VIS-NIR.

Inside the electromagnetic range, the most well-known wavelengths that are utilized for crop monitoring applications are green, red and NIR. Plant pigments overwhelm the reflectance of vegetation in the VIS wavelengths. Chlorophyll absorbs red and blue energy and reflects green. Leaves cell structure and canopy structure overwhelm the reflectance of NIR, with healthier vegetation producing high reflectance. Hypothetically, NIR and red reflected energy could be evaluated to gauge plants development improvement, the quantity of vegetation biomass, or even their health status. Be that as it may, because of a few factors such a measurement isn't precise enough to screen the vegetation state.

The utilization of Vegetation Indices (VIs, for example, the NDVI and the Enhanced Vegetation Index (EVI) could be an answer for this issue, by consolidating at least two groups into a solitary condition so as to precisely depict, screen and measure the condition of vegetation. Hyperspectral remote sensing offers fine spectral goals resolution over different wavelengths of the VIS-NIR extend, and can give improved estimation of a wide variety of plant biophysical properties (Dorigo et al., 2007). Thermal infrared is additionally helpful to recognize alterations due to abiotic stress (water scarcity) or biotic (infection). In certain territories with permanent cloud cover, optical and thermal sensors can't be utilized. Synthetic Aperture Radar (SAR) isn't influenced by sun light, cloud, haze and rain, in this way could be utilized in a more extensive scope of conditions and can get information during day and night.

The most Frequently utilized spatial resolution in PA applications are those offered by high resolution satellites or UAVs, which permit mapping the spatial variation inside a field. Since the dispatch of IKONOS satellite in 1999, information at spatial goals superior to 1 m are made accessible by a few VHR satellites, for example, Quickbird, GeoEye-1, WorldView 1, 2, and 3, among others. Pleiades satellite constellation and the most recent SPOT satellites offer spatial resolution of a couple of meters, which are satisfactory for PA applications. ESA's Sentinel satellites offer coarser resolution pictures beginning from 10 m, which probably won't be helpful to catch the spatial variability of the majority of the fields. Mapping with planes is additionally not popular in PA applications since they are not as cost-effective, adaptable and accessible as VHR satellites and UAVs.

Later innovative advancements consist of advantages such as imaging with enhanced spatial resolution (circa 5cm), on-demand accessibility, and the potential to get information during cloudy conditions. The most widely recognized platforms utilized are multicopters fit for vertical take-off and landing and the target an, which is helpful to limit obscuring. Little, fixed-wing air ships are progressively utilized, having circumvent the necessity for a take-off and runway, which was hard to avoid in numerous agrarian areas. They offer the benefit of longer flight length and therefore more further coverage in a solitary flight, as they are less energy consuming than multi- copters. The drawbacks of UAV acquisitions, when contrasted with the VHR satellite attained pictures, are: constrained geographic inclusion, little payload that confines the sensors performance that are mounded on the platform and regulations that may block flying at certain regions over the farmland.

A few sensors can be carried on UAV platforms, such as the RGB camera, multispectral, hyperspectral, and LiDAR based sensors. Multispectral cameras are accessible in a wide scope of spatial resolutions and band selections, as off-the-self arrangements and value for money. These sensors cover the VIS-NIR ranges of the spectrum in wide and overlapping bands. Smaller than usual hyperspectral broom or snapshot sensors have been created for UAVs, anyway their expense is still high. Low weight laser scanners, or LiDAR have additionally been grown as of late for UAVs, and can reproduce the three-dimensional (3D) space for different urban examinations, mining estimations, and woodland industry.

The fundamental standard of LiDAR functioning is the estimation of the distance between the sensor and the object, through deciding the passed time between a laser beam outflow and the arrival to the sensors' receiver. Increasing this time with the speed of light and dividing by two to represent the doubled distance voyaged, gives an exact gauge of the distance between the sensor and the object. At the point when the sensor is loaded on an airborne vehicle, it gives a transect of detailed data acquisitions concerning the canopy and ground, where gaps appear (Lefsky, Cohen, Parker, & Harding, 2002).

Preparing these estimations creates an exact depiction of the canopy stature, an Digital Surface model (DSM). An issue identified with canopy stature estimations utilizing LiDAR incorporate the powerlessness to decide the ground surface height in thick or complex canopies or the highest point of the canopy in sparse vegetation. An elective strategy for evaluating an DSM from UAV picture stereopairs is utilizing the structure from movement (SfM) calculation, which is a photogrammetric method that was at first produced for archeological locations (Verhoeven, Doneus, Briese, & Vermeulen, 2012).

The calculation evaluates the unidentified camera directions through correlation of numerous acquired image feature points in different pictures. It at that point delivers the 3D structure-defining model of a scene from the covering 2D picture groupings taken from the overlapping image arrangements and directions. SfM is for the most part recommended when pictures were procured by value for money cameras on board UAVs, as opposed to costly metric cameras, which follow photogrammetric strategies and require high ability (Küng et al., 2011).

3.1.3.1 Remote earth observation for crop diseases identification

Crop biotic stresses can emerge from weeds, insects or infections. Biotic stress cause serious monetary harm to crops when limits for occurrence of the stress are exceeded. Early recognition of biotic stress is basic for control through practices that incorporate tilling, spraying, or IPM. In any case, early discovery of biotic stresses is demanding (Pinter Jr et al., 2003). At the initial stage of infection, weeds are small and hard to recognize from crops. Insects and sicknesses may cause no discernible marks of crop damage at beginning periods of invasion. At later phases of expansion a crop harm can show up as spots or stripes of varying hues (e.g., yellow, dark, or white). Also, leaves can twist or shrivel and crop biomass can be decreased.

In regular agribusiness, crop exploring is utilized to survey the rate and occurrence of the biotic stresses. Domain specialists regularly scout fields for biotic stresses at fixed growth crop stages dependent on historical records of developing degree days, wind and moistness patterns that are known to advance development of weeds, insects or diseases. Exploring happens in constrained zones of the field, ordinarily along edges and in w or z patterns that can be effectively crossed by strolling. Subsequently, exploring may miss viewing clusters of a biotic stress in remote zones from these pathways.

These bunches of pressure can quickly grow if not identified, causing expansive crop harm.

Remote detecting offers the likelihood of quickly studying huge regions of a field for biotic stresses dependent on images gathered by utilizing satellites, planes, or UAVs. Every one of these platforms has clear pros and cons (Mulla, 2013). Be that as it may, recognition of biotic stresses at beginning periods of occurrence, makes an especially significant requirement for high resolution imagery. Indeed, even with high resolution imagery, biotic stress detection has demonstrated considerably more difficult than abiotic stress detection, as biotic stress indicators may not be identified at the leaf area or canopy until seriousness of disease raises beyond the threshold levels.

3.1.3.2 Weed detection

Early identification of weeds utilizing remote detecting regularly depends on segmentation of images to separate weeds from soil or from growing crops. Segmentation of weeds from crops has proven to be feasible utilizing high goals remote sensing imagery (Onyango & Marchant, 2003). Segmentation attempts to initially distinguish the area of harvest columns dependent on the assumption that crop lines are linear. As a subsequent advance, green vegetation varying in area from crop lines is delegated a weed. Segmentation alone can't separate between various weed species. Precision of weed segmentation techniques varies, relying upon the degree of shadows in crop rows and overlap between leaves of crop and weeds. Weeds patches inside a crop row or in crops that are not cultivated in columns like the little grain crops, can't be separated from crops through employing segmentations techniques.

Crop row detection through utilizing segmentation regularly depends on analytical strategies, for example, the basic Hough transform (Montalvo et al., 2012). The Hough transform requires that panchromatic pictures be changed over into grayscale ones, which are then renamed into binary classes, one corresponding to vegetation and the other to soil. Straightforward Hough transforms depend on iteratively assessing the angle and the distance from origin of all lines going through a plant in a crop row, and repeating this procedure with every other plant. The line that goes through most of part the plants in the column is assigned as the line. Clearly, this methodology works best for straight columns. Recognition of bended columns is conceivable utilizing a Generalized Hough Transform (Mukhopadhyay & Chaudhuri, 2015). The Hough transforms and their variations are generally used to recognize crop rows in PA because of their sturdiness to planting skips and failed development. Be that as it may, Hough transform, regularly do not manage to precisely recognize crop rows in the events of high weed weight, where overlapping between weeds and crop occur.

Peña, Torres-Sánchez, de Castro, Kelly, and López-Granados (2013) utilized a UAV equipped with a multispectral TetraCam with 2 cm spatial resolution to label different growth stages (V4-V6) at a spanish maize field into areas with high (>20%), medium (5–20%) and low (<20%) weed occurrence. Segmentation strategies dependent on NDVI >0.2 and object oriented order were utilized to detect yield and weeds from soil, and after that to recognize weeds from crop. Weed inclusion was then mapped in pixels of measurement of 1 m × 0.7 m.

Brilliant understanding ($R^2 = 0.89$) was gotten between groundtruth observations and UAV based estimations of weed occurrence Tellaeche, Pajares, Burgos-Artizzu, and Ribeiro (2011) utilized segmentation and SVM calculations to effectively recognize regions with over the top abundant wilds oat weed presence in a small field of cereal. SVM method is an administered grouping approach for separating data that possess two conditions (e.g., weed inclusion and weed weight) just as a binary decision label (e.g., spray weeds or don't spray weeds) into two classes having the best separation across a hyperplane separating the two classes from each other (Behmann, Mahlein, Rumpf, Römer, & Plümer, 2015).

The high resolution RGB images were obtained from the field through utilizing a UAV were first segmented into a binary images denoting soil (dark) and crop or weeds (white). Crop rows were recognized utilizing a Hough transform, and lines opposite to crop lines were built up on imagery at fixed distance interims to characterize cells where the choice to spray or not to spray was assessed utilizing groundtruth data. Groundtruth data for estimated weed coverage and weed pressure, alongside the groundtruth data for sspray/no spray choices were utilized to train the SVM utilizing 20 out of 86 pictures obtained from the field area on bright and cloudy days. A nonlinear Gaussian radial basis function kernel demonstrated the best separation between weed presence and weed pressure in the two choice classes separating cells that ought to be sprayed or not sprayed with herbicide. Support vectors from the training data were then used to make spray/no spray choices in the rest of the 66 pictures with a general precision above 73%.

Satellite images is not of sufficiently high spatial resolution to recognize individual weeds (Lopez-Granados, 2011). However, there have been fruitful investigations dependent on satellite images including identification of locations inside a field that are invaded by weeds. Casady, Hanley, and

Seelan (2005) utilized Ikonos satellite images to find areas invaded by leafy spurge at two study fields in North Dakota. In this investigation, Ikonos multispectral images at 4 m spatial resolution was acquired at different dates for two years. After radiometric and geometric correction of imagery, supervised classification on all images was utilized to isolate the investigation destinations into areas that had less than 26%, 26–50%, 51–75% or >75% weed cover calibrated with ground truth data. Spectral signatures of leafy spurge and different classes of vegetative spread were then created from Ikonos images from individual dates at every one of the training sites.

Maximum Likelihood classification produced from spectral signatures was precise at recognizing leafy spurge from green vegetation at the principal study site, however it failed at recognizing leafy spurge from different weeds and forage crops at the second study field. Ikonos images was not capable of recognizing areas at either study field that included leafy spurge patches that of 200 m^2 size or with a weed spread less than 30%. Higher goals satellite images, for example, images attained through employing the GeoEye-1 (with a spatial resolution of 1.84 m) or WorldView-2 (with a spatial resolution of 0.46 m) satellites would hypothetically permit weed patches having a size of 16 m^2 or larger to be identified (Lopez-Granados, 2011).

3.1.3.3 Detection of insects

Biotic stress caused by insects or disease presence utilizing spectral indices got from remote sensing images has to a great extent not been economically reasonable. The main reason behind this is side effects of insects or disease occurrence are ordinarily not detected by remote sensing images at the beginning of presence. At beginning periods of populace development, insects may dwell on the lower sides of leaves or in crop roots, where they are not noticeable utilizing remote sensing imaging. A wide scope of remote detecting methodologies have been utilized to identify insects and diseases at later phases of pervasion under wisely controlled conditions when just one kind of insect or infection harm appears (Pinter Jr et al., 2003).

Insects and diseases harms in these cases can be recognized through utilizing different spectral indices dependent on VIS-NIR reflectance, combined with Partial Least Squares Regression (PLSR) analysis or AI models. Insects and diseases presence normally cause higher reflectance, most of the times due to lower absorption at blue and red wavelengths because of lower concentrations of chlorophyll pigments, and lower reflectance at NIR wavelengths because of lower biomass. Green and red-edge wavelengths likewise may react to an assortment of biotic crop stresses. Generally utilized spectral indices for biotic stress involve NDVI, Green NDVI, Red Edge NDVI, and Soil Adjusted VI (SAVI). In any case, these spectral indices are not indicative of a particular insect or disease presence, and frequently react similarly also to other biotic or abiotic stresses, including nitrogen (N) insufficiency (Pinter Jr et al., 2003).

Some examinations employing satellite remote sensing have been carried out aimg to recognize insect harm in crops (Hunt Jr & Daughtry, 2018). Satellite remote sensing of insects presence is fundamentally used to recognize the area and greatness of woodland extent (Rullan-Silva, Olthoff, de la Mata, & Alonso, 2013). Normally, changes in NDVI values from consecutive satellite pictures gathered crosswise over enormous forested regions at moderate to low spatial resolution are utilized to find the trees defoliation due to insects presence. UAVs have the capability of insect effect recognition compared to satellites due to their spatial resolution, less effect because of cloud cover and an expanded capacity to get images at basic phases of insects invasion and development.

Hunt and Rondon (2017) demonstrated that harms to potato crops from Colorado potato beetles could be distinguished with UAV images inside one day of an expansion in invasion. Recognition of harm depended on changes in crop height from one day to another, utilizing potato canopy cover remaking with Structure from Motion (SfM) and Object Based Image Analysis (OBIA). Identification of insects invasion was more exact with this method than with one that is dependent on recognizing when UAV based vegetation indices lowered under a basic limit esteem.

Puig, Gonzalez, Hamilton, and Grundy (2015) utilized K-means grouping of high resolution RGB imaging acquired though utilizing a UAV in order to characterize white grub harm in a 6 ha Australian field, cultivated with sorghum. Groundtruth perceptions were utilized to distinguish regions of the field with heavy, moderate and no insect harm. Before K-means clustering, images was smoothed for high frequency noise utilizing a Gaussian convolution kernel applied to the red, green and blue parts of images separately. K-means clustering effectively classified insects invasion in the field, heavy on 1.7 ha, moderate on 1.1 ha and no insect harmed on 3.2 ha. It ought to be noticed that there were no yield stresses other than white grubs in the investigated field area. In this manner, it is vague whether the methodology introduced by Puig et al. (2015) could be spreadable to different fields where changes in crop biomass occur because of stressors other than white grubs.

3.1.3.4 Crop disease detection

Yuan, Pu, Zhang, Wang, and Yang (2016) utilized multispectral images with the SPOT-6 satellite at 6 m spatial resolution for regional mapping of powdery mildew of wheat crop, in China. Red and green satellite bands just as the NDVI, Triangular Vegetation Index (TVI), and Atmospherically Resistant Vegetation Index (ARVI) spectral indices were consolidated utilizing Spectral Angle Mapper (SAM) to appraise whether certain farmer fields were invaded or not. SAM includes pairwise examination of two reflectance bands or vegetation indices for an undiseased reference area in respect to a second (uninvestigated) area. Red and green reflectance reacts straightforwardly to fine buildup, while NDVI demonstrates the effect of disease on crop biomass. TVI is an indicator of the infection effects on chlorophyll, while ARVI filters out environmental disturbances in satellite images. SAM satellite based model assessments of infection propagation were correlated ($R^2 = 0.78$) with groundtruth disease evaluations, dependent on hyperspectral images gathered with an ASD FieldSpec spectrometer.

Yellow rust wheat infection in China were precisely identified utilizing airborne hyperspectral remote sensing with a 1 m spatial resolution and a photochemical reflectance index (PRI) (Huang et al., 2007). Estimations of PRI were assessed utilizing reflectance values at 531 and 570 nm. Groundtruth assessments of the percent of territory with yellow rust were estimated more than four growth stages for three wheat varieties. Airborne estimations of PRI were strongly correlated ($R^2 = 0.91$) with estimated disease rate. Linear regression based on airborne estimates of PRI in one year was utilized to partition wheat fields into classes with no, 1–10%, 10–45%, 45–80% or >80% disease occurrence expected for the next year, demonstrating that the factual connection among PRI and infection occurrence was steady over years. The PRI based appraisals of disease occurrence were not influenced by either water pressure or N stress treatments applied at the experimental site.

UAVs have been progressively used to recognize areas inside horticultural fields influenced by crop disease (Thomasson et al., 2018). Duarte-Carvajalino et al. (2018) precisely recognized potato blight invasions utilizing images gathered through utilizing an UAV. Imagery was gathered in a period of almost three months utilizing an digital camera with a blue-green-NIR channel having a spatial resolution of 0.8 cm. The trial site included three potato varieties including 14 diverse potato assortments, with treatments incorporating a control with no fungicide, and two fungicide treatments followed by a fixed schedule or a schedule followed by IPM.

Images were collected into plots of 0.4×0.32 m. Groundtruth information for disease seriousness was assessed by testing four potato plants from each test plot on each date of image acquisition. A RF classifier dependent on spectral differences (green and blue or NIR and green) was highly correlated ($R^2 = 0.75$) with ground truth information for severity of potato blight occurrence.

Conventional methodologies for the biotic crop stress recognition are regularly founded on multispectral and hyperspectral remote sensing and the utilization of spectral indices, for example, NDVI. Spectral indices are to a great extent incapable of identifying biotic stresses at beginning periods of invasion, when treatment with the help of spraying the infected crop is regarded the best solution. These records are additionally frequently not capable to separate between one kind of biotic stress and another, even at later phases of invasion, or among biotic and abiotic yield stresses. They are most helpful when a field is known to be swarmed with a specific kind of biotic pressure, and the goal is to measure the spatial degree and seriousness of invasion.

Considerably more development is required by specialists to introduce to the market financially reasonable methodologies for identifying insects and diseases invasion utilizing spectral indices got from multispectral or hyperspectral remote crop sensing. The fundamental issues that must be concerned to incorporate precision under field conditions where various stressors concurrently, speed of analysis, scaleability of algorithms, usability, and lower expenses.

3.1.3.5 Remote sensing for crop abiotic stresses

A noteworthy job of remote sensing information in the field of PA is to survey crop condition in the field, to delineate homogeneous zones, and describe and examine them to create recommendation maps for variable rate application (VRA) of inputs, for example, fertilizer, pesticide and water. Remote sensing data information regarding crop variability, plant health status and yield constraining factors that can have a positive impact on the crop productivity. While mapping crop variations requires basic remote sensing information at several times in a season, basic decision making requires increasingly advanced remote sensing at high temporal and spatial resolutions.

3.1.3.6 Remote sensing for prediction of yield productivity

There are some parameters utilized in the reference to gauge crop development and yield. These incorporate among others crop size, crop biomass, LAI, NDVI, crop volume and thickness.

Yield and biomass prediction

One of the primary targets of PA is the precise and effective expectation of crop development and yield. This practice is mainly followed by farmers and leaders to agricultural consultants during the crop growth season. Notwithstanding accessibility of soil supplements, crop genotype and farming management practices, crop improvement is a function of meteorological conditions, for example, temperature, daylight and precipitation (Rembold, Atzberger, Savin, & Rojas, 2013). Harvest crop growth models recreate these connections to anticipate the expected yield and biomass improvement. It is basic to take note of that crop yield prediction or yield potential will be profitable to farmers for raising the expected yield.

Crop models are arranged in two principle classes, in view of their structure; to be specific, powerful models and observational models. Dynamic models, likewise called procedure based models, which recreate the improvement of the yield utilizing differential conditions (Rauff & Bello, 2015). These models require countless info parameters to assimilate the natural conditions (Lobell & Burke, 2010). Empirical models, otherwise called measurable models, are presented as relapse conditions with one or couple of parameters. These models require less information than dynamic models. Empirical models are helpful to gauge harvest yield in specific situations to give valuable information to policymakers. A constraint of experimental models is that they can't gauge crop yield without historical data and they can't be generalize to different areas than to those that have been initially used for calibration. Contingent upon the kind of model and the application, dynamic crop models may require a lot of information, which at times may confine their usability. Each crop model requires at least soil data, climate conditions, initial conditions (e.g., cultivar type, planting date, previous yield) and the management practices.

Since the early advancement of crop models, agrarian researchers have exploited the accessible remote sensing images to improve models' performance, screen crop biomass, phenology and crop yield at scales ranging from field or sub-field level to territorial and national. The primary benefit of fusing remote sensing information with crop models are the accurate depiction of the crop's condition during the growing season and the augmentation of the missing spatial data. Among all the accessible remotely detected parameters, LAI, NDVI and portion of ingested photosynthetically dynamic radiation (fAPAR) are the most ordinarily utilized for assessing crop yield.

Lambert, Blaes, Traoré, and Defourny (2017) considered the utilization of remote sensing to gauge and map yield in sub-Sahara Africa. They utilized Sentinel-2 time-series information to map crop types and to gauge the potential yield dependent on linear regresssions between VIs or LAI. There have been also considered ground truth reference data on yield in 105 fields where different crops including of cotton, maize, millet and sorghum were cultivated. The outcomes of the investigation demonstrated that the maximum LAI during the growth season is a decent factor for yield estimation as it comes to $R^2 = 0.61$ as a mean value. Rice yield forecast is demanding, since water background at early growth phases of rice prevents the utilization of remote sensors for mapping crop condition. Along these lines, UAV images have been utilized after complete canopy closure to appraise NDVI maps of trial fields with variable N administration in Thailand (Swain, Thomson, & Jayasuriya, 2010). A linear model ($R^2 = 0.760$, RMSE $=$ $0.598\,\mathrm{Mg\,ha^{-1}}$) was utilized to anticipate rice yield at the panicle initial establishment. Delivering yield variability maps at that beginning time could be valuable to characterize phytosanitary practices inside a same growing season, for example, site-explicit N treatment.

AI is likewise utilized for yield prediction, while fusing remote sensing information. The arised models created through these architectures permit the investigation of variables that influence crop growth in an unsupervised manner (Elavarasan, Vincent, Sharma, Zomaya, & Srinivasan, 2018). They give a viable technique to look at the huge datasets and to extract insights from the acquired data so as to give a progressively significant understanding into the procedures impacting crop yield. Pantazi et al. (2016) employed three SOMs for wheat yield prediction utilizing NDVI information obtained from the Disaster Monitoring Constellation for International Imaging (DMCii) satellite pictures, just as VIS-NIR spectra were obtained from an on-line soil scanner (Mouazen, 2006). The performance of the employed SOM models exceeded 90% of accuracy, demonstrating that the models potential of predicting wheat yield and of labeling correctly the management zones.

Albeit a few effective models have been proposed to gauge yield for cereal crops with remote sensing, research oriented to fruit and high profit crops has given a variety of performances for citrus (Ye, Sakai, Manago, Asada, & Sasao, 2007), avocado (Robson, Rahman, & Muir, 2017), and pears (Van Beek et al., 2015). The principle reasons are the other yield bearing and individual ranch the board rehearses, for example, flower thinning. Direct recognizable proof of natural products with remote sensing has been conceivable with high resolutions UAV pictures, for example, counting tomato crops in a field (Senthilnath et al., 2016).

A case of the need of combining crop models with remote sensing information for improved yield prediction is cotton crop since its yield can't be directly correlated to the proportions of biomass (Leon et al., 2003). Recently, Haghverdi, Washington-Allen, and Leib (2018) examined which of the Landsat 8 inferred VIs give the best forecast of cotton yield at the subfield level, employing ANN models in a 73-ha irrigated field of cotton, situated in western Tennessee (USA). Results demonstrated that the Greenness Index (GI) and Wetness Index (WI) were the more powerful VIs for evaluating cotton yield with high performance. Moreover, Cheng, Meng, and Wang (2016) managed to estimate yield through employing the WOFOST model and utilizing remote sensing information. Time series of Huan Jing (HJ-CCD) data has been employed to figure LAI utilizing an exact regression model, which was adjusted to the WOFOST model.

It has been proved that this methodology can upgrade the model's yield prediction ability and give improved spatial simulation capacity. Basso, Ritchie, Pierce, Braga, and Jones (2001) applied the CROPGRO-Soybean model with airborne remote sensing information on uniformly characterized defined areas or the management zones, in light of NDVI input information. The model has been proved capable of gauging crop yield with the assistance of remote sensing information, which upgraded its capacity to discover spatial patterns over the field and to give spatial inputs to the model. This was demonstrated valuable to recognize management zones and the reasons of yield fluctuation, which are key components in PA.

Vegetation height

Vegetation height is regarded a significant structural parameter, characterizing the growth status of crops, feed and forestry. It is a straight-forward parameter for observing vegetation growth, health condition, and estimating its over the ground biomass. Harvest height is associated with grain yield of cereal crops (e.g., wheat, grain, oat, rye) (Bendig et al., 2014). In spite of the fact that it is a simple to quantify parameter in situ, high fluctuation between plants renders it hard to appraise with remote sensing for wide regions. Besides, the required redundancies along the cropping season to screen the vegetation condition, raises the expense of field-based reviews. Various methodologies and sensor techniques are utilized to gauge crop height. For PA applications, these sensors are mounted to UAVs and VHR satellites.

Despite the fact that NIR reflectance is delicate to plant structure, its accuracy in mapping plant height is very constrained. This is on the grounds that reflectance information are for the most part delicate to plant pigments and vegetation thickness instead of height; therefore canopy height can be determined just in an indirect manner (Chopping et al., 2008). Also, vegetation indices saturation for the most part happens at high plant densities, in this manner decreasing the contribution of VIS-NIR information to plant height estimation. Consequently, different techniques are preferred for evaluating canopy structure parameters, e.g., covering height, for example, photogrammetric strategies to appraise computerized surface models, and dynamic sensors, for example, LiDARs.

RGB pictures acquired through a cheap UAV digital camera were utilized to assess different vegetation indices, which, in fusion with plant heights got from SfM, delivered a stepwise linear regression model for assessing maize canopy height (Li et al., 2016). Basic and structural information combination was viewed as crucial for accomplishing high precision of height estimation (0.11 m ME). In another exploratory plot in Texas (USA), maize height was assessed utilizing SfM examination of RGB pictures procured from different UAV platforms (Chu, Starek, Brewer, Murray, & Pruter, 2018). They tried different picture information extraction configurations on estimation precision. The most noteworthy accuracies concerning maize height estimation were accomplished with a width of testing along a maize developing line of 0.25 m, utilizing the 99th percentile of height measurements and the canopy surface model rather than direct point cloud information, and a DSM spatial resolution superior to 0.12 m. High-goals RGB UAV pictures were utilized to assess grain tallness in a trial plot in Germany along a growing season (Bendig et al., 2014).

They assessed barley plant height from crop surface model got from SfM examination of UAV pictures with spatial resolution of 0.01 m. Contrasted with in situ height measurements, the chosen linear model accomplished high precision ($R^2 = 0.92$). Subsequently, arable crop height can be estimated with an extremely high performance for various growth stages utilizing UAV-based high-resolution pictures. A VIS-NIR camera on board a UAV was utilized to gauge tree heights over olive plantations in southern Spain (Zarco-Tejada, Diaz-Varela, Angileri, & Loudjani, 2014). SfM was utilized to evaluate the DSM and from this the tree height. Evaluation against in situ tree height estimations yield satisfactory performance

(RMSE between 0.33 and 0.4m). Lowering spatial resolution yielded reduced performance in the occasion that the pixel size goes lower than 0.35 m, meaning that higher flying elevations or lower resolutions could be utilized for raising productivity of mapping in future applications.

Airborne LiDARs have been tried in evaluating crop height. Maize height was assessed from a LiDAR mounted on a plane in a test site, situated in northwest China (Li et al., 2015). Among the assessed parameters, most extreme height of maize had the highest performance with a RMSE and R^2 equal to 18.48 cm and 0.79 respectively, when contrasted to field estimations. A small scale UAV with a mounted LiDAR was utilized to gauge crop height in an area cultivated with maize (Anthony, Elbaum, Lorenz, & Detweiler, 2014). The miniaturized scale UAV was flown in close proximity to yields to accomplish exceptionally high resolution and to consider the canopy gaps crated by the structural sparsity and give estimations of crop height. Such low flights elevate efficiently the spatial resolution, giving the opportunity to lower cost, of-the-self UAV frameworks to be utilized in yield height mapping applications.

Leaf area index (LAI) parameter

A significant parameter to describe vegetation biomass is called LAI, which is characterized as the uneven green leaf region per unit ground area (estimated in m^2/m^2). LAI is a biophysical parameter, and the most widely recognized remotely detected parameter absorbed in crop growth estimation models. Nonetheless, VIs including NDVI have been utilized in experimental models. Information of LAI is essential to describe vegetation yield growth, gauge water loss through transpiration, and assess different stressfactors.

Warren and Metternicht (2005) stated that LAI is the essential canopy parameter involved in two fundamental physiological procedures, such as photosynthesis and evapotranspiration, both very reliant on sun based radiation. Moreover, they showed that for by far most of procedure based crop development and crop models an estimate of LAI for the crop canopies is needed. Wu, Wang, and Bauer (2007) concluded that LAI is an essential parameter for the linkage of multispectral remote sensing to crop growth and condition for biological estimations. For PA applications, LAI is related with agronomic, natural, ecological, and physiologic procedures (e.g., development examination, photosynthesis, transpiration and energy balance).

A few strategies for the estimation of LAI at field scales have been created and effectively evaluated. VIs, for example, NDVI, EVI, and so on, have been utilized to infer LAI values, with the focal point of late research focusing on constructing consistent models. Information for LAI estimation valuable for PA applications can be reached with the help of multispectral cameras on board UAVs, from commercial VHR satellite pictures (WorldView2, Quickbird, etc.), and from commercial high resolution satellites (Sentinel-2, SPOT, Landsat, and so forth.) for fields of great extent. Gevaert, Tang, García-Haro, Suomalainen, and Kooistra (2014) built up a data fusion technique for crop monitoring in PA applications utilizing hyperspectral-UAV and Formosat-2 images in field cultivated with potato, situated in Netherlands. Their technique demonstrated high connection to the Weighted Difference Vegetation Index (WDVI) ($r = 0.969$), LAI ($r = 0.896$) and canopy chlorophyll ($r = 0.788$).

Richter, Atzberger, Vuolo, and D'Urso (2011) utilized spectral inspecting methods on Sentinel-2 information to gauge the LAI for wheat, sugar beet, and maize. Their intended accuracy of 10%, as per the Global Monitoring for Environment and Security (GMES) for green LAI accuracy achieved for sugar beet (8%), less satisfactory for wheat (11%) and was much worse for maize (19%). For these crops, the RMSE fluctuated between 0.4 and 0.6. Notwithstanding, they function as indicators, proving that the previously mentioned accuracy ought to be conceived cautiously, considering the possibility of bias due to the acquisition of the reference data set. Warren and Metternicht (2005) investigated the relationship of airborne advanced multispectral images data with 1 m spatial resolution and determined yield parameters for canola fields.

Significant associations of LAI with the blue, red and NIR regions ($r > 0.82$), showed that they are efficient for timely detection of canola growth fluctuations. SfM has additionally been utilized to evaluate LAI. Mathews and Jensen (2013) utilized SfM to make a 3D vineyard point cloud and envision vegetation and furthermore endeavored to foresee vine-LAI dependent on data got from the produced SfM point cloud. They determined various indicators (point heights, amount of points, etc.) from the SfM point cloud and contrasted them with LAI estimations, so as to find possible correlations. Results demonstrated that SfM could be utilized to foresee LAI for vineyards, with only a moderate portion of variety of field-estimated LAI being clarified ($R^2 = 0.567$).

Li, Niu, Chen, and Li (2017) modeled the basic multifaceted nature of maize canopies utilizing point cloud got from airborne LiDAR and UAV imaging. They tried different measurements for the estimation of maize LAI with their outcomes demonstrating critical relationships with in situ estimated LAI (r > 0.60). Forget about Leave one- out cross validation for LAI estimation demonstrated that the most noteworthy strength (RMSE = 0.16) was gotten by the general informational collection model, clarifying 75% of the LAI variation. Li et al. (2015) inspected the capability of airborne LiDAR and GF-1 satellite information for the estimation of maize LAI and biomass during the pinnacle of the growing season. They found that both maize biophysical parameters inspected indicated high Pearson's connection to the remote detecting measurements utilized with the highest relationship found among LAI and NDVI (r = 0.77) and pursued by that among biomass and mean crop height (r = 0.72).

Verger et al. (2014) created and assessed a strategy for LAI estimation for wheat and rapeseed crops, utilizing UAV reflectance estimations and radiative transfer inversion system. Evaluation with in situ LAI gave RMSE of 0.17 and R^2 of 0.97. Richter et al. (2011) proposed coupling the PROSPECT SAILH model with a look-up table calculation and observed the approach to be valuable for supporting the handling of Sentinel-2 information for the high value crop production. They simulated design with a RMSE of 0.4–0.6, considering the biophysical parameter of LAI however estimation failures must be also considered. They additionally proposed the utilization of sophisticated radiation propagation models in instances of not yet grown crops.

3.1.4 Spectral and thermal properties of plants in a nutshell

The VIS-NIR range (400–2500 nm): Green vegetation has very prominent spectral features in the VIS spectral portion: two chlorophyll pigment absorptions in the blue (450 nm) and red (680 nm) regions that bound a reflection in the green region (550 nm).

This explanation behind the ability of the human eye seeing healthy vegetation as green. At the point when the plant is stressed that subsequently blocks ordinary growth and chlorophyll generation, there is less absorption in the red and blue bands and the measure of appearance in the red waveband raises. These bands were generally used to screen N content. The spectral reflectance signature in the VIS-NIR range implies an elevation in the reflection for healthy vegetation at about 700 nm. In the NIR range between 750 and 1300 nm, a plant leaf normally reflects between 40% and 80%, of the incident radiation; the rest is transmitted, with just about 5% being absorbed. For correlation, the reflectance in the green range reaches 15–20% of the incident radiation. This high reflectance in the NIR is attributed to the light in the intercellular volume of the leaves' mesophyll. At wavelengths longer than 1300 nm, water vapor absorption increases significantly at the reflectance of 2500 nm. Water absorption bands are for the most part noted at around 760, 970, 1200, 1470, 1900, and 2500 nm. These groups were accordingly generally used to screen water content.

Far or Thermal Infrared (TIR) Range (3–14μ), plants assimilate efficiently the far infrared radiation at wavelengths greater than 2500 nm (2.5 μ) (Gates, Keegan, Schleter, & Weidner, 1965). Leaf temperature can be detected by estimating the far-infrared or thermal infrared (8–14 μm) radiation they transmit. Evapotranspiration is the procedure where water buffered in the soil or vegetation is converted over from the fluid into the vapor stage and is released to the air (Maes & Steppe, 2012). Water evaporation is an energy requesting process, elevates evapotranspiration rates, lowering consequently the surface temperature of leaves and plants. As plant stomata close, evapotranspiration rate diminishes; the energy heat balance between the vegetation and its status alters and leaf temperature rises. First approaches concerning the employment of canopy temperature for characterizing the plant water status and plant health condition, were developed in the 1960s (Fuchs & Tanner, 1966). The accessibility of thermal cameras provoked a critical advancement of the thermal remote sensing during 2000s (Maes & Steppe, 2012).

3.1.4.1 Spectral analysis methods to estimate N and water status in crop

Spectral information in the VIS-NIR-short wavelength infrared (SWIR) range, split into multispectral and hyperspectral information. Multispectral imagery for the most part corresponds from 3 to 5 bands, which have a width of 20–100 nm. Hyperspectral imaging comprises of much smaller bands (1–5 nm), having hundreds of bands. As of late, the term super-spectral images was introduced, referring to satellites pictures that give more than 10–15 bands, some generally limited groups, particularly in the red-edge and water absorption areas. Basically, there are three fundamental strategies for spectral investigation: (a) spectralfiles, (b) band selection, and (c) linear and nonlinear multivariate statistics and data mining. The primary strategy is generally utilized for each of the three image types while the other two are for the most part used to break down hyperspectral pictures to address dimensionality complexity. An extensive amount of reference material was composed on spectral image investigation (Thenkabail, Lyon, & Huete, 2019), and the reader has the opportunity to get introduced to further information.

3.1.4.2 Nitrogen crop status

Nitrogen deficiency belongs to one of the most significant crop abiotic stress to be identified, since it influences directly the yield profitability. An extra reason that makes the recognition of N deficiency significant is the way that N leaches below the root zone in the occasion that irrigation system or water administration is not proper, yielding conditions that are problematic for crop development. Numerous VIs have been created to gauge crop N status at leaf and canopy levels. A list of indices can be found in different works, for instance, in Tian et al. (2011). Most of the indices depend on surrogate markers generally of chlorophyll content, which is demonstrated to be physiologically connected to N concentration. Best-performing VIs incorporate common ratios in the red edge area, in the blue band and in the SWIR bands (1200–2500 nm), especially the 1510 nm band (Herrmann, Karnieli, Bonfil, Cohen, & Alchanatis, 2010).

The abilities of the VIs to foresee N content are influenced by different components such as crop type, growth stage, site, year, and spectral, spatial and temporal resolutions of the sensor. This fluctuation prompted the improvement of many indices, and numerous investigations appear to rehash a similar scenario: looking at all indices created until this point, one next to the other upcoming combinations of 2–3 bands that performed some way or another better with the new data sets gathered in the new feed campaign. Along these lines, new indices are continually created. The accompanying sequence of recent studies quickly shows that apparently endless improvement or fitting of indices. The Medium Resolution Imaging Spectrometer (MERIS) terrestrial chlorophyll index (MTCI) (Dash & Curran, 2004) is regarded as the best spectral index to be utilized for variable rate N supplementations in two fields cultivated with potato (USA) (Nigon et al., 2015). In an examination in some fields, cultivated with rice coming from various locations in China during four seasons the MTCI performed essentially more poor than another 3-band spectral index presented by Tian et al. (2011). The new index performed well utilizing ground spectra just as utilizing collected Hyperion satellite cube.

Moharana and Dutta (2016) investigated the spatial fluctuation of N content of rice from Hyperion cube in India, has presumed that in spite of the fact that the 3-band index proposed, gave great N estimation abilities, pursuing a totally different relationship. Along these lines, the new adjusted connections were better for mapping the spatial fluctuation of N content than the connections demonstrated by Tian et al. (2011). This succession of studies, demonstrated the significant impacts of crop type and sites.

The impact of growth stage was exhibited by a few investigations (e.g., Cohen et al., 2010).

In Li et al. (2010) the red edge and NIR bands were increasingly efficient for N estimation at early growth stage, yet VIS bands, particularly the blue oned, were more delicate at later growth stage. Tian et al. (2011) indicated impacts presented by various sensors. In their investigation they have demonstrated the significance of the blue bands to predict N content in rice crops utilizing ground spectral information, while for Hyperion satellite pictures, they found that the red edge and NIR were delicate.

The watched intercorrelations between wavelengths and indices along growth stages and cultivars might be assessed by multivariate techniques, which utilize the entire range and not just chosen wavelengths. For example, multivariate strategies were utilized to gauge leaf N content dependent on narrow-band spectral information in potatoes. PLSR investigation has brought about a more higher correlation among's predicted and estimated leaf N content with a R^2 equal to 0.95 than the N-specific changed chlorophyll absorption reflectance index (TCARI) where R^2 was found equal to 0.82 (Cohen et al., 2010). Additionally, with PLSR the improved correlation was accomplished with a single model for both the vegetative and the tuber-bulking periods, while the TCARI provided an alternate model corresponding to every period. In a similar report, when reenacted Venus satellite spectra were utilized, the accuracy of the spectral model was lowered reaching an R^2 equal to 0.78, yet still included both vegetative and tuber-bulking periods. This implied the fact that while a single 3-band record was affected by the impacts of growth stage, the spectra of just 11 level bands were increasingly sturdy.

Strategies identified with machine or deep learning ideas, for example, RF and ANNs, were acquainted with the examination of remotely sensed information to appraise N levels. This happens because of the potential of these strategies to process an enormous number of data sources and handle nondirect assignments. The utilization of these procedures for harvest yield and N estimation was as of late investigated by Chlingaryan, Sukkarieh, and Whelan (2018).

3.1.4.3 Canopy/leaf relative water content (CRWC/LRWC)

Spectral features of water can be utilized to evaluate the leaf and canopy relative water content (LRWC/CRWC). For wavelengths easily affected to water retention, leaf reflectance diminishes as water content increases. Various examinations have demonstrated the capacity of spectral indices to define LWC, for instance, the early work of Hunt and Rock (1989), the investigation of Ceccato, Gobron, Flasse, Pinty, and Tarantola (2002), and later examinations for crops (Zhang & Zhou, 2015). While a different scope of vegetation indices have prior been produced for the remote estimation of CRWC. The majority of them are characterized for specific crop types and zones, making them less generally relevant (Pasqualotto et al., 2018). In a couple of studies, endeavors to utilize indices as derivations of reflectance values for individual wavelengths did not yield huge correlations regarding the canopy level (Rodríguez-Pérez, Riaño, Carlisle, Ustin, & Smart, 2007). It is particularly valid for early phenological stages, described by low fractional vegetation cover. However, strategies that utilize the entire range performed superior to commonly utilized water index indicators like in Clevers, Kooistra, & Schaepman, 2008). An ongoing report proposed an index that depends on a spectral area difference that was appropriate for a wide assortment of crop types, having a R^2 equal to 0.8, got utilizing an exponential regression algorithm, contrasted with less performing ($R^2 = 0.6$) CRWC spectral indices (Pasqualotto et al., 2018).

3.1.4.4 Thermal image processing to predict water content in crops

Until now, thermal cameras give either panchromatic or multispectral pictures in the an area of 3–14 μm. However, the extensive research for assessing crop status in agribusiness was led with panchromatic images, which means there is no spectral dimensionality associated with the investigation of thermal pictures. Thus, the center of the thermal image processing is to change over the surface temperature to horticulturally important yield water status indices.

Leaf Water Potential (LWP) in crops and Stem Water Potential (SWP) in plantations are significant biophysical parameters that demonstrate the capacity of the crop to move water from soil to the environment by the leaf (Jarvis, 1976). The parameters LWP and SWP are utilized for water supply administration, since the water content isn't an indicator on whether the plant is capable of utilizing the available water. Literature provides constrained knowledge regarding to their remote estimation through applying hyperspectral sensing in the VIS-NIR area because they express the affection of water in the plant tissue (Rapaport, Hochberg, Shoshany, Karnieli, & Rachmilevitch, 2015). An recent investigation acquired average relationships between spectral indices features in the VIS-NIR area and SWP in vineyards. For this reason, they took advantage of the daily revisit time of the Planet-Labs satellites and high volume data corresponding to seasonal SWP estimations from dozens of vineyard fields (Helman et al., 2018).

L/SWP do influence the leaves' stomata condition, which are responsible for the evapotranspiration procedure and influence leaf temperature. A significant result of the stomatal closure, happening when plants are exposed to water stress is that energy scattering is lowered, so leaf temperature will in general ascent (Jones, 1999). The utilization of canopy temperature as a marker of plant water status was advanced by Idso and associates (Jackson, Idso, Reginato, & Pinter Jr, 1981). Since canopy temperature is influenced by both plant water status and natural conditions, water stress indices that align the ecological conditions were created. The crop water pressure index (CWSI) in view of canopy temperature (Jackson et al., 1981) has turned into an efficient index to delineate field fluctuation of crop water condition utilizing thermal images. CWSI is characterized as a small amount of the canopy temperature between dry (upper) and wet (lower) baselines under field conditions. It is determined by the accompanying condition:

$$CWMSI = \frac{T_{canopy} - T_{wet}}{T_{dry} - T_{wet}} \tag{3.1}$$

where T_{canopy} denotes the canope temperature, T_{wet} is the temperature of a leaf that transpires and T_{dry} symbolizes the temperature of a leaf that does not transpire. Thermal crop sensing detecting techniques have been generally utilized as sensors for observing and mapping crop water status in different orchards (Gonzalez-Dugo et al., 2013), olives (Agam et al., 2013) almonds (Gonzalez-Dugo et al., 2012), and grapevines (Möller et al., 2007).

3.1.4.5 Decision support for fertilization and irrigation regimes based on augmentation on remote sensed data

Studies have demonstrated that actual N treatment isn't really effective (Cohen et al., 2010) and that coordinating the planning and rate of N application with the N crop needs during various growth stages may expand N Use Efficiency (NUE) and limit N losses (Errebhi, Rosen, Gupta, & Birong, 1998). As appeared above, N content can be observed by spectral indices obtained got from multispectral and hyperspectral satellite images and like water supply system management, the utilization of remote-sensing stages to screen N content can be supported by the utilization of the successive and generally high spatial resolution of the Sentinel-2 images (Delloye, Weiss, & Defourny, 2018). However, in spite of similarities mentioned in literature, chosen bands were regarded not representative for various crops or even to a similar cultivated area in various regions. In addition, apart from dissimilarities among crops, regions, and years, growth level likewise has been appeared to affect the performance of various VIs for assessing N. Further apart issues related with model construction, accuracies and functional qualities of the sensors, other crop biophysical and biochemical properties affect canopy reflectance.

That is to say, for instance, that while N content denotes the ideal property, data given by processed spectral estimations would be biased by other variables apart from N, for example, water capacity, stand density, and pesticides. In perspective on this, we question the utility of seeking after the best arrangement of spectral bands or the best spectral index. As such, it appears that there is any global set for bands for foreseeing any crop property including N. Rather, a Normalized Sufficiency Index (NSI) approach was proposed to address the intercorrelation between wavelengths and indices along crops, growth stages, years, and regions. NSI can be connected to spectral imagery to standardize the estimations to a non-limiting N reference region with the goal that the capacity of a producer to produce N administration decision can be optimized (Peterson, Blackmer, Francis, & Schepers, 1993). NSI with regards to spectral based N estimation, is determined by the accompanying expression:

$$NSI = \frac{N_i}{N_{ref}} \tag{3.2}$$

where Ni denotes the deliberate estimation of the spectral index; and Nref is the spectral index value of the non-limiting N reference region. NSI was correlated to leaf N fixation and spectral indices/models to standardize them by relative purposes between spectral indices and PLSR prediction models (Nigon et al., 2015). Applying the NSI equation to spectral information made it more robust towards external variables, for example, cultivar and growth stage. Practically, it implies that for an effective utilization of multi-or hyperspectral information the farmer should keep N-rich cultivation areas inside the field (Samborski, Tremblay, & Fallon, 2009) to be utilized as Nref for NSI estimation and N prescription maps for variable N application rate.

Irrigation system basic decision support likewise profits by utilizing devices that screen water crop needs all through the season. Regarding irrigation system, the crop coefficient (Kc) strategy is a pragmatic and solid procedure for assessing evapotranspiration coefficient (ETc) (Allen, Pereira,

Raes, & Smith, 1998), and has been employed by the farmers in different rural districts globally. Keeping that in mind, digital devices are accessible for farmer that keep (or help with storing) accumulated water needed for irrigation as directed by the crops type, growth stage and climate conditions. Be that as it may, Kc has been appeared to change among different regions within the field and between seasons (Kumar, Udeigwe, Clawson, Rohli, & Miller, 2015), creating biased ETc estimations due to climate oddities (Hunsaker, Pinter, Barnes, & Kimball, 2003). One way to deal with this need is by utilizing satellite remote sensing imaging. This methodology is at present supported by the no- charge Sentinel-2 imaging that gives pictures in spatial and temporal resolutions reasonable for precision farming (Rozenstein, Haymann, Kaplan, & Tanny, 2018).

While the Kc approach, utilizes estimations from continuous (or practically ongoing) remote sensing images can give crop growth prediction, it can't gauge crop water status, which is required either to recognize moderate water supply availability or to apply controlled deficit irrigation regime, carried out in different crops including cotton, vineyards and. Keeping that in mind, thermal imaging can be introduced and has been utilized in a few reported cases for uniform and VRI administration (Prenger, Ling, Hansen, & Keener, 2005; O'Shaughnessy, Evett, & Colaizzi, 2015). Cohen, Alchanatis et al. (2017) have demonstrated that for upscaling thermal imaging to gauge and map water status commercially, a well-watered reference is required to be kept up or the virtual reference approach ought to be used.

From the abovementioned, it is inferred that for remotely-sensing supported uniform or VRN fertilizer application, the NSI is proposed to be applied through utilizing multi-otherworldly satellite imagery. Hyperspectral sensing can conceivably give extra data over multispectral sensing. Nonetheless, because of the absence of hyperspectral satellite frameworks with high spatial and temporal resolution, these tools are not yet utilized in currently followed cultivating practices (Hank et al., 2019). Additionally, contrary to the multispectral satellite frameworks, the hyperspectral imaging ones don't have Application Programing Interfaces (APIs) that empower their standard usage that is required for business use. Due the APIs that no-cost satellite pictures like Landsat and Sentinel have, their images are accessible nearly continuously.

They can be envisioned and further processed on clouds given by Google, or big data computational models are needed to be used into by the user. For instance, Google-Earth-Engine (GEE) offers "a multi-petabyte repository of satellite images and geospatial datasets with planetary-scale processing capacities and makes it accessible for researchers, analysts, and engineers to detect changes, map trends, and evaluate anomalies on the Earth's surface". Numerous destinations like https://clim-engine.appspot.com/ provide free of charge and fast examination of time-series of satellite images for different purposes including farming. It implies that the data volume can be visualized and processed on the web and can be more effectively than any time in recent memory augmented in decision support platforms.

Additionally, multispectral satellite platforms are capable of improving uniform and VRI administration. On contrary to the N management, the multispectral images to this end are not used to screen water status, yet rather to screen the crop growth, which thus alters the crop coefficient. Crop water status can be assessed through thermal imaging to further improve water supply effectiveness. Be that as it may, a trade off exists between satellite and airborne thermal imaging regarding spatial coverage and spatio-temporal precision. Satellites give images covering immense regions with limited cost per hectar and have therefore turned into a typical instrument for precision agriculture. Recently, satellite-based images in the region of VIS-NIR have a generally high spatiotemporal resolution, however in the thermal range their best resolutions (90m in Landsat-TM satellite and 120m in ASTER satellite) and their long revisit time are not proper for water supply management. Then again, airborne based thermal images, which, hypothetically, can be obtained every day, have high spatial resolution.

However, they are restricted by their spatial coverage, and are therefore costly per area unit. A trend of thermal imaging application for water status mapping has occurred recently with the expanding accessibility of small size, flexible and uncooled microbolometer based designs. These cameras can be effectively mounted on UAVs. In any case, with a lack of a method for internal reference, they can't be utilized for dependable evaluation of canopy temperature and subsequently solid estimation of the crop water status. At present it appears that they are used incorrectly. New research outcomes have suggested the possibility for retroactive tuning of such cameras (Ribeiro-Gomes et al., 2017) however their use in practice is still under testing.

Improvement of the spatial resolution of the airborne thermal images can likewise be accomplished by utilizing resolution that take advantage of the extensive overlap between consecutive pictures (Cohen, Agam et al., 2017). In the event that pertinent, airborne thermal images might be obtained from higher elevations and coverage.

3.2 Data fusion background

The human brain is incredibly proficient at learning from numerous sources. As depicted in Fig. 3.3, data flows from a variety of sensors are fused and organized by complex processing stages utilizing biochemical energy at the cerebrum. These kinds of coordination and arrangement are predominantly adjusted to the surroundings and incoming information. For instance, someone in the assembly hall is hearing to a speech, the most significant data originates from the visual and hearing related sensory systems.

Despite the fact that at the exact instant the cerebrum is likewise accepting contributions from different senses (e.g., the temperature, the smell, the taste), it flawlessly suppresses these less applicable senses and keeps the focus on the most significant data. This prioritization additionally happens in the senses of a similar classification. For example, some parts of the body (e.g., fingertips, toes, lips) have a lot stronger representations than different less sensitive regions. For human, a few capacities of multisource learning are congenital, though some others are built up by training.

Hall and Llinas (1997) defined data fusion as a combination of data derived from multiple sources including sensors and correlated information from a plethora of relevant data streams. These data streams include Internet of Things and social networks that are used in order to reach more reliable

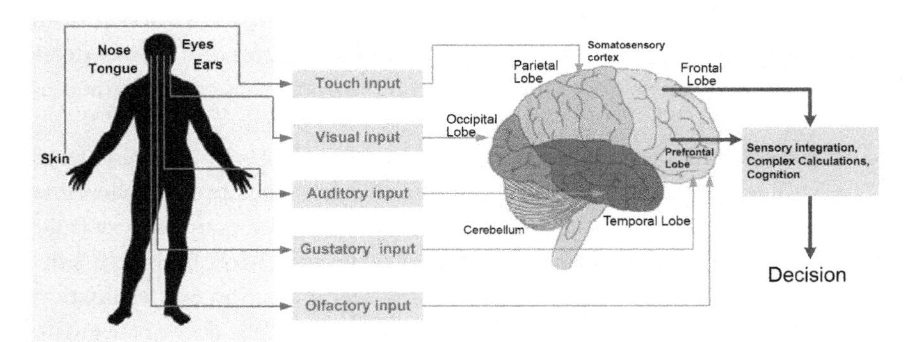

Fig. 3.3 The human decision depends on fusion of different senses. Data attained by the eyes is sent to the occipital projections of the cerebrum. Sound data is processed by the sound-related cortex in the prefrontal lobes. Smell and taste are dissected in the olfactory bulb contained in prefrontal lobes. Tough related data goes to the somatosensory cortex spreading out along the mind surface. Data originates from various sensors is incorporated and processed at the frontal and prefrontal lobes of the brain, where the most perplexing estimations and perceptions happen. (http://www.bhrhospitals.nhs.uk).

conclusions and actionable insights compared to a single source (Klein, 1993). The multiple senses of both humans and animals have contributed to the improvement of their survival capability. For instance, in fact the sense of vision is not reliable regarding the edibility of food but it has to be integrated with other senses like touch, smell, and taste. In the case of animals, a possible prey may not see their forthcoming predator due low visibility caused by high vegetation, but the sense of hearing can save their life functioning as an advance warning. Consequently, both animals and humans employ multisensory data fusion aiming to enhance environment awareness in order to extend their life expectancy.

Multisensor data fusion techniques have been applied extensively for several purposes. Primarily, they have been used in Department of Defence (DoD) applications for example automated threat assessment (e.g., identification-friendfoe-neutral (IFFN) systems), vehicles and generally early defense systems (Hall, Linn, & Llinas, 1991). Data fusion techniques have been further expanded so as to offer solutions to nonmilitary cases, including health monitoring, complex machinery management, safety and operations, robotics (Abidi & Gonzalez, 1992), crop monitoring, remote sensing and medical applications.

3.2.1 Taxonomy of fusion architectures

Sensor data can be assimilated, or fused, in a plethora of ways starting from sensor output to a processed level as state vector of features or fusion of decisions from each sensor (decision level). The fusion levels are described as follows:

1. Raw sensor data fusion concerns data that can be directly combined and is feasible only in the case that the sensor data originate from the same physical manifestations of a natural phenomenon, for example data from a number of cameras or an array of acoustic sensors. Most of the raw sensor data fusion algorithms are based on classic detection and estimation principles. In the opposite case when the sensor data do not originate from the same physical manifestation, the data fusion procedure will need to operate at feature/state vector level or decision level.
2. Feature-level data fusion concerns the production of relevant characteristics from raw sensor data that are correlated to the observed phenomenon. An illustration of feature generation is those physiognomical features of a person to produce a representation of them. This method

is employed often by specialists like artists aiming to bring easily to mind well-known figures. A similar strategy is followed by the human brain in order to identify objects by performing a hierarchical self-organizing fusion. To perform feature-level fusion, the features are produced by processing several raw sensor data streams, and assimilated in a vector format by direct concatenation of feature values to produce one combined state vector that feeds into machine learning algorithms like neural networks either supervised or unsupervised to produce the fusion decision.
3. The decision level fusion concerns the augmentation of sensor based decisions, following the process of determining the identity and other semantic or categorical data about the observed phenomenon. Representative instances that illustrate the function of decision level fusion concern the voting method that assigns weights to the fused decisions, inferential procedures, inference based on prior probabilities and their combinations through Bayesian augmentation and the method of Dempster–Shafer.

Apart from the different data fusion architectures that concerned the level of processing that is applies on the raw sensor data from none to decision based there are other taxonomies base on the semantic information level which are presented below:
1. The centralised architectures comprised of plain algorithms which are cannot follow possible sensor alterations.
2. The hierarchical architectures are based on cooperative processing of sensor data and require bidirectional communication.
3. The decentralized architectures are characterized by robustness to sensor faults and operational regime alterations but a significant drawback of them is that require complex algorithms to function properly.

3.2.2 Data fusion advantages

The combination of data obtained by multisensory networks with data fusion framework enables a faster and lower cost processing, additionally to lowering the level of uncertainty hence guaranteeing higher reliability. These data can be fused in a variety of ways, for instance: linear combiner, combination of posteriors and product of posteriors. The advantages of data fusion compared to classical algorithms include:
1. Increased confidence that is a consequence of the complementary nature of the antecedent sources of information;

2. Reliable navigation information regarding position and state estimation in noisy and rapidly changing environments (limited visibility, overlapping objects);
3. Elevated spatial and temporal covering of important regions of interest and effective tackling of the dimensionality of the input space;
4. Improved separability when comparing hypotheses thanks to fuller and more relevant information availability;
5. Increased system resilience with self-healing capability regarding operation in the event that one or more sensors are in a fault condition;
6. An effective solution for handling the big data that become available from sensor information and other sources like social media, remote sensing and open data repositories.

The main criterion that concerns the optimization of fusion function is the minimal error of identification of the fusion decision compared to actual situation.

3.3 Data fusion applications in precision agriculture

Precision Agriculture depends in a variety of sensors in order to assess the spatial variability in crop and soil characteristics that are needed for site-specific management. The different sensors carry complementary information regarding the crops, soil and subsequently this complementarity calls for a combination of sensor information in order to estimate accurately the crop and soil parameters. This combination of sensor information is the so-called **sensor fusion.**

The approach used by novel fusion methods involves high-level fusion techniques accuracy enhancement in precision agriculture. High-level fusion techniques refer to feature based fusion or decision level fusion. They are necessary for attaining combination features such as spectral signatures, context awareness decisions and texture features in a more precise way. Features combination enhance the precision of image classification and information acquisition. The current algorithms of high-level fusion techniques include the Bayesian theory, fuzzy logic, artificial neural networks etc.

Examples of high-level fusion techniques from different sensors (Hall and Raper, 2005) were employed to classify land use and more precisely vegetation monitoring. Sankey et al. (2018) combined an octocopter UAV with an embedded lidar scanner and hyperspectral imager, yielding a big dataset with augmentation potential for vegetation mapping. The additional advantage of the fusion with respect to the hyperspectral data alone, proves that the lidar fusion and hyperspectral data increase by more than 30% the species identification. This is achieved by the synergy of hyperspectral with the plant height spatial variations mapped by lidar. The fusion method reached an accuracy of 84–89% concerning vegetation classification. On the other hand, the acquired hyperspectral features contributed to an accuracy of 72–76%. Consequently, all the above mentioned methods applied, contributed highly on vegetation monitoring and the detection of ecological alterations.

3.3.1 Fusion of optical images and synthetic aperture radar (SAR) approaches

SAR-based approaches are regarded as an effective research tool for assessing spatial variation and monitoring a crop-planting area. As regards dryland crops, radar reflection complexity is increased due to dependence on the dielectric variability of canopy, its geometric characteristics and the dielectric soil properties and surface texture. The afore mentioned characteristics make the crop identification more complex when SAR-based methods are applied. On the contrary, have indicated that the SAR data features can provide useful information for dryland crops recognition. However, in the case of individual crops, reflection complexity patterns demonstrate reflection variability.

Forkuor (2014) increased the recognition for maize crop by employing a fusion of optical data derived from RapidEye and TerraSAR-X radar data. COSMO-SkyMed (CSK) and Radarsat-2 data combined fused with optical data have been utilized by Sukawattanavijit and Chen (2015), resulting to results of high precision. However, the occasional lack of access to optical data circumstances blocks the function of data fusion. On the contrary, UAV approaches provide optical observations without any interruptions compared to satellite optical remote sensing.

Due to the fact that they are susceptible to common weather phenomena (rain, snow, clouds) that reduce visibility. This results into the limitation of availability of optical remote sensing data in certain periods during crop growth. In the case of SAR, the visibility is independent of cloud existence and lighting. Consequently, SAR data offer several advantages for crop variation assessment (Wang, Skidmore, Wang, Darvishzadeh, & Hearne, 2015).

3.3.2 Fusion of light detection and ranging (LiDAR) data and images

LiDAR is an active sensor that is used routinely for modeling the stereo structure of crops and terrain mapping. Digital Elevation Models (DEM) resulting from LiDAR (Sithole & Vosselman, 2004) and Canopy Height Models (CHM) (Sankey, Glenn, Ehinger, Boehm, & Hardegree, 2010) are possible to be produced with higher accuracy due to the resolution of UAV sensors (James & Robson, 2014). The elevated resolution can make possible the higher definition of individual canopies and their structure together with microtopographic characteristics of the terrain.

A LiDAR sensor produces 3D point clouds taking into consideration the strength of the reflected signals. In certain cases, a burst of pulses can be produced by specific hardware devices (Wagner et al., 2006). Like in the case of optical sensors mounted on the aerial platforms or satellite optical imagers, the raw data need a fair amount of processing to extract the information about the terrain from the from the point cloud of LiDAR.

The combination of LiDAR data and camera produced imaging data has been widely applied in many cases including Digital Surface Model (DSM)/Digital Elevation Model (DEM) production, 3D object detection and simulation to land use identification. The success of this combination is due to the fact that the image obtained from the UAV has higher grander resolution than the resolution of the LiDAR data and the morphological image features are retained better than the similar features obtained from the LiDAR data.

Laser scanning data make possible the determination of the type of land use/land cover and on top of that the morphology of the canopy can be deduced from height variations (Morsdorf et al., 2004). Park et al. (2001) combined LiDAR data and RGB images for land cover identification. The augmentation results have led to the conclusions that different types of imaging sensors provide synergistic content which can increase precision of recognition and lead to a more consistent identification of land cover categories. In case that a burst of pulses is generated or the full wave of the signal are available, then find a resolution components can be revealed that can lead to a better analysis of the vegetation because the emitted light can penetrate the canopies and acquire the reflected signal which originates from branches, trucks and ground. By using this technique, the variability of crop species can be defined at a high resolution (Verrelst et al., 2009).

3.3.3 Fusion of optical images and GIS data

GIS spatial data, including topographic features, land use, streets and demographic data, can be fused with data acquired from remote sensing to enhance precision of image categorization, object identification, alteration detection 3D GIS. The combination of remote sensor and GIS data has contributed significantly to geospatial visualization (Yang, Liu, & Zhang, 2008). However, the different perspectives between them inhibit a direct matching. Hence, the fusion of data from various applications requires the consideration of the deviation of the fusion object representation and the object's category (Weis et al., 2005).

Remote sensing images are formed by a matrix of pixels denoting the color levels in the RGB-domain, while GIS data encloses regions with label metadata, which carry semantic information concerning the region of the objects. It is possible that pixel values are not associated to objects, but signify radiometric features. To include GIS features as supplementary data in the training region determination and the classification result, an effective solution is to use the supplementary data as extra bands. The different characteristics between data and semantics, the comparison of imaging and GIS data cannot be accompliced at the level of a single pixel, but should take place on a level where augmentation of low-level image items to significant objects occurs (Weis et al., 2005).

The object recognition and alteration detection technique represents a way of fusing these two heterogeneous data sources in an effortless way because a default feature of this technique is the image pixels augmentation to label semantic polygons to allow overlay and further analysis with vectorized GIS data.

3.3.4 Data fusion of satellite, aerial and close-range images

Thanks to the emergence of multiple aspect and variable resolution remote sensing, image combination of satellites, airborne platforms and proximal sensing is needed for certain applications, including environmental surveillance, street mapping, traffic mapping, mapping of ancient artifacts, constructions detection, etc. Due to the different resolutions of images ranging from very high resolution to coarse, the augmentation of those images is characterized from the features of the sensors in this resolution. Regarding satellite images, automated geo-referencing can be implemented through the fusion of already existent digitized orthophotos acquired from aerial platforms. Unmanned aerial vehicles (UAVs) represent a novel way for acquiring high resolution and detailed mapping and forming a spatial database with low cost and minimal risk (Anderson & Gaston, 2013) in greater spatial and temporal resolution. However, UAV imagery usually achieve reduced coverage due to their limited autonomy.

Recent progress led to elevated use of UAVs in sensor technology (Valavanis & Vachtsevanos, 2015). Current UAV sensors can acquire image with a precision of 1 cm at hourly intervals (Javernick, Brasington, & Caruso, 2014). Their high temporal and spatial resolution capability at a range between 1 and 100 hectares, characterize the UAVs ideal for accurate and repetitive flight campaigns. UAV-obtained information have shown potential in in biodiversity monitoring and alterations detection concerning land use (Koh & Wich, 2012), precision agriculture (Hunt et al., 2010), water management (DeBell, Anderson, Brazier, King, & Jones, 2015), and biotic and abiotic stress detection (Zarco-Tejada, Gonzalez-Dugo, & Berni, 2012). UAV extracted vegetation indices and more notably the Normalized Difference Vegetation Index (NDVI) have been associated with biophysical features like the leaf area index (LAI) and nitrogen consumption (Lelong et al., 2008).

The fusion of airborne sensors and vehicle-mounted sensors can be employed for enhanced semantic annotation (Zhang, Zhao, Horn, & Baumgartl, 2001). The vehicle mounted sensor data are acquired by ground platform with a mounted calibrated camera, differential Global Positioning System (GPS) and laser scanner. The semantic data are applied to the local and global subdivision of an aerial image into segments. After correlating the vehicle images to the characteristics of an airborne image, the data are applied as parameters in a segmentation algorithm to be used for classification and 3D modeling.

Guan et al. (2017) introduced a couple of important synergistic effects concerning of satellite features for prediction crop yield. One of the satellites features serves as the surrogate of the canopy biomass, as has been measured by the reflection in the GCVI that was used as indicator of the presented study. The other satellite features indicated the environmental stress which was useful since the climatic parameters could not be predicted. Hence, the fusion of multiple sensors can achieve the discrimination of stresses based on their type either biotic or abiotic stresses.

3.3.5 Optical and fluorescence imaging

There is substantial evidence concerning the detectability of crop infections and the potential to automate it (Moshou et al., 2004). The indicators of disease occurrence cited by this work are alterations in electromagnetic spectra. The main assumption is that plants in healthy condition respond to incident electromagnetic waves differently in comparison to diseased plants. Some plants' infection symptoms are often related to different plants' optical properties that can be indicated either by human observation, or detected by utilizing advanced instruments. Fluorescence is regarded as the most reliable approach for fungal disease detection at an early stage, since it assesses the health condition regarding how efficient the phytosynthesis process is (Lee et al., 2010).

The leaf pigments absorb partially the incident visible light as a driver of photosynthetic activity and another part is redirected to the production of fluorescence. The fluorescence emission appears at VIS-NIR waveband demonstrating a peak emission at 690 nm while a part of energy is dissipated as heat radiation appearing in the thermal infrared region (TIR). Fluorescence by itself is not able to reach a conclusion regarding plant stress detection, reaching an average certainty of detection, due to light level sensitivity. At an early stage of infection, metabolic alterations take place, while the fungus activity extends radially around the infection area. The center of infected area shows signs of a necrosis, there is loss of pigmentation, the photosynthetic system is damages as the cell walls break down, allowing the infection symptoms to become visible. The credibility of disease detection is relies on the utilized algorithms. Every case of crop and disease, spectral imaging techniques are available to make user friendly and automate infection assessment, by employing Data mining algorithms like ANNs and Machine Learning, or other AI techniques.

Lee et al. (2010) suggests that the light reflection analysis can contribute to detect possible infections. The pathogen infected areas can be spotted in the visible spectrum range from 400 to 700 nm. The pathogen appearance is highly related to its nature; the levels of chlorophyll alterations in the visible spectral bands and red edge bands. Moreover leaf browning symptoms are related to senescence in the visible spectral bands and short-wave near infrared range from 680 to 800 nm while near-infrared range from 1400 to 1600 nm and 1900–2100 nm are highly associated to water deficiency, several canopy alterations and to the general leaf area.

Consequently as the disease spreads on the entire plant, a general closure of the stomata occurs aiming to prevent further water losses. This stage can be detected and monitored by thermal infra-red imaging sensors. Due to the rapid changes of leaf temperature and the dependence on ambient temperature along with illumination and wind, thermography provides a poor estimation of temperature distribution if it is used as component proximal sensing platforms.

Fluorescence is regarded as a useful technique for Early stage disease identification even before leaf alterations take place. However, is not capable of identifying the type of stress factor and the level of infection providing an average detection accuracy. Moreover, the accuracy fluctuates with light intensity. Consequently, it can be used as an early alarm for the anticipating of future health plant conditions and is considered as ideal for laboratory use or in absence of light.

References

Abidi, M. A., & Gonzalez, R. C. (1992). *Data Fusion in Robotics and Machine Intelligence.* Boston, MA: Academic.

Adao, T., Hruska, J., Padua, L., Bessa, J., Peres, E., Morais, R., et al. (2017). Hyperspectral imaging: a review on UAV-based sensors, data processing and applications for agriculture and forestry. *Remote Sensing, 9*(11), 30.

Agam, N., Segal, E., Peeters, A., Levi, A., Dag, A., Yermiyahu, U., et al. (2013). Spatial distribution of water status in irrigated olive orchards by thermal imaging. *Precision Agriculture, 15*, 346–359.

Alchanatis, V., Cohen, Y., Cohen, S., Moller, M., Sprinstin, M., Meron, M., et al. (2010). Evaluation of different approaches for estimating and mapping crop water status in cotton with thermal imaging. *Precision Agriculture, 11*(1), 27–41.

Allen, R. G., Pereira, L. S., Raes, D., & Smith, M. (1998). Crop evapotranspiration: Guidelines for computing crop water requirements. *FAO Irrigation and Drainage Paper, 56*, 300.

Anderson, K., & Gaston, K. J. (2013). Lightweight unmanned aerial vehicles will revolutionize spatial ecology. *Frontiers in Ecology and the Environment, 11*, 138–146.

Anthony, D., Elbaum, S., Lorenz, A., & Detweiler, C. (2014). On crop height estimation with UAVs. In *Paper presented at the Intelligent Robots and Systems (IROS 2014), 2014 IEEE/RSJ International Conference* (pp. 4805–4812).

Apostol, S., Viau, A. A., Tremblay, N., Briantais, J. -M., Prasher, S., Parent, L. -E., et al. (2003). Laser-induced fluorescence signatures as a tool for remote monitoring of water and nitrogen stresses in plants. *Canadian Journal of Remote Sensing, 29*, 57–65.

Bakhshipour, A., Jafari, A., Nassiri, S. M., & Zare, D. (2017). Weed segmentation using texture features extracted from wavelet sub-images. *Biosystems Engineering, 157*, 1–12.

Basso, B., Ritchie, J., Pierce, F., Braga, R., & Jones, J. (2001). Spatial validation of crop models for precision agriculture. *Agricultural Systems, 68*(2), 97–112.

Behmann, J., Mahlein, A. K., Rumpf, T., Römer, C., & Plümer, L. (2015). A review of advanced machine learning methods for the detection of biotic stress in precision crop protection. *Precision Agriculture, 16*(3), 239–260.

Belasque, J., Jr., Gasparoto, M., & Marcassa, L. G. (2008). Detection of mechanical and disease stresses in citrus plants by fluorescence spectroscopy. *Applied Optics*, *47*(11), 1922–1926.

Bellon-Maurel, V., & McBratney, A. (2011). Near-infrared (NIR) and mid-infrared (MIR) spectroscopic techniques for assessing the amount of carbon stock in soils—critical review and research perspectives. *Soil Biology and Biochemistry*, *43*, 1398–1410.

Bendig, J., Bolten, A., Bennertz, S., Broscheit, J., Eichfuss, S., & Bareth, G. (2014). Estimating biomass of barley using crop surface models (CSMs) derived from UAV-based RGB imaging. *Remote Sensing*, *6*(11), 10395–10412.

Ben-Dor, E., Chabrillat, S., Demattê, J. A. M., Taylor, G. R., Hill, J., Whiting, M. L., et al. (2009). Using imaging spectroscopy to study soil properties. *Remote Sensing of Environment*, *113*, S38–S55.

Borregaard, T., Nielsen, H., Nørgaard, L., & Have, H. (2000). Crop-weed discrimination by line imaging spectroscopy. *Journal of Agricultural Engineering Research*, *75*(4), 389–400.

Bosco, G. L. (2013). Development and application of portable, hand-held X-ray fluorescence spectrometers. *TrAC Trends in Analytical Chemistry*, *45*, 121–134.

Bravo, C., Moshou, D., West, J., McCartney, A., & Ramon, H. (2003). Detailed spectral reflection information for early disease detection in wheat fields. *Biosystems Engineering*, *84*(2), 137–145.

Bregaglio, S., Frasso, N., Pagani, V., Stella, T., Francone, C., Cappelli, G., et al. (2015). New multi-model approach gives good estimations of wheat yield under semi-arid climate in Morocco. *Agronomy for Sustainable Development*, *35*(1), 157–167.

Brown, D. J., Shepherd, K. D., Walsh, M. G., Dewayne Mays, M., & Reinsch, T. G. (2006). Global soil characterization with VNIR diffuse reflectance spectroscopy. *Geoderma*, *132*, 273–290.

Casady, G. M., Hanley, R. S., & Seelan, S. K. (2005). Detection of leafy spurge (*Euphorbia esula*) using multidate high-resolution satellite imagery. *Weed Technology*, *19*(2), 462–467.

Castrignanò, A., Wong, M. T. F., Stelluti, M., De Benedetto, D., & Sollitto, D. (2012). Use of EMI, gamma-ray emission and GPS height as multi-sensor data for soil characterisation. *Geoderma*, *175–176*, 78–89.

Ceccato, P., Gobron, N., Flasse, S., Pinty, B., & Tarantola, S. (2002). Designing a spectral index to estimate vegetation water content from remote sensing data: Part 1—Theoretical approach. *Remote Sensing of Environment*, *82*, 188–197.

Cheng, Z., Meng, J., & Wang, Y. (2016). Improving spring maize yield estimation at field scale by assimilating time-series HJ-1 CCD data into the WOFOST model using a new method with fast algorithms. *Remote Sensing*, *8*(4), 303.

Chlingaryan, A., Sukkarieh, S., & Whelan, B. (2018). Machine learning approaches for crop yield prediction and nitrogen status estimation in precision agriculture: A review. *Computers and Electronics in Agriculture*, *151*, 61–69.

Chopping, M., Moisen, G. G., Su, L., Laliberte, A., Rango, A., Martonchik, J. V., et al. (2008). Large area mapping of southwestern forest crown cover, canopy height, and biomass using the NASA multiangle imaging spectro-radiometer. *Remote Sensing of Environment*, *112*(5), 2051–2063.

Chu, T., Starek, M. J., Brewer, M. J., Murray, S. C., & Pruter, L. S. (2018). Characterizing canopy height with UAS structure-from-motion photogrammetry—results analysis of a maize field trial with respect to multiple factors. *Remote Sensing Letters*, *9*(8), 753–762.

Clevers, J. G. P. W., Kooistra, L., & Schaepman, M. E. (2008). Using spectral information from the NIR water absorption features for the retrieval of canopy water content. *International Journal of Applied Earth Observation and Geoinformation*, *10*, 388–397.

Cohen, Y., Agam, N., Klapp, I., Karnieli, A., Beeri, O., Alchanatis, V., et al. (2017). Future approaches to facilitate large-scale adoption of thermal based images as key input in the production of dynamic irrigation management zones. *Advances in Animal Biosciences*, *8*, 546–550.

Cohen, Y., Alchanatis, V., Meron, M., Saranga, Y., & Tsipris, J. (2005). Estimation of leaf water potential by thermal imagery and spatial analysis. *Journal of Experimental Botany*, *56*, 1843–1852.

Cohen, Y., Alchanatis, V., Zusman, Y., Dar, Z., Bonfil, D. J., Karnieli, A., et al. (2010). Leaf nitrogen estimation in potato based on spectral data and on simulated bands of the VENμS satellite. *Precision Agriculture*, *11*, 520–537.

Cohen, Y., Alchanatis, V., Saranga, Y., Rosenberg, O., Sela, E., & Bosak, A. (2017). Mapping water status based on aerial thermal imagery: Comparison of methodologies for upscaling from a single leaf to commercial fields. *Precision Agriculture*, *18*, 801–822.

Cook, S. E., Corner, R. J., Groves, P. R., & Grealish, G. J. (1996). Use of airborne gamma radiometric data for soil mapping. *Australian Journal of Soil Research*, *34*(1), 183–194.

Danielsen, S., & Munk, L. (2004). Evaluation of disease assessment methods in quinoa for their ability to predict yield loss caused by downy mildew. *Crop Protection*, *23*(3), 219–228.

Dash, J., & Curran, P. J. (2004). The MERIS terrestrial chlorophyll index. *International Journal of Remote Sensing*, *25*, 5403–5413.

De Groot, A. V., van der Graaf, E. R., de Meijer, R. J., & Maučec, M. (2009). Sensitivity of in-situ γ-ray spectra to soil density and water content. *Nuclear Instruments and Methods in Physics Research Section A: Accelerators, Spectrometers, Detectors and Associated Equipment*, *600*(2), 519–523.

DeBell, L., Anderson, K., Brazier, R. E., King, N., & Jones, L. (2015). Water resource management at catchment scales using lightweight UAVs: current capabilities and future perspectives. *Journal of Unmanned Vehicle Systems*, *3*, 1–24.

Delloye, C., Weiss, M., & Defourny, P. (2018). Retrieval of the canopy chlorophyll content from Sentinel-2 spectral bands to estimate nitrogen uptake in intensive winter wheat cropping systems. *Remote Sensing of Environment*, *216*, 245–261.

Dennerley, C., Huang, J., Nielson, R., Sefton, M., & Triantafilis, J. (2018). Identifying soil management zones in a sugarcane field using proximal sensed electromagnetic induction and gamma-ray spectrometry data. *Soil Use and Management*, *34*, 219–235.

Devadas, R., Lamb, D., Simpfendorfer, S., & Backhouse, D. (2009). Evaluating ten spectral vegetation indices for identifying rust infection in individual wheat leaves. *Precision Agriculture*, *10*(6), 459–470.

Dhawale, N. M., Adamchuk, V. I., Prasher, S. O., Viscarra Rossel, R. A., Ismail, A. A., & Kaur, J. (2015). Proximal soil sensing of soil texture and organic matter with a prototype portable mid-infrared spectrometer. *European Journal of Soil Science*, *66*, 661–669.

Dickson, B. L., & Scott, K. M. (1997). Interpretation of aerial gamma-ray surveys? Adding the geochemical factors. *AGSO Journal of Australian Geology and Geophysics*, *17*, 187–200.

Dierke, C., & Werban, U. (2013). Relationships between gamma-ray data and soil properties at an agricultural test site. *Geoderma*, *199*, 90–98.

Dorigo, W. A., Zurita-Milla, R., de Wit, A. J. W., Brazile, J., Singh, R., & Schaepman, M. E. (2007). A review on reflective remote sensing and data assimilation techniques for enhanced agroecosystem modeling. *International Journal of Applied Earth Observation and Geoinformation*, *9*(2), 165–193.

Douglas, R. K., Nawar, S., Alamar, M. C., Coulon, F., & Mouazen, A. M. (2019). Rapid detection of alkanes and polycyclic aromatic hydrocarbons in oil-contaminated soil with visible near-infrared spectroscopy. *European Journal of Soil Sciences*, *70*(1), 14–150.

Duarte-Carvajalino, J., Alzate, D., Ramirez, A., Santa-Sepulveda, J., Fajardo-Rojas, A., & Soto-Suárez, M. (2018). Evaluating late blight severity in potato crops using unmanned aerial vehicles and machine learning algorithms. *Remote Sensing*, *10*(10), 1513.

Dudka, M., Langton, S., Shuler, R., Kurle, J., & Grau, C. (1999). Use of digital imagery to evaluate disease incidence and yield loss caused by sclerotinia stem rot of soybeans. *Precision Agriculture*, 1549–1558.

Elavarasan, D., Vincent, D. R., Sharma, V., Zomaya, A. Y., & Srinivasan, K. (2018). Forecasting yield by integrating agrarian factors and machine learning models: A survey. *Computers and Electronics in Agriculture, 155*, 257–282.

Errebhi, M., Rosen, C. J., Gupta, S. C., & Birong, D. E. (1998). Potato yield response and nitrate leaching as influenced by nitrogen management. *Agronomy Journal, 90*, 10–15.

Forkuor, G. (2014). Agricultural land use mapping in West Africa using multi-sensor satellite imagery, University of Wuerzburg, Wuerzburg, Germany.

Fuchs, M., & Tanner, C. B. (1966). Infrared thermometry of vegetation 1. *Agronomy Journal, 58*, 597–601.

Gates, W. P. (2006). X-ray absorption spectroscopy. In F. Bergaya, B. K. G. Theng, & G. Lagaly (Eds.), *Developments in Clay Science: vol. 1. Handbook of clay science* (pp. 789–864): Elsevier Ltd.

Gates, D. M., Keegan, H. J., Schleter, J. C., & Weidner, V. R. (1965). Spectral properties of plants. *Applied Optics, 4*, 11–20.

Gevaert, C., Tang, J., García-Haro, F., Suomalainen, J., & Kooistra, L. (2014). Combining hyperspectral UAV and multispectral Formosat-2 imagery for precision agriculture applications. In *Paper presented at the 2014 6th Workshop on Hyperspectral Image and Signal Processing: Evolution in Remote Sensing (WHISPERS), IEEE* (pp. 1–4).

Goel, P., Prasher, S., Patel, R., Smith, D., & DiTommaso, A. (2002). Use of airborne multispectral imagery for weed detection in field crops. *Transactions of ASAE, 45*(2), 443.

Gonzalez-Dugo, V., Zarco-Tejada, P., Berni, J. A. J., Suárez, L., Goldhamer, D., & Fereres, E. (2012). Almond tree canopy temperature reveals intra-crown variability that is water stress-dependent. *Agricultural and Forest Meteorology, 154–155*, 156–165.

Gonzalez-Dugo, V., Zarco-Tejada, P., Nicolás, E., Nortes, P. A., Alarcón, J. J., Intrigliolo, D. S., et al. (2013). Using high resolution UAV thermal imagery to assess the variability in the water status of five fruit tree species within a commercial orchard. *Precision Agriculture, 14*, 660–678.

Grasty, R. L. (1997). Applications of gamma radiation in remote sensing. In E. Schanda (Ed.), *Remote Sensing for Environmental Science* (pp. 257–276). New York: Springer-Verlag.

Green, D., Burpee, L., & Stevenson, K. (1998). Canopy reflectance as a measure of disease in tall fescue. *Crop Science, 38*, 1603–1613.

Guan, J., & Nutter, F., Jr. (2002). Relationships between defoliation, leaf area index, canopy reflectance, and forage yield in the alfalfa-leaf spot pathosystem. *Computers and Electronics in Agriculture, 37*(1–3), 97–112.

Guan, K., Wu, J., Kimball, J. S., Anderson, M. C., Frolking, S., Li, B., et al. (2017). The shared and unique values of optical, fluorescence, thermal and microwave satellite data for estimating large-scale crop yields. *Remote Sensing of Environment, 199*, 333–349.

Haghverdi, A., Washington-Allen, R. A., & Leib, B. G. (2018). Prediction of cotton lint yield from phenology of crop indices using artificial neural networks. *Computers and Electronics in Agriculture, 152*, 186–197.

Hall, D. L., Linn, R. J., & Llinas, J. (1991). A survey of data fusion systems. In *vol. 1470. Proceedings of SPIE conference on data structure and target classification* (pp. 13–136). Orlando, FL, Apr. 1991.

Hall, D. L., & Llinas, J. (1997). An introduction to multisensor data fusion. *Proceedings of the IEEE, 85*(1), 6–23.

Hall, H. E., & Raper, R. L. (2005). Development and concept evaluation of an on-the-go soil strength measurement system. *Transactions of the ASAE, 48*(2), 469–477.

Hank, T. B., Berger, K., Bach, H., Clevers, J. G., Gitelson, A., Zarco-Tejada, P., et al. (2019). Spaceborne imaging spectroscopy for sustainable agriculture: Contributions and challenges. *Surveys in Geophysics*, *40*(3), 515–551.

Hanks, J. E., & Beck, J. L. (1998). Sensor-controlled hooded sprayer for row crops. *Weed Technology*, 308–314.

Heggemann, T., Welp, G., Amelung, W., Angst, G., Franz, S. O., Koszinski, S., et al. (2017). Proximal gamma-ray spectrometry for site-independent in situ prediction of soil texture on ten heterogeneous fields in Germany using support vector machines. *Soil & Tillage Research*, *168*, 99–109.

Helman, D., Bahat, I., Netzer, Y., Ben-Gal, A., Alchanatis, V., Peeters, A., et al. (2018). Using time series of high-resolution planet satellite images to monitor grapevine stem water potential in commercial vineyards. *Remote Sensing*, *10*, 1615.

Herrmann, I., Karnieli, A., Bonfil, D. J., Cohen, Y., & Alchanatis, V. (2010). SWIR-based spectral indices for assessing nitrogen content in potato fields. *International Journal of Remote Sensing*, *31*, 5127–5143.

Horta, A., Malone, B., Stockmann, U., Minasny, B., Bishop, T. F. A., McBratney, A. B., et al. (2015). Potential of integrated field spectroscopy and spatial analysis for enhanced assessment of soil contamination: A prospective review. *Geoderma*, *241–242*, 180–209.

Huang, W., Lamb, D. W., Niu, Z., Zhang, Y., Liu, L., & Wang, J. (2007). Identification of yellow rust in wheat using in-situ spectral reflectance measurements and airborne hyperspectral imaging. *Precision Agriculture*, *8*(4–5), 187–197.

Huang, J., Lark, R. M., Robinson, D. A., Lebron, I., Keith, A. M., Rawlins, B., et al. (2014). Scope to predict soil properties at within-field scale from small samples using proximally sensed γ-ray spectrometer and EM induction data. *Geoderma*, *232–234*, 69–80.

Hunsaker, D. J., Pinter, P. J., Barnes, E. M., & Kimball, B. A. (2003). Estimating cotton evapotranspiration crop coefficients with a multispectral vegetation index. *Irrigation Science*, *22*, 95–104.

Hunt, E. R., Hively, W. D., Fujikawa, S. J., Linden, D. S., Daughtry, C. S., & McCarty, G. W. (2010). Acquisition of NIR green-blue digital photographs from unmanned aircraft for crop monitoring. *Remote Sensing*, *2*, 290–305.

Hunt, E. R., Jr., & Daughtry, C. S. (2018). What good are unmanned aircraft systems for agricultural remote sensing and precision agriculture? *International Journal of Remote Sensing*, *39*(15-16), 5345–5376.

Hunt, E. R., & Rock, B. N. (1989). Detection of changes in leaf water-content using near infrared and middle-infrared reflectances. *Remote Sensing of Environment*, *30*, 43–54.

Hunt, E. R., & Rondon, S. I. (2017). Detection of potato beetle damage using remote sensing from small unmanned aircraft systems. *Journal of Applied Remote Sensing*, *11*(2), 26013.

Jackson, R. D., Idso, S., Reginato, R., & Pinter, P., Jr. (1981). Canopy temperature as a crop water stress indicator. *Water Resources Research*, *17*(4), 1133–1138.

Jacquemoud, S., & Ustin, S. L. (2001). Leaf optical properties: A state of the art. In *Paper presented at the 8th International Symposium of Physical Measurements & Signatures in Remote Sensing* (pp. 223–332). Aussois France: CNES.

James, M. R., & Robson, S. (2014). Mitigating systematic error in topographic models derived from UAV and ground-based image networks. *Earth Surface Processes and Landforms*, *39*, 1413–1420.

Jarvis, P. G. (1976). The interpretation of the variations in leaf water potential and stomatal conductance found in canopies in the field. *Philosophical Transactions of the Royal Society of London. Series B, Biological Sciences*, *273*, 593–610.

Javernick, L., Brasington, J., & Caruso, B. (2014). Modeling the topography of shallow braided rivers using Structure-from-Motion photogrammetry. *Geomorphology*, *213*, 166–182.

Ji, W., Adamchuk, V. I., Biswas, A., Dhawale, N. M., Sudarsan, B., Zhang, Y., et al. (2016). Assessment of soil properties in situ using a prototype portable MIR spectrometer in two agricultural fields. *Biosystems Engineering, 152*, 14–27.

Jones, H. G. (1999). Use of infrared thermometry for estimation of stomatal conductance as a possible aid to irrigation scheduling. *Agricultural and Forest Meteorology, 95*, 139–149.

Kaniu, M. I., Angeyo, K. H., Mwala, A. K., & Mangala, M. J. (2012). Direct rapid analysis of trace bioavailable soil macronutrients by chemometrics-assisted energy dispersive X-ray fluorescence and scattering spectrometry. *Analytica Chimica Acta, 729*, 21–25.

Kaniu, M. I., Angeyo, K. H., Mwala, A. K., & Mwangi, F. K. (2012). Energy dispersive X-ray fluorescence and scattering assessment of soil quality via partial least squares and artificial neural networks analytical modeling approaches. *Talanta, 98*, 236–240.

Kawamura, K., Tsujimoto, Y., Rabenarivo, M., Asai, H., Andriamananjara, A., & Rakotoson, T. (2017). Vis-NIR spectroscopy and PLS regression with waveband selection for estimating the total C and N of paddy soils in Madagascar. *Remote Sensing, 9*(10), 1081.

Khanal, S., Fulton, J., & Shearer, S. (2017). An overview of current and potential applications of thermal remote sensing in precision agriculture. *Computers and Electronics in Agriculture, 139*, 22–32.

Klein, L. A. (1993). *Sensor and data fusion concepts and applications. vol. 14.* Bellingham, Washington, United States: SPIE Opt. Engineering Press, Tutorial Texts.

Koh, L. P., & Wich, S. A. (2012). Dawn of drone ecology: Low-cost autonomous aerial vehicles for conservation. *Tropical Conservation Science, 5*, 121–132.

Kuang, B., Mahmood, H. S., Quraishi, M. Z., Hoogmoed, W. B., Mouazen, A. M., & van Henten, E. J. (2012). Sensing soil properties in the laboratory, in situ, and on-line: A review. *Advances in Agronomy, 114*, 155–223.

Kumar, V., Udeigwe, T. K., Clawson, E. L., Rohli, R. V., & Miller, D. K. (2015). Crop water use and stage-specific crop coefficients for irrigated cotton in the mid-south, United States. *Agricultural Water Management, 156*, 63–69.

Küng, O., Strecha, C., Beyeler, A., Zufferey, J. -C., Floreano, D., Fua, P., et al. (2011). The accuracy of automatic photogrammetric techniques on ultra-light UAV imagery. In *Paper presented at the UAV-g 2011-Unmanned Aerial Vehicle in Geomatics.*

Lambert, M. -J., Blaes, X., Traoré, P. S., & Defourny, P. (2017). Estimate yield at parcel level from S2 time series in sub-Saharan smallholder farming systems. In *Paper presented at the Analysis of Multitemporal Remote Sensing Images (MultiTemp), 2017 9th International Workshop on the, IEEE* (pp. 1–7).

Lee, W. S., Alchanatis, V., Yang, C., Hirafuji, M., Moshou, D., & Li, C. (2010). Sensing technologies for precision specialty crop production. *Computers and Electronics in Agriculture, 74*(1), 2–33.

Lee, W. S., Slaughter, D., & Giles, D. (1999). Robotic weed control system for tomatoes. *Precision Agriculture, 1*(1), 95–113.

Lefsky, M. A., Cohen, W. B., Parker, G. G., & Harding, D. J. (2002). Lidar remote sensing for ecosystem studies. *BioScience, 52*(1), 19–30.

Lelong, C. C., Burger, P., Jubelin, G., Roux, B., Labbe, S., & Baret, F. (2008). Assessment of unmanned aerial vehicles imagery for quantitative monitoring of wheat crop in small plots. *Sensors, 8*, 3557–3585.

Leon, C. T., Shaw, D. R., Cox, M. S., Abshire, M. J., Ward, B., Wardlaw, M. C., et al. (2003). Utility of remote sensing in predicting crop and soil characteristics. *Precision Agriculture, 4*(4), 359–384.

Li, F., Miao, Y. X., Hennig, S. D., Gnyp, M. L., Chen, X. P., Jia, L. L., et al. (2010). Evaluating hyperspectral vegetation indices for estimating nitrogen concentration of winter wheat at different growth stages. *Precision Agriculture, 11*, 335–357.

Li, W., Niu, Z., Chen, H., & Li, D. (2017). Characterizing canopy structural complexity for the estimation of maize LAI based on ALS data and UAV stereo images. *International Journal of Remote Sensing*, *38*(8–10), 2106–2116.

Li, W., Niu, Z., Chen, H., Li, D., Wu, M., & Zhao, W. (2016). Remote estimation of canopy height and aboveground biomass of maize using high-resolution stereo images from a low-cost unmanned aerial vehicle system. *Ecological Indicators*, *67*, 637–648.

Li, W., Niu, Z., Huang, N., Wang, C., Gao, S., & Wu, C. (2015). Airborne LiDAR technique for estimating biomass components of maize: A case study in Zhangye City, Northwest China. *Ecological Indicators*, *57*, 486–496.

Li, W., Niu, Z., Wang, C., Huang, W., Chen, H., Gao, S., et al. (2015). Combined use of airborne LiDAR and satellite GF-1 data to estimate leaf area index, height, and aboveground biomass of maize during peak growing season. *IEEE Journal of Selected Topics in Applied Earth Observations and Remote Sensing*, *8*(9), 4489–4501.

Lobell, D. B., & Burke, M. B. (2010). On the use of statistical models to predict crop yield responses to climate change. *Agricultural and Forest Meteorology*, *150*(11), 1443–1452.

Lopez-Granados, F. (2011). Weed detection for site-specific weed management: Mapping and real-time approaches. *Weed Research*, *51*(1), 1–11.

Logan, E., Lee, J., Landis, E., Custer, S., Bennett, A., Fulton, J., et al. (2018). *Crop protection apps*. Columbus, OH: Ohio State Univ. Extension. Service Bull. FABE-552.03.

Lottes, P., Hörferlin, M., Sander, S., & Stachniss, C. (2017). Effective vision-based classification for separating sugar beets and weeds for precision farming. *Journal of Field Robotics*, *34*(6), 1160–1178.

Maes, W. H., & Steppe, K. (2012). Estimating evapotranspiration and drought stress with ground-based thermal remote sensing in agriculture: A review. *Journal of Experimental Botany*, *63*, 4671–4712.

Mahlein, A. -K. (2016). Plant disease detection by imaging sensors—parallels and specific demands for precision agriculture and plant phenotyping. *Plant Disease*, *100*(2), 241–251.

Mahmood, H. S., Hoogmoed, W. B., & van Henten, E. J. (2013). Proximal gamma-ray spectroscopy to predict soil properties using windows and full-spectrum analysis methods. *Sensors*, *13*, 16263–16280.

Malley, D., Martin, P., & Ben-Dor, E. (2004). Application in analysis of soils. In C. A. Roberts, J. Workman, & J. B. Reeves III, (Eds.), *Near-infrared spectroscopy in agriculture* (pp. 729–784). Madison, USA: A Three Society Monograph (ASA, SSSA, CSSA).

Malthus, T. J., & Madeira, A. C. (1993). High resolution spectroradiometry: Spectral reflectance of field bean leaves infected by Botrytis fabae. *Remote Sensing of Environment*, *45*(1), 107–116.

Mathews, A. J., & Jensen, J. L. (2013). Visualizing and quantifying vineyard canopy LAI using an unmanned aerial vehicle (UAV) collected high density structure from motion point cloud. *Remote Sensing*, *5*(5), 2164–2183.

McClure, W. F. (2003). 204 Years of near Infrared Technology: 1800–2003. *Journal of Near Infrared Spectroscopy*, *11*, 487–518.

Meng, J., & Bradley, K. (2016). *ID Weeds App*. Columbus, MO: Univ. Missouri Extension Service.

Minasny, B., McBratney, A. B., Bellon-Maurel, V., Roger, J. -M., Gobrecht, A., Ferrand, L., et al. (2011). Removing the effect of soil moisture from NIR diffuse reflectance spectra for the prediction of soil organic carbon. *Geoderma*, *167*, 118–124.

Moharana, S., & Dutta, S. (2016). Spatial variability of chlorophyll and nitrogen content of rice from hyperspectral imagery. *ISPRS Journal of Photogrammetry and Remote Sensing*, *122*, 17–29.

Möller, M., Alchanatis, V., Cohen, Y., Meron, M., Tsipris, J., Naor, A., et al. (2007). Use of thermal and visible imagery for estimating crop water status of irrigated grapevine. *Journal of Experimental Botany*, *58*, 827–838.

Montalvo, M., Pajares, G., Guerrero, J. M., Romeo, J., Guijarro, M., Ribeiro, A., et al. (2012). Automatic detection of crop rows in maize fields with high weeds pressure. *Expert Systems with Applications, 39*(15), 11889–11897.

Moran, M. S., Inoue, Y., & Barnes, E. M. (1997). Opportunities and limitations for image-based remote sensing in precision crop management. *Remote Sensing of Environment, 61*, 319–346.

Morsdorf, F., et al. (2004). LiDAR-based geometric reconstruction of boreal type forest stands at single tree level for forest and wildland fire management. *Remote Sensing of Environment, 92*, 353–362.

Moshou, D., Bravo, C., Oberti, R., West, J., Bodria, L., McCartney, A., et al. (2005). Plant disease detection based on data fusion of hyper-spectral and multi-spectral fluorescence imaging using Kohonen maps. *Real-Time Imaging, 11*(2), 75–83.

Moshou, D., Bravo, C., Oberti, R., West, J., Ramon, H., Vougioukas, S., et al. (2011). Intelligent multi-sensor system for the detection and treatment of fungal diseases in arable crops. *Biosystems Engineering, 108*(4), 311–321.

Moshou, D., Bravo, C., Wahlen, S., West, J., McCartney, A., De Baerdemaeker, J., et al. (2006). Simultaneous identification of plant stresses and diseases in arable crops using proximal optical sensing and self-organising maps. *Precision Agriculture*, 7(3), 149–164.

Moshou, D., Bravo, C., West, J., McCartney, A., & Ramon, H. (2004). Automatic detection of 'yellow rust' in wheat using reflectance measurements and neural networks. *Computers and Electronics in Agriculture, 44*(3), 173–188.

Mouazen, A. M. (2006). *Soil Survey Device. International publication published under the patent cooperation treaty (PCT).* World Intellectual Property Organization, International Bureau. International Publication Number: WO2006/015463; PCT/BE2005/000129; IPC: G01N21/00; G01N21/00.

Mouazen, A. M., & Kuang, B. (2016). On-line visible and near infrared spectroscopy for in-field phosphorous management. *Soil & Tillage Research, 155*, 471–477.

Mouazen, A. M., Kuang, B., De Baerdemaeker, J., & Ramon, H. (2010). Comparison between principal component, partial least squares and artificial neural network analyses for accuracy of measurement of selected soil properties with visible and near infrared spectroscopy. *Geoderma, 158*, 23–31.

Muhammed, H. H., & Larsolle, A. (2003). Feature vector based analysis of hyperspectral crop reflectance data for discrimination and quantification of fungal disease severity in wheat. *Biosystems Engineering, 86*(2), 125–134.

Mukhopadhyay, P., & Chaudhuri, B. B. (2015). A survey of Hough Transform. *Pattern Recognition, 48*(3), 993–1010.

Mulla, D. (2013). Twenty five years of remote sensing in precision agriculture: Key advances and remaining knowledge gaps. *Biosystems Engineering, 114*, 358–371.

Naidu, R. A., Perry, E. M., Pierce, F. J., & Mekuria, T. (2009). The potential of spectral reflectance technique for the detection of Grapevine leafroll-associated virus-3 in two red-berried wine grape cultivars. *Computers and Electronics in Agriculture, 66*(1), 38–45.

Nawar, S., Buddenbaum, H., Hill, J., Kozak, J., & Mouazen, A. M. (2016). Estimating the soil clay content and organic matter by means of different calibration methods of vis-NIR diffuse reflectance spectroscopy. *Soil & Tillage Research, 155*, 510–522.

Nawar, S., & Mouazen, A. M. (2017). Comparison between random forests, artificial neural networks and gradient boosted machines methods of on-line Vis-NIR spectroscopy measurements of soil total nitrogen and total carbon. *Sensors, 17*, 2428.

Nigon, T. J., Mulla, D. J., Rosen, C. J., Cohen, Y., Alchanatis, V., Knight, J., et al. (2015). Hyperspectral aerial imagery for detecting nitrogen stress in two potato cultivars. *Computers and Electronics in Agriculture, 112*, 36–46.

Ohnesorg, W. J., Hunt, T. E., & Wright, R. J. (2011). *Soybean aphid speed scouting spreadsheet.* Lincoln, NE: Univ. Nebraska Extension Service.

Okparanma, R. N., Coulon, F., & Mouazen, A. M. (2014). Analysis of petroleum-contaminated soils by diffuse reflectance spectroscopy and sequential ultrasonic solvent extraction-gas chromatography. *Environmental Pollution, 184*, 298–305.

Onyango, C. M., & Marchant, J. A. (2003). Segmentation of row crop plants from weeds using colour and morphology. *Computers Electronics Agriculture, 39*(3), 141–155.

Osborne, L. E., & Deneke, D. (2010). *Soybean diseases: A pictorial guide for South Dakota.* Brookings, SD: South Dakota State University Extension Service, Bull. EC932.

O'Shaughnessy, S. A., Evett, S. R., & Colaizzi, P. D. (2015). Dynamic prescription maps for site-specific variable rate irrigation of cotton. *Agricultural Water Management, 159*, 123–138.

Pantazi, X. E., Moshou, D., Alexandridis, T., Whetton, R. L., & Mouazen, A. M. (2016). Wheat yield prediction using machine learning and advanced sensing techniques. *Computers and Electronics in Agriculture, 121*, 57–65.

Pantazi, X. E., Moshou, D., Oberti, R., West, J., Mouazen, A. M., & Bochtis, D. (2017). Detection of biotic and abiotic stresses in crops by using hierarchical self organizing classifiers. *Precision Agriculture, 18*(3), 383–393.

Pantazi, X., Moshou, D., & Tamouridou, A. (2019). Automated leaf disease detection in different crop species through image features analysis and One Class Classifiers. *Computers and Electronics in Agriculture, 156*, 96–104.

Pantazi, X. E., Tamouridou, A. A., Alexandridis, T. K., Lagopodi, A. L., Kashefi, J., & Moshou, D. (2017a). Evaluation of hierarchical self-organising maps for weed mapping using UAS multispectral imagery. *Computers and Electronics in Agriculture, 130*, 224–230.

Pantazi, X. E., Tamouridou, A. A., Alexandridis, T. K., Lagopodi, A. L., Kontouris, G., & Moshou, D. (2017b). Detection of Silybum marianum infection with Microbotryum silybum using VNIR field spectroscopy. *Computers and Electronics in Agriculture, 137*, 130–137. https://doi.org/10.1016/j.compag.2017.03.01.

Park, J. Y., et al. (2001). Land-cover classification using combined ALSM (LiDAR) and color digital photography. In *Presented at ASPRS conference*. St. Louis, MI: ASPRS. April, 23–27.

Parsons, C., Margui Grabulosa, E., Pili, E., Floor, G. H., Roman-Ross, G., & Charlet, L. (2013). Quantification of trace arsenic in soils by field-portable X-ray fluorescence spectrometry: Considerations for sample preparation and measurement conditions. *Journal of Hazardous Materials, 262*, 1213–1222.

Pasqualotto, N., Delegido, J., Van Wittenberghe, S., Verrelst, J., Rivera, J. P., & Moreno, J. (2018). Retrieval of canopy water content of different crop types with two new hyperspectral indices: Water absorption area index and depth water index. *International Journal of Applied Earth Observation and Geoinformation, 67*, 69–78.

Patel, D., & Patel, H. (2016). Survey of android apps for agriculture sector. *International Journal of Information Sciences and Techniques, 6*, 61–67.

Peña, J. M., Torres-Sánchez, J., de Castro, A. I., Kelly, M., & López-Granados, F. (2013). Weed mapping in early-season maize fields using object-based analysis of unmanned aerial vehicle (UAV) images. *PLoS One, 8*(10) e77151.

Peterson, T. A., Blackmer, T. M., Francis, D. D., & Schepers, J. S. (1993). *Using a chlorophyll meter to improve N management.* Historical Materials from University of Nebraska-Lincoln Extension, G93-1171.

Pinter, P. J., Jr., Hatfield, J. L., Schepers, J. S., Barnes, E. M., Moran, M. S., Daughtry, C. S., et al. (2003). Remote sensing for crop management. *Photogrammetric Engineering & Remote Sensing, 69*(6), 647–664.

Polischuk, V., Shadchina, T., Kompanetz, T., Budzanivskaya, I., Boyko, A., & Sozinov, A. (1997). Changes in reflectance spectrum characteristic of Nicotiana debneyi plant under the influence of viral infection. *Archives of Phytopathology and Plant Protection, 31*(1), 115–119.

Pongnumkul, S., Chaovalit, P., & Surasvadi, N. (2015). Applications of smartphone-based sensors in agriculture: A systematic review of research. *Journal of Sensors*, *2015*, 1–18.
Prenger, J. J., Ling, P. P., Hansen, R. C., & Keener, H. M. (2005). Plant response-based irrigation control system in a greenhouse: System evaluation. *Transactions of the ASAE*, *48*, 1175–1183.
Puig, E., Gonzalez, F., Hamilton, G., & Grundy, P. (2015). Assessment of crop insect damage using unmanned aerial systems: a machine learning approach. In *21st International Congress on Modelling and Simulation, Gold Coast, Australia, 29 Nov to 4 Dec 2015*.
Rampant, P., & Abuzar, M. (2004). Geophysical tools and digital elevation models: tools for understanding crop yield and soil variability. *Super Soil 2004 3rd Aust. New Zeal. Soils Conf. 5–9 December 2004*. Aust: Univ. Sydney.
Rapaport, T., Hochberg, U., Shoshany, M., Karnieli, A., & Rachmilevitch, S. (2015). Combining leaf physiology, hyperspectral imaging and partial least squares-regression (PLS-R) for grapevine water status assessment. *ISPRS Journal of Photogrammetry and Remote Sensing*, *109*, 88–97.
Rauff, K. O., & Bello, R. (2015). A review of crop growth simulation models as tools for agricultural meteorology. *Agricultural Sciences*, *6*(09), 1098.
Rembold, F., Atzberger, C., Savin, I., & Rojas, O. (2013). Using low resolution satellite imagery for yield prediction and yield anomaly detection. *Remote Sensing*, *5*(4), 1704–1733.
Ribeiro-Gomes, K., Hernández-López, D., Ortega, J., Ballesteros, R., Poblete, T., & Moreno, M. (2017). Uncooled thermal camera calibration and optimization of the photogrammetry process for UAV applications in agriculture. *Sensors*, *17*(10), 2173.
Richter, K., Atzberger, C., Vuolo, F., & D'Urso, G. (2011). Evaluation of sentinel-2 spectral sampling for radiative transfer model based LAI estimation of wheat, sugar beet, and maize. *IEEE Journal of Selected Topics in Applied Earth Observations and Remote Sensing*, *4*(2), 458–464.
Riedell, W. E., & Blackmer, T. M. (1999). Leaf reflectance spectra of cereal aphid-damaged wheat. *Crop Science*, *39*(6), 1835–1840.
Robson, A., Rahman, M., & Muir, J. (2017). Using Worldview Satellite Imagery to Map Yield in Avocado (*Persea americana*): A case study in Bundaberg, Australia. *Remote Sensing*, *9*(12), 1223.
Rodríguez-Pérez, J. R., Riaño, D., Carlisle, E., Ustin, S., & Smart, D. R. (2007). Evaluation of hyperspectral reflectance indexes to detect grapevine water status in vineyards. *American Journal of Enology and Viticulture*, *58*, 302.
Rozenstein, O., Haymann, N., Kaplan, G., & Tanny, J. (2018). Estimating cotton water consumption using a time series of Sentinel-2 imagery. *Agricultural Water Management*, *207*, 44–52.
Rullan-Silva, C. D., Olthoff, A. E., de la Mata, J. A. D., & Alonso, A. P. (2013). Remote monitoring of forest insect defoliation: A review. *Forest Systems*, *22*(3), 377–391.
Samborski, S. M., Tremblay, N., & Fallon, E. (2009). Strategies to Make Use of Plant Sensors-Based Diagnostic Information for Nitrogen Recommendations All rights reserved. No part of this periodical may be reproduced or transmitted in any form or by any means, electronic or mechanical, including photocopying, recording, or any information storage and retrieval system, without permission in writing from the publisher. *Agronomy Journal*, *101*, 800–816.
Sankey, T. T., Glenn, N., Ehinger, S., Boehm, A., & Hardegree, S. (2010). Characterizing western juniper expansion via a fusion of Landsat 5 Thematic Mapper and lidar data. *Rangeland Ecology & Management*, *63*(5), 514–523.
Sankey, T. T., McVay, J., Swetnam, T. L., McClaran, M. P., Heilman, P., & Nichols, M. (2018). UAV hyperspectral and lidar data and their fusion for arid and semi-arid land vegetation monitoring. *Remote Sensing in Ecology and Conservation*, *4*(1), 20–33.

Sasaki, Y., Okamoto, T., Imou, K., & Torii, T. (1998). Automatic diagnosis of plant disease-Spectral reflectance of healthy and diseased leaves. *IFAC Proceedings Volumes*, *31*(5), 145–150.

Senthilnath, J., Dokania, A., Kandukuri, M., Ramesh, K. N., Anand, G., & Omkar, S. N. (2016). Detection of tomatoes using spectral-spatial methods in remotely sensed RGB images captured by UAV. *Biosystems Engineering*, *146*, 16–32.

Sharma, A., Weindorf, D. C., Man, T., Aldabaa, A. A. A., & Chakraborty, S. (2014). Characterizing soils via portable X-ray fluorescence spectrometer: 3. Soil reaction (pH). *Geoderma*, *232*, 141–147.

Sharma, A., Weindorf, D. C., Wang, D. D., & Chakraborty, S. (2015). Characterizing soils via portable X-ray fluorescence spectrometer: 4. Cation exchange capacity (CEC). *Geoderma*, *239*, 130–134.

Shi, T., Cui, L., Wang, J., Fei, T., Chen, Y., & Wu, G. (2012). Comparison of multivariate methods for estimating soil total nitrogen with visible/near-infrared spectroscopy. *Plant & Soil*, *366*, 363–375.

Shirzadifar, A., Bajwa, S., Mireei, S. A., Howatt, K., & Nowatzki, J. (2018). Weed species discrimination based on SIMCA analysis of plant canopy spectral data. *Biosystems Engineering*, *171*, 143–154.

Shrestha, D., Steward, B., & Birrell, S. (2004). Video processing for early stage maize plant detection. *Biosystems Engineering*, *89*(2), 119–129.

Sithole, G., & Vosselman, G. (2004). Experimental comparison of filter algorithms for bare-Earth extraction from airborne laser scanning point clouds. *ISPRS Journal of Photogrammetry and Remote Sensing*, *59*(1–2), 85–101.

Soodan, R. K., Pakade, Y. B., Nagpal, A., & Katnoria, J. K. (2014). Analytical techniques for estimation of heavy metals in soil ecosystem: A tabulated review. *Talanta*, *125*, 405–410.

Soriano-Disla, J. M., Janik, L. J., Viscarra Rossel, R. A., Macdonald, L. M., & McLaughlin, M. J. (2014). The performance of visible, near-, and mid-infrared reflectance spectroscopy for prediction of soil physical, chemical, and biological properties. *Applied Spectroscopy Reviews*, *49*, 139–186.

Stenberg, B., Viscarra Rossel, R. A., Mouazen, A. M., & Wetterlind, J. (2010). Visible and Near Infrared Spectroscopy in Soil Science. *Advances in Agronomy*, *107*, 163–215.

Sui, R., Thomasson, J. A., Hanks, J., & Wooten, J. (2008). Ground-based sensing system for weed mapping in cotton. *Computers and Electronics in Agriculture*, *60*(1), 31–38.

Sukawattanavijit, C., & Chen, J. (2015). Fusion of RADARSAT-2 imagery with LANDSAT-8 multispectral data for improving land cover classification performance using SVM. In *2015 IEEE 5th Asia-Pacific Conference on Synthetic Aperture Radar (APSAR) (pp. 567–572)*, IEEE.

Sun, H., Slaughter, D., Ruiz, M. P., Gliever, C., Upadhyaya, S., & Smith, R. (2009). Development of an RTK GPS plant mapping system for transplanted vegetable crops. In *Paper presented at the 2009 Reno, Nevada, June 21–June 24, pp. 1*.

Swain, K. C., Thomson, S. J., & Jayasuriya, H. P. (2010). Adoption of an unmanned helicopter for low-altitude remote sensing to estimate yield and total biomass of a rice crop. *Transactions of the ASABE*, *53*(1), 21–27.

Swanhart, S., Weindorf, D. C., Acree, A., Bakr, N., Zhu, Y., Nelson, C., et al. (2013). Soil salinity assessment via portable X-ray fluorescence spectrometry. In *ASA-CSSA-SSSA National Meetings*. Madison, WI: Soil Science Society of America. 3–6 November, Tampa, FL.

Taubenhaus, J., Ezekiel, W., & Neblette, C. (1929). Airplane photography in the study of cotton root rot. *Phytopathology*, *19*(11), 1025–1029.

Taylor, M. J., Smettem, K., Pracilio, G., & Verboom, W. (2002). Relationships between soil properties and high-resolution radiometrics, central eastern Wheatbelt, western Australia. *Exploration Geophysics*, *33*, 95–102.

Tellaeche, A., Pajares, G., Burgos-Artizzu, X. P., & Ribeiro, A. (2011). A computer vision approach for weeds identification through Support Vector Machines. *Applied Soft Computing, 11*(1), 908–915.

Thenkabail, P. S., Lyon, G. J., & Huete, A. (Eds.), (2019). *Hyperspectral remote sensing of vegetation* (2 ed.). Boca Raton, London, New York: CRC Press- Taylor and Francis group. four-volume-set.

Thomasson, J. A., Wang, T., Wang, X., Collett, R., Yang, C., & Nichols, R. L. (2018 May). Disease detection and mitigation in a cotton crop with UAV remote sensing. In *Proc. SPIE, 10664*. 106640L.

Tian, Y. C., Yao, X., Yang, J., Cao, W. X., Hannaway, D. B., & Zhu, Y. (2011). Assessing newly developed and published vegetation indices for estimating rice leaf nitrogen concentration with ground- and space-based hyperspectral reflectance. *Field Crops Research, 120*, 299–310.

Tillett, N., Hague, T., Grundy, A., & Dedousis, A. (2008). Mechanical within-row weed control for transplanted crops using computer vision. *Biosystems Engineering, 99*(2), 171–178.

Valavanis, K. P., & Vachtsevanos, G. J. (Eds.), (2015). *Handbook of unmanned aerial vehicles*. Netherlands: Springer.

Van Beek, J., Tits, L., Somers, B., Deckers, T., Verjans, W., Bylemans, D., et al. (2015). Temporal dependency of yield and quality estimation through spectral vegetation indices in Pear Orchards. *Remote Sensing*, 7(8), 9886.

Van der Weide, R., Bleeker, P., Achten, V., Lotz, L., Fogelberg, F., & Melander, B. (2008). Innovation in mechanical weed control in crop rows. *Weed Research, 48*(3), 215–224.

Van Egmond, F. M., Loonstra, E. H., & Limburg, J. (2010). Gamma ray sensor for topsoil mapping: The mole. In R. A. V. Rossel, A. B. McBratney, & B. Minasny (Eds.), *Proximal soil sensing. Progress in soil science* (pp. 323–332). Dordrecht, Heidelberg, London, New York: Springer Science and Business Media B.V.

Verger, A., Vigneau, N., Chéron, C., Gilliot, J. -M., Comar, A., & Baret, F. (2014). Green area index from an unmanned aerial system over wheat and rapeseed crops. *Remote Sensing of Environment, 152*, 654–664.

Verhoeven, G., Doneus, M., Briese, C., & Vermeulen, F. (2012). Mapping by matching: A computer vision-based approach to fast and accurate georeferencing of archaeological aerial photographs. *Journal of Archaeological Science, 39*(7), 2060–2070.

Verrelst, J., et al. (2009). Mapping of aggregated floodplain plant communities using image fusion of CASI and LiDAR data. *International Journal of Applied Earth Observation and Geoinformation, 11*(1), 83–94.

Viscarra Rossel, R. A., Adamchuk, V. I., Sudduth, K. A., McKenzie, N. J., & Lobsey, C. (2011). Proximal soil sensing: an effective approach for soil measurements in space and time. *Advances in Agronomy, 113*, 243–291.

Viscarra Rossel, R. A., & Behrens, T. (2010). Using data mining to model and interpret soil diffuse reflectance spectra. *Geoderma, 158*, 46–54.

Viscarra Rossel, R. A., Taylor, H. J., & McBratney, A. B. (2007). Multivariate calibration of hyperspectral γ-ray energy spectra for proximal soil sensing. *European Journal of Soil Science, 58*, 343–353.

Viscarra Rossel, R. A., Walvoort, D. J. J., McBratney, A. B., Janik, L. J., & Skjemstad, J. O. (2006). Visible, near infrared, mid infrared or combined diffuse reflectance spectroscopy for simultaneous assessment of various soil properties. *Geoderma, 131*, 59–75.

Vohland, M., Besold, J., Hill, J., & Fründ, H. -C. (2011). Comparing different multivariate calibration methods for the determination of soil organic carbon pools with visible to near infrared spectroscopy. *Geoderma, 166*, 198–205.

Vohland, M., & Emmerling, C. (2011). Determination of total soil organic C and hot water-extractable C from VIS-NIR soil reflectance with partial least squares regression and spectral feature selection techniques. *European Journal of Soil Science, 62*, 598–606.

Wagner, W., et al. (2006). Gaussian decomposition and calibration of a novel small-footprint full wave form digitising airborne laser scanner. *ISPRS Journal of Photogrammetry and Remote Sensing*, *60*(2), 100–112.
Wang, Z., Skidmore, A. K., Wang, T., Darvishzadeh, R., & Hearne, J. (2015). Applicability of the PROSPECT model for estimating protein and cellulose+lignin in fresh leaves. *Remote Sensing of Environment*, *168*, 205–218.
Warren, G., & Metternicht, G. (2005). Agricultural applications of high-resolution digital multispectral imagery. *Photogrammetric Engineering & Remote Sensing*, *71*(5), 595–602.
Weindorf, D. C., Bakr, N., & Zhu, Y. (2014). Advances in portable X-ray fluorescence (PXRF) for environmental, pedological, and agronomic applications. *Advances in Agronomy*, *128*, 1–45.
Weindorf, D. C., Herrero, J., Castañeda, C., Bakr, N., & Swanhart, S. (2013). Direct soil gypsum quantification via portable X-ray fluorescence spectrometry. *Soil Science Society of America Journal*, 77, 2071–2077.
Weis, M., et al. (2005). A framework for GIS and imagery data fusion in support of cartographic updating. *Information Fusion*, *6*(4), 311–317.
West, J. S., Bravo, C., Oberti, R., Lemaire, D., Moshou, D., & McCartney, H. A. (2003). The potential of optical canopy measurement for targeted control of field crop diseases. *Annual Review of Phytopathology*, *41*(1), 593–614.
Wong, M. T. F., & Harper, R. J. (1999). Use of on-ground gamma-ray spectrometry to measure plant-available potassium and other topsoil attributes. *Australian Journal of Soil Research*, *37*, 267–277.
Wu, J., Wang, D., & Bauer, M. E. (2007). Assessing broadband vegetation indices and QuickBird data in estimating leaf area index of corn and potato canopies. *Field Crops Research*, *102*(1), 33–42.
Yang, C. -M., & Cheng, C. -H. (2001). Spectral characteristics of rice plants infested by brown planthoppers. *Proceedings of the National Science Council, Republic of China. Part B, Life Sciences*, *25*(3), 180–186.
Yang, G. J., Liu, Q. H., & Zhang, J. X. (2008). Automatic land cover change detection based on image analysis and quantitative methods. *The International Archives of the Photogrammetry, Remote Sensing and Spatial Information Sciences, XXXVII*, *B*7, 1555–1558.
Ye, X., Sakai, K., Manago, M., Asada, S. -I., & Sasao, A. (2007). Prediction of citrus yield from airborne hyperspectral imagery. *Precision Agriculture*, *8*(3), 111–125.
Yuan, L., Pu, R., Zhang, J., Wang, J., & Yang, H. (2016). Using high spatial resolution satellite imagery for mapping powdery mildew at a regional scale. *Precision Agriculture*, *17*(3), 332–348.
Zarco-Tejada, P. J., Diaz-Varela, R., Angileri, V., & Loudjani, P. (2014). Tree height quantification using very high resolution imagery acquired from an unmanned aerial vehicle (UAV) and automatic 3D photo-reconstruction methods. *European Journal of Agronomy*, *55*, 89–99.
Zarco-Tejada, P. J., Gonzalez-Dugo, V., & Berni, J. A. (2012). Fluorescence, temperature and narrow-band indices, acquired from a UAV platform for water stress detection using a micro-hyperspectral imager and a thermal camera. *Remote Sensing of Environment*, *117*, 322–337.
Zhang, F., & Zhou, G. S. (2015). Estimation of canopy water content by means of hyperspectral indices based on drought stress gradient experiments of maize in the North Plain China. *Remote Sensing*, 7, 15203–15223.
Zhang, B., Zhao, Q. G., Horn, R., & Baumgartl, T. (2001). Shear strength of surface soil as affected by soil BD and soil water content. *Soil & Tillage Research*, *59*(3–4), 97–106.
Zhu, Y., Weindorf, D. C., & Zhang, W. (2011). Characterizing soils using a portable X-ray fluorescence spectrometer: 1. Soil texture. *Geoderma*, *167–168*, 167–177.
Zwiggelaar, R. (1998). A review of spectral properties of plants and their potential use for crop/weed discrimination in row-crops. *Crop Protection*, *17*(3), 189–206.

Further reading

Barry, D.J.: Design of and studies with a novel one meter multi-element spectroscopic telescope. Ph.D dissertation, University of Cornell (1995).

Cohen, Y., & Alchanatis, V. (2019). Spectral and spatial methods for hyperspectral and thermal image analysis to estimate biophysical and biochemical properties of agricultural crops. In P. S. Thenkabail, G. J. Lyon, & A. Huete (Eds.), *Hyperspectral remote sensing of vegetation* (2nd ed.). Boca Raton, London, New York: CRC Press-Taylor and Francis Group. Four-Volume-Set.

Taton, R. (1966). La premire note mathmatique de Gaspard Monge (juin 1769). *Revue d'Histoire des Sciences et de leurs Applications*, *19*, 143–149.

CHAPTER 4

Tutorial I: Weed detection

Contents

4.1 Introduction

The field distribution of weeds relies on the weed and crop genotype, ambient conditions and the applied crop management practices. Most of the times, weeds grow in the form of patches and their presence differentiates among different types of fields. However, in every field the patches pattern that is created during the years, tends to demonstrate great characteristics. The different patterns of weed appearance within the field, require subsequently customized crop management for targeted treatment application. The patchy weed appearance implies that the field is partially weed free, so it has to be treated locally. In terms of site-specific crop management, the weeds locations are obtained by employing various techniques including using GNSS tools while weed mapping is performed by portable instruments, due harvesting or through remote sensing (Whelan & Taylor, 2013).

Intelligent Data Mining and Fusion Systems in Agriculture
https://doi.org/10.1016/B978-0-12-814391-9.00004-2

Weed control is mostly based chemical application which leads to environmental degradation. For this reason, novel weed management solutions are high on demand. Harker and O'donovan (2013) refers to Integrated Weed Management (IWM) as the combination of various weed management methods depending on the weed appearance with respect to the crop life cycle.

Hence, IWM concerns the use of different techniques without exclusions of certain techniques. The repetition of a specific weed treatment leads to population adaptation since some weeds would become resistant to this specific treatment. For example, the hand-weeding removal of barnyard grass (*Echinochloa crus-galli* (L.) Beauv.) in rice fields (*Oryza sativa L.*) triggered the emergence of biotypes that evated hand-weeding removal approaches (Barrett, 1983). Therefore, weeds tend to adapt any repetitive of a specific weed management technique for this reason it is more preferable to apply different methods than persisting to a specific one. Spatial and temporal fluctuations are regarded major constraints for applying successful integration of weed management practices.

Targeting the reduction of herbicide use, site-specific treatment aims on the targeted identification of plants for spraying application. To achieve individual target treatment, it is crucial to develop weed sensing for detection and machine learning for weed recognition. Sensing equipment and AI are under development for real-time identification of weeds hence enabling site-specific treatment with high accuracy and (Tellaeche, Pajares, Burgos-Artizzu, & Ribeiro, 2011). The identification of the weed species from sensor signatures (Moshou, Ramon, & De Baerdemaeker, 2002) is crucial for determining the type of chemical and its exact dosage for spraying application. For applying the most appropriate dosage or treat mechanically the detected weeds, it is necessary to perform weed mapping prior to treatment application in order to access weed distribution and the appearance of weed patches. The term "weed mapping" refers to represents all synergies of spatial information for weed identification, including data acquisition, crop classification and the results visualization.

The efficient combination of autonomous ground vehicles or unmanned aerial vehicles (UAVs) with multi-sensory systems have offered to the market novel solutions for weed mapping (Fernández-Quintanilla et al., 2018). More precisely, the employment of several Machine Learning techniques has substantially enhanced the automatic recognition of weeds during the last decade (De Castro et al., 2018). Multispectral imagery enable the successful discrimination between weeds and crops with similar appearance like green grass versus rice (Barrero & Perdomo, 2018) or black grass versus winter wheat (Lambert, Hicks, Childs, & Freckleton, 2018). One further asset of automated weed mapping instruments is their rapid operation and the fast determination of weed appearance within the field (Laursen, Jørgensen, Midtiby, Mortensen, & Baby, 2017).

The reflectance of leaves and the resultant spectral signatures have been used for discriminating between weeds and crops. Under laboratory conditions, the feasibility of using the spectral reflectance for recognizing crops versus weeds and the weed species has been proven (Borregaard, Nielsen, Nørgaard, & Have, 2000).

There have been various approaches targeting on weed identification that employed artificial intelligence algorithms in literature. Specific weed recognition methods rely on analyzing the morphological characteristics of plant leaves of crops or weeds. Søgaard (2005) presented a machine learning approach aiming to classify different types of weeds by applying active shape modeling (ASM). Other research contacted by Moshou, Vrindts, De Ketelaere, De Baerdemaeker, and Ramon (2001) demonstrated that different spectral behavior in specific spectral regions is sufficient for classifying plants species.

Wang, Zhang, Zhu, and Geng (2008) used a portable spectrometer and hyperspectral images acquired from a UAV in order to map the extent of invasion caused by the weed Sericea lespedeza in Mid-Missouri, USA. A weed species classification approach based on hyperspectral signatures has been demonstrated by Moshou et al. (2002), employing a self-organizing map (SOM) which is enriched with local linear layers aiming to optimize classification.

The Self-Organizing Maps (SOMs) belong to the most prominent Artificial Neural Networks techniques in literature (Kohonen, 1988). Their wide range of applicability to several scientific fields until recently made them principal machine learning tools (Marini, 2009). They are capable of learning the training set characteristics without the need of target data, which means they do not require any supervision. The necessity of training unsupervised models so as to function in a supervised manner has led to the adaptation of unsupervised architectures. Such examples include the counter propagation Artificial Neural Networks (CP-ANNs), which are SOMs variant, which results from a further addition to the SOM layer (Zupan, Novic, & Gasteiger, 1995).

More precisely, the CP-ANNs are capable of solving non-linear classification problems. Common modifications of CP-ANNs produced novel supervised neural network methods and associated training methods including Supervised Kohonen Networks (SKNs) and XY-fused Networks (XY-Fs) (Melssen, Wehrens, & Buydens, 2006).

Novelty Detection (ND) concerns the procedure of representing a known target class for detecting a novelty testing data samples as outliers. It can be defined as a specific case of classification task where the existent information concerns only one type of class and the training samples come from this target class. In ND, the classification accuracy s characterized according to the capability of the classifier to characterize samples of a reference class (Clifton et al., 2014; Pimentel, Clifton, Clifton, & Tarassenko, 2014). On the contrary, the outlier class is not used for training. In certain cases, it is almost impossible to obtain outlier samples especially in the case of safety critical systems. Simulating abnormal situations in such systems would imply the intentional damaging of such systems to produce a variety of faults. Such an approach would result in negative impact in terms of safety and financial losses. Novelty detection has been used in agricultural applications, where the presence of biotic stresses are treated as an outliers detection problem.

AlSuwaidi, Grieve, and Yin (2018) have proposed a novelty detector by SVM (ND-SVM) which has been employed to construct a prediction model for differentiating control and *Cercospora* infected or cold stressed sugarbeet plants based on hyperspectral images. The presented chapter demonstrates various Self Organizing Map derived models using supervised learning and novelty detection methodology in order to classify reflectance data for the identification of different weed species and their discrimination from *Zea mays* crop.

4.2 Materials and methods

4.2.1 Defining hyperspectral versus multispectral imaging

Multispectral imaging can function as the basis for the development of hyperspectral imaging, which captures images at various wavebands in the electromagnetic spectrum and associating the spectral signatures with the chemical compounds that produce them, by absorbing the light frequencies that resonate with the chemical bonds. The multispectral imaging usually concerns the acquisition of images in a few wavelengths usually up to six spectral bands between visible and near-infrared (NIR) (Jensen, Apan, Young, & Zeller, 2007). Image acquisition in such a narrow wavelength range yields gaps in the spectrum, resulting in a loss of information and lack of exploitation of the spectral signature. On the contrary hyperspectral imaging enables image acquisition in a wider spectral range including hundreds of 200 spectral bands between visible and NIR regions. Hyperspectral imaging is considered an innovative method for assessing quality in agricultural industry, combining standard imaging, spectroscopy and radiometric principles.

Radiometry is the estimation of the level of electromagnetic energy (Watt) that resides in spectral range. Common radiometer is constructed with a single sensor augmented with a filter in order to isolate a targeted spectral range. Spectrometry denotes the intensity of light in a spectral region (W/m^2). Compared to radiometer principles, spectrometers split the spectral region into several wavebands based on diffraction grating or prisms. Goetz, Vane, Solomon, and Rock (1985) coined the term of 'hyperspectral imaging' for earth remote sensing by applying spectroscopic imaging methods.

The term 'hyper', for image processing denotes the imagery acquisition at multiple regions across the electromagnetic spectrum. Chemicals can be identified by their specific combination of spectral peaks that correspond to the absorbance bands that result from the chemical bonds resonance with the incident light (Goetz, 2009). The chemical compounds that constitute every biomaterials display interactions with the incident light. By employing image transformations simultaneously with spectral processing it is possible to identify the regions of interest and determine the chemical composition of the material. Moreover, hyperspectral imaging is regarded as an effective tool for quantity and quality assessment, due to its capability to detect the occurrence of the material simultaneously with their position.

Hyperspectral imaging is an suitable method for a variety of applications by producing the spatial distribution of spectral signatures (ElMasry & Sun, 2010). The terms 'hypercubes' or 'datacubes' refer to the amount of data obtained by hyperspectral imagery form 3D structures consisting of the spatial variation corresponding to each of the wavebands of the hyperspectral sensor (Mehl, Chen, Kim, & Chan, 2004). A hypercube contains a spatial arrangement at each wavelength corresponding to the pixels of an image. Individual pixels includes the spectrum, which corresponds the sample's chemical compounds that correspond to specific pixel.

4.2.2 Advantages of hyperspectral imaging

Hyperspectral imaging is characterized as a pioneer tool for quality control in agri-products. Its main advantages regarding quality control are presented as follows:

1. No sample preparation is required.
2. It is a non-destructive method.

3. By the time the model is formed, trained and tested, application of model is easy.
4. Eco friendly and cost effective technique, since no further inputs (such as chemical substances, waste management, further chemical analysis) are needed for the model application.
5. The capability of storing a great amount of spectral information corresponding to every pixel enables more precise knowledge concerning the sample chemical composition.
6. The regions of interest are defined based on the number of spectral bands for each pixel, which runs through the hypercube as a column defined by a group of pixels.
7. Qualitative and quantitative estimations can be performed by using the same hyperspectral images.
8. Several substances can be analyzed simultaneously from the same hyperspectral images.
9. Additional analysis can reveal the chemical composition of the examined material and produce two dimensional map of chemical concentrations which is called chemical imaging (ElMasry & Sun, 2010).

4.2.3 Disadvantages of hyperspectral imaging

1. Hyperspectral imaging system require high expenditure compared to other image processing tools.
2. Due to the hyperspectral data volume, the requirements for data storage and speed of processing are very demanding.
3. The amount of the collected images covering the whole spectral region lead to longer acquisition times compared to traditional digital imaging devices.
4. They demand effective and accurate prediction algorithms to estimate the concentration of chemical components.
5. They are not appropriate for continuous measurement systems since the image acquisition and processing requires extended periods of time.
6. Imaging results are dependent on several ambient illumination factors caused by coverage, scattering angle of light incidence, shadows and cloudiness resulting in noisy images. The effect of the external factors can be mitigated by spectral transformations that alleviate the impact of these factors of image integrity.

7. For an effective data processing, accurate pre-processing and predictive model generation is needed since row hyperspectral imaging offer only qualitative insights (ElMasry & Sun, 2010).

4.2.4 Hyperspectral imaging applications in agriculture

An initial attempt of applying hyperspectral imaging techniques has been made by Goetz et al. (1985) oriented to remote sensing. Further expansions of his study were oriented towards agriculture (Sánchez & Pérez-Marín, 2011), and more precisely crop disease detection (Kumar et al., 2012; Kuska et al., 2015; Xie, Shao, Li, & Hea, 2015).

Bauriegel, Giebel, Geyer, Schmidt, and Herppich (2011) presented a hyperspectral imaging approach aiming to identify Fusarium infection at an early stage in wheat crops. A PCA method has been applied so as to discriminate infected from healthy plants. Both healthy and diseased plant were recognized reaching a accuracy level of 87% after the data have been subjected to the Spectral Angle Mapper image analysis method.

Zhang, Paliwal, Jayas, and White (2007) utilized NIR hyperspectral imaging to recognize 3 different types of storage fungi infections in wheat kernels. The PCA was employed for dimensionality reduction. An SVM was used to build classifier which had a high performance reaching accuracies of 100%, 87.2%, 92.9% and 99.3% were achieved for the detection of healthy condition and the three different types of fungi, respectively.

4.2.5 Experimental setup

In the current case study, hyperspectral images were acquired with a gray scale camera. Where a spectrograph was integrated with it. A slim linear stripe allows the light to pass through, and then to a grated prism which produce a spectrum, corresponding to each point on the linear stripe (Herrala, Okkonen, Hyvarinen, Aikio, & Lammasniemi, 1994). In other words, the linear strip scans the field of interest and acts like a scanner. A common hyperspectral camera consists of three components: a camera, an objective lens and an imaging spectrograph. The working principle of an common spectrograph is illustrated in Fig. 4.1.

The objective lens formulate the light passing through the slit aperture to from a field patch. The light that passes from the slit is collimated by the optics inside the spectrograph. The light is separated into spectral regions by passing through a diffraction grating. After that, the diffracted light falls into monochromatic camera that registers the spectra. This provides the

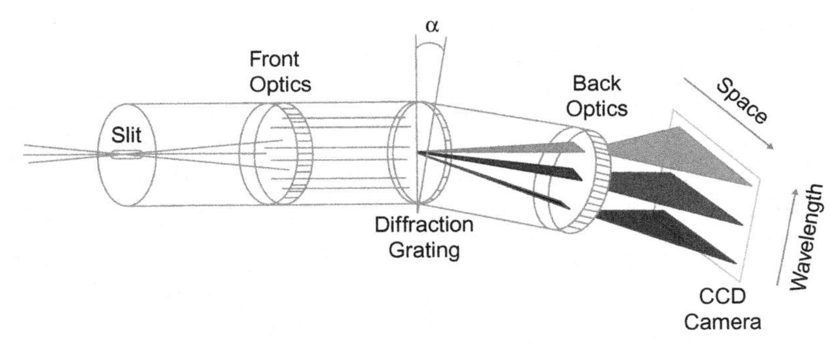

Fig. 4.1 Illustrative depiction of a spectrograph operation (Moshou et al., 2002).

camera with two types of spectral axes: one spatial and one orthogonal. The hyperspectral camera system was manufactured and developed by Specim (Oulu, Finland). To overcome the illumination variability the reflected light was normalized by using the ambient light which was obtained by a 25% Spectralon panel. This kind of normalization is useful for maintaining the same relation between the magnitudes of the incident reflected light. By applying this technique, the spectral reflectance appears illumination invariant.

The device characteristics of the spectral camera included spectral resolution 1.5–5 nm, spectral range 435–855 nm, slit dimensions 80 μm × 8.8 mm, CCD specifications ½" (4.8 × 6.4 mm) and the number of narrow band spectral bands was two hundred. The artificial lighting that was used as illumination source for the measurements included a halogen lamp (100 W) (Fig. 4.2).

Experimental trials in corn (*Zea mays*) plants were carried out greenhouse environment at Biologische Bundesanstalt, Bundessortenamt und Chemische Industrie (BBCH) 12–14 (Meier, 2001). The main crop of the field was mixed in equal proportions with different weed species. Including: *Cirsium arvense*, *Medicago lupulina*, *Oxalis europaea*, *Poa annua*, *Sinapis arvensis*, *Tarraxacum officinale*, *Stellaria media*, *Poligonum persicaria*, *Ranunculus repens*, and *Urtica dioica*.

For the current case study, the spectral information was highly intercorrelated due to multiple peaks of chemical bonds. This led to redundant information that affected the computation time and the accuracy of estimation. A reduced number of wavelengths can lead to the same information

Fig. 4.2 Ispector V10 spectrograph manufactured by Specim, Finland.

regarding plant recognition. The optimal set of wavelengths corresponds to the higher discriminative ability. Eq. (4.1) is used for extracting the wavelengths that can lead to maximization of the class separability function and is described as follows:

$$F(\lambda) = |X(\lambda) - Y(\lambda)| / \sqrt{\sigma_X^2(\lambda) + \sigma_Y^2(\lambda)} \quad (4.1)$$

where $X(\lambda)$ and $Y(\lambda)$ denote the mean reflected light corresponding to X and Y class values at an λ wavelength. The $\sigma_X^2(\lambda)$ and $\sigma_Y^2(\lambda)$ symbolize the reflected light standard deviations at wavelength λ, concerning X and Y class respectively.

Local peaks correspond to wavelengths for which the reparability is high so they are considered more suitable for classifying the different plant species. Owning to the objective of multiple class classification the wavelength selection aims to increase the separability between crop and individual weed species. This procedure produces a pairwise wavelength selection, resulting in an automatic selection for each pair of classes. In the event that the wavelengths are not in close proximity to those already picked, lower ranking wavelengths are appended. When adding crop-weed pair the procedure continues to include lower ranking wavelengths until the fifth rank is reached.

The whole procedure yielded 17 wavebands of 20 nm centered at the following wavelengths: 539, 540, 542, 545, 549, 557, 565, 578, 585, 596, 605, 639, 675, 687, 703, 814 and 840 nm.

4.3 Explanation of experiments

4.3.1 Plant selection

In order to separate the plant regions, is important to define a signature that can identify canopy spectra from non-plant spectra. The Normalized Difference Vegetation Index (NDVI) is considered a suitable indicator that can be employed for leaf spectra detection (Rouse, Haas, Schell, & Deering, 1974) and its NDVI is given as follows:

$$\mathrm{NDVI} = \frac{R_{\mathrm{NIR}} - R_{\mathrm{R}}}{R_{\mathrm{NIR}} + R_{\mathrm{R}}} \tag{4.2}$$

where R_{NIR} represents the near infrared reflectance (NIR) between 740 and 760 nm, and R_{R} denotes the red reflectance between 620 and 640 nm. A threshold of 0.4 has been used for separating different types of vegetation from soil. An illustration of NDVI indicator employed for the soil–plant separation is given in Fig. 4.3.

4.3.2 NDVI and spatial resolution

A more scholastic examination of the NDVI images, it is evident that the vegetation appears quite variable over the same plant. As we come closer

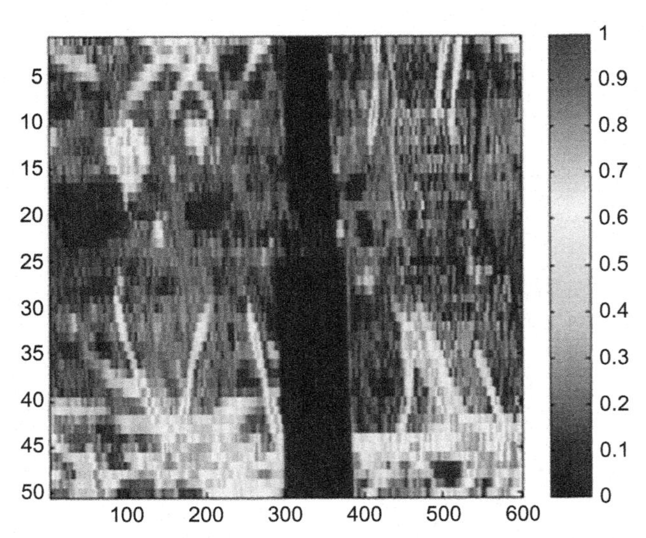

Fig. 4.3 NDVI depiction employed for the soil–plant separation. The bright areas where NDVI is higher than 0.5 denote plants presence while the dark areas with NDVI values lower than 0.3 denote soil. The dark blue column in the middle of the picture denotes the white Teflon plate presence.

to the center of the plant (or leaf), it is indicated that the NDVI value tends to be higher. The surface of vegetation demonstrates a consistently high NDVI. In the current application, in order to assure that only plant areas will be examined, only areas with NDVI value greater than 0.7 were retained.

4.4 Results and discussion for weed detection

For the current case study three different SOMs including CPANN, XYF and SKN and One Class Classifiers such as MOG, SVM, SOM and Auto encoder have been employed for recognizing *Zea mays* plants from ten different types of weed species. The theoretical scientific background of the utilized classifiers has been described in Chapter 2 in Sections 2.9.3, 2.9.4, 2.9.5 and 2.8.2, 2.8.3, 2.8.5.

4.4.1 Results for hierarchical maps

The training procedure is described as follows:

(1) 1210 samples corresponding to 17 features have been used for training. Each of them has been derived from 110 spectral signatures corresponding to corn plants and 110 to ten different weed species.

(2) The Supervised SOM classifiers were calibrated by using the training sample spectra.

(3) To test generalization the trained networks were validated with 54 unknown samples acquired from *Zea mays* crops and 54 samples from individual weed species. The efficiency of the presented technique is highly correlated to the capability of the each of the three SOM classifiers to recognize *Zea mays* and each of the 10 weed plant species and a successful and precise way.

The Table 4.1 demonstrates the classification accuracy for *Zea mays* crop and the 10 different weed species as follows:

As it can be seen in Table 4.1, the CPANN classifier classifies perfectly *Zea mays* crop. But indicates lower classification capability in comparison to SKN and XYF. More precisely, the CPANN classifier achieved an 8 out of 10 recognition accuracy. On the whole, SKN achieves an equally high classification accuracy concerning all the 10 different crop species and the *Zea mays* crop.

To assess which spectral characteristics are more important and how they have influenced the classification outcome of employed hierarchical SOM architectures, an illustration of the SOM weights and output layer weights are depicted in Figs. 4.4–4.9. Each figure couple shows the layers of the

Table 4.1 Classification accuracies for the *Zea mays* crop recognition versus different weed species which each of them is presented with an identification number (*Ranunculus repens* = w1, *Cirsium arvense* = w2, *Sinapis arvensis* = w3, *Stellaria media* = w4, Tarraxacum officinale = w5, *Poa annua* = w6, Poligonum persicaria = w7, *Urtica dioica* = w8, Oxalis europea = w9 and *Medicago lupulina* = w10)

	Successful detection (%)										
Method	**Z.M**	**w1**	**w2**	**w3**	**w4**	**w5**	**w6**	**w7**	**w8**	**w9**	**w10**
skn	94.44	98.15	90.74	98.15	92.60	100	100	96.30	92.59	90.74	90.74
xyf	90.74	100	94.44	100	98.14	87.04	96.30	85.19	88.89	100	100
cpann	100	100	100	94.44	83.33	94.44	83.33	77.78	83.33	92.60	83.33

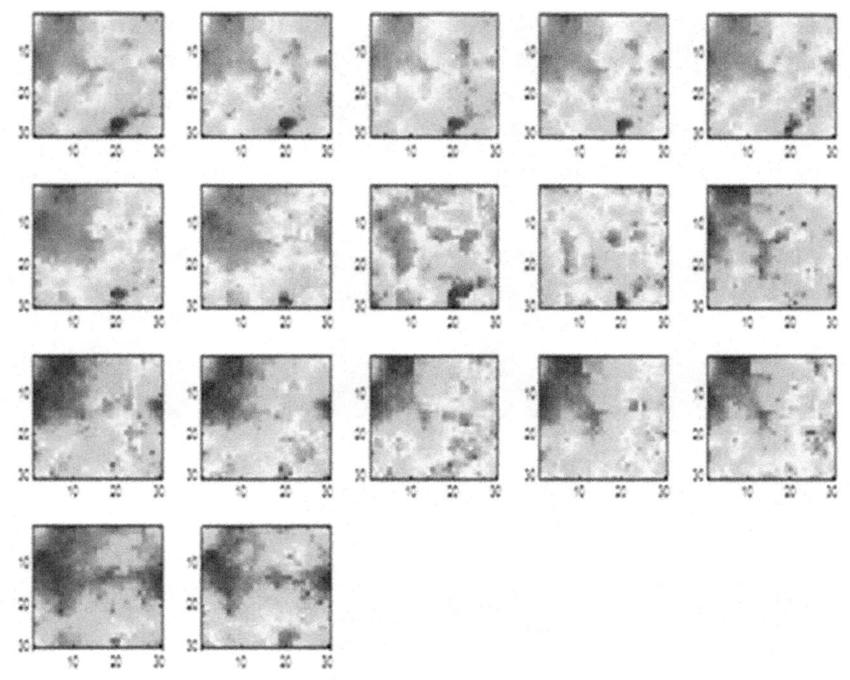

Fig. 4.4 Weight plot of SKN classifier related to the spectral features attained from 17 spectral signatures. The spectral features vary from lower values (denoted with blue color, dark gray in monochrome) to higher values (denoted with red color, light gray in monochrome).

hierarchical SOM architectures. The weights visualization may unveil relevant correlations among the features and the category. This can be explained by the preservation of the topological features that are inherent to the SOM learning algorithm. The SOM training algorithm exploits the topology connections between the data features in a way that similar data vectors correspond to proximal neurons in the SOM network. The color variation map emanating from the codebook vector components leads to identification of possible relations existing between the component layers of the SOM since the values in each component layer exhibits similarities when the components are uncorrelated substantially.

As it is demonstrated above (Figs. 4.5, 4.7, and 4.9), it is observable that in SKN and XYF classifiers the output weights are collocated in the same cluster that corresponds to the input components for each weed species. Every weed corresponds to a compact cluster in the output domain except the CPANN classifier for which the weights of the output layer demonstrate

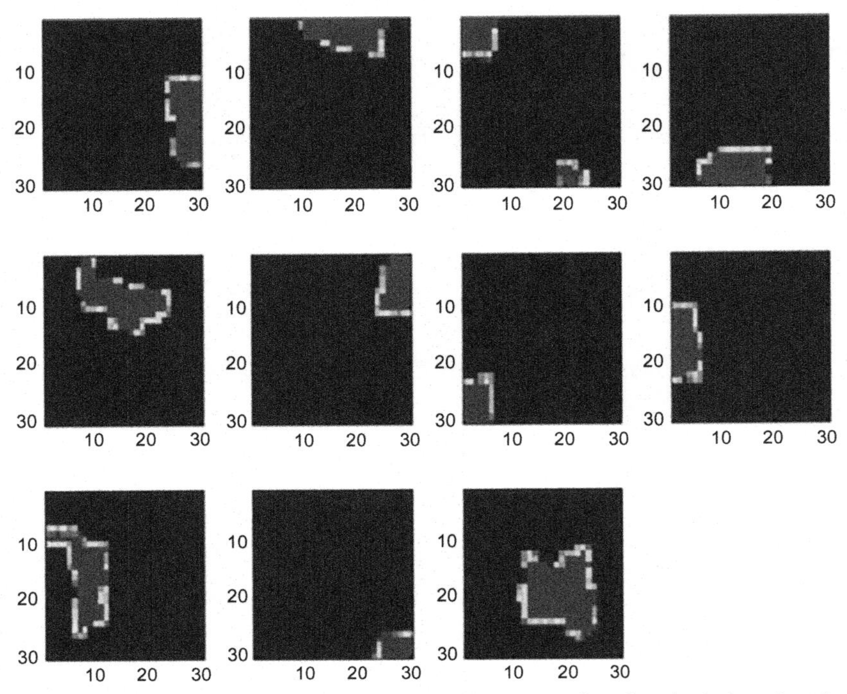

Fig. 4.5 Weights of target layer corresponding to SKN classifier depicting the class labels of *Zea mays* crop and the 10 different investigated weed species. The areas with blue (light gray in monochrome) color equals to 0 while red (dark gray in monochrome) areas denote 1.

some sparsity and are not compact like the SKN and XYF classifiers. This behavior is the reason for the reduced classification accuracy (Table 4.1) which is visible with the results of the CPANN classifier with respect to the other two employed classifiers SKN and XYF. SKN appears to be the most performant network which is probably to the capability of the classifier to perform a simultaneous clustering of both input and output layer.

The comparison of the clusters formed by the input weights in Figs. 4.4, 4.6, and 4.8 leads to the conclusion that the component layers corresponding to the spectral bands of 539, 540, 542, 545, 549, 557, 565 nm demonstrated higher correlation. The wavebands situated at 578, 585 nm demonstrated high independence to other spectral bands. The wavebands situated at 596, 605, 639, 675, 687, 703, 814, and 840 nm show high correlation between them. The most representative waveband for *Zea mays* as the clusters of SKN indicate it is centered at 703 nm symbolized by the red colored cluster.

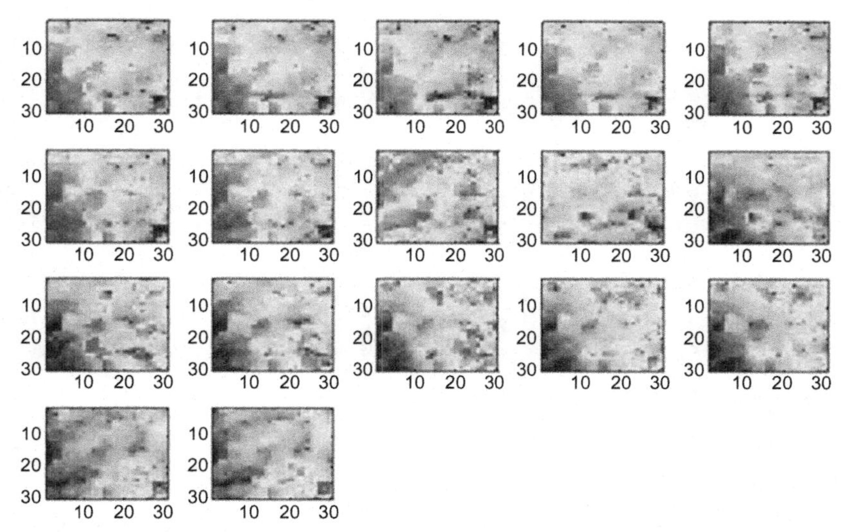

Fig. 4.6 Weight plot of XYF classifier related to the spectral features attained from 17 spectral signatures. The spectral features vary from lower values (denoted with blue color, dark gray in monochrome) to higher values (denoted with red color, light gray in monochrome).

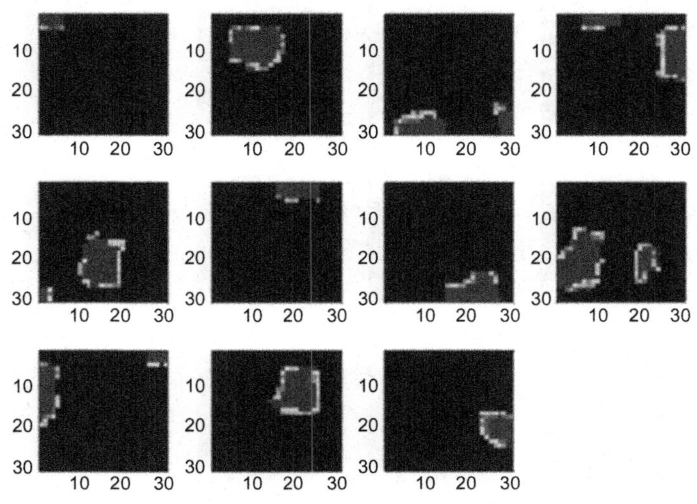

Fig. 4.7 Weights of target layer corresponding to XYF classifier depicting the class labels of *Zea mays* crop and the 10 different investigated weed species. The areas with blue (light gray in monochrome) color equals to 0 while red (dark gray in monochrome) areas denote 1.

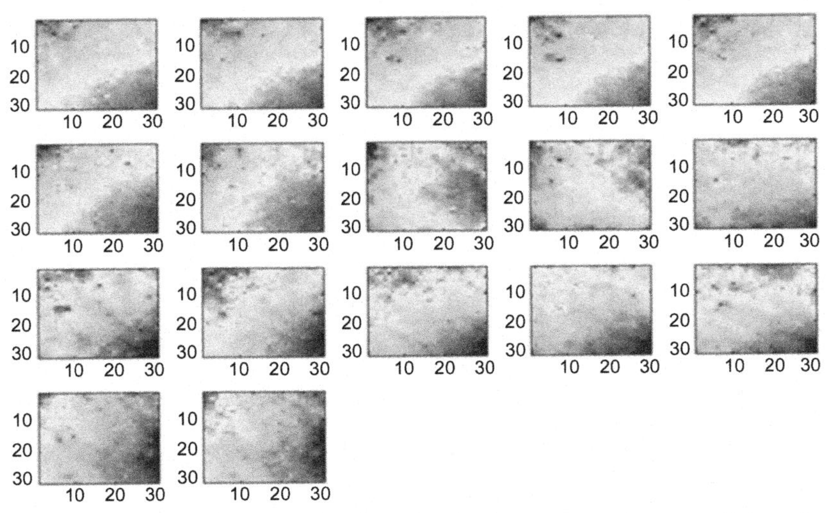

Fig. 4.8 Weight plot of CPANN classifier related to the spectral features attained from 17 spectral signatures. The spectral features vary from lower values (denoted with blue color, light gray in monochrome) to higher values (denoted with red color, dark gray in monochrome).

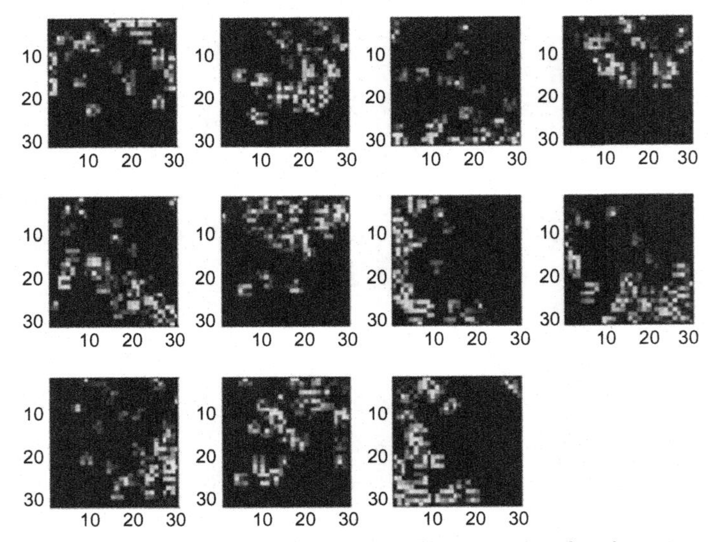

Fig. 4.9 Weights of target layer corresponding to CPANN classifier depicting the class labels of *Zea mays* crop and the 10 different investigated weed species. The areas with blue (light gray in monochrome) color equals to 0 while red (dark gray in monochrome) areas denote 1.

As it is stated by the above observations and visualizations (Figs. 4.5–4.9), the topographic mapping preservation property enables the assessment of significance of certain characteristics that carry novel information and deduce the redundant nature of other characteristics due to their intercorrelation. The intercorrelation of features occurs when the visual appearance show that that behavior has the same trend in the colormap visualization of the component layer of this feature.

4.4.2 Results for active learning

The procedure to realize the active learning algorithm for the identification of different weed species was implemented as follows:

(1) The target set was defined by Maize plants which consisted of 110 samples comprising of 17 features emanating from feature selection.

(2) Each of the one-class classifiers was calibrated by using the target datasets from step 1.

(3) The calibrated one-class classifiers underwent testing with 54 unseen samples which came from Maize plants and additional 54 originating from a specific weed species. The evaluation metric of classifier performance was the ability to classify unknown species as an outlier in comparison to those crop plants comprising the baseline set.

(4) The crop combined with the outlier spectra originating from a single weed species as were the new default baseline set. The afore mentioned process concerning outlier discovery was iterated (i) by incorporating crop data samples with the initial weed species, (ii) by confronting the one-class classifiers to an unknown weed species and simultaneously with data from the known and stored weed and crop species. The evaluation metric of the classification performance was the capability to recognize unknown weed species as outliers in comparison with the baseline set that has resulted from the augmentation from crop and weed. In the case that the unknown sample was classified as the baseline set it would belonged to one of the two classes that comprised the augmented dataset.

It is notable that the proposed process does not need any external expert information to operate and can progress autonomously. The functionality of the procedure is mainly attributable on an iteration of outlier identification and subsequent augmentation cycles which are functioning in an automatic manner. The implementation software for constructing the One Class Classifiers was DDtools for Matlab (Tax, 2015).

Several one-class classifiers have been evaluated, however the most promising results for the active learning have been produced by SOM, MOG, SVM and the Autoencoder classifier. Regarding the one-class MOG classifier identification accuracy for crop identification were up to 100%. At the same time, the accuracy of identification of different weed species ranged from 31.48% to 98.15%. The one-class SOM crop identification accuracy reached 100% and with respect to other weed species identification performance varied between 53.70% and 94.44% (Fig. 4.6). By looking at Table 4.1, it can be observed that positive recognition concerning *Zea mays* is greater that concerning distinct weed species some classifiers as SOM and MOG. On the other hand, specific weed species achieved higher performance in identification regarding SOM by attaining performance exceeding 80%. The SOM classifier was not able to identify successfully *Ranunculus repens* which is annotated with weed index 1 and reached an identification accuracy of 53.7%. For the MOG classifier, an identification rate exceeding 80% was attained for *Cirsium arvense, Stellaria media, Urtica dioica* and *Medicago lupulina* (denoted with weed indices 2, 4, 8 and 10 respectively.

4.4.3 Discussion on SOM based models and active learning

Various models of one-class algorithms were developed to test the process of active learning. The most significant results were attained by SOM, MOG, SVM and the Autoencoder classifier. The one-class MOG classifier achieved an accuracy of 100% for crop type identification. The weed type identification fluctuated from as low as 31.48% as high as 98.15%. The one-class SOM classifier crop identification accuracy reached 100%. The identification accuracy for various weed species, varied from 53.70% to 94.44% (Fig. 4.10). As it can be seen in Table 4.1 the correct identification concerning *Zea mays* appears compared to that concerning specific weed species both for SOM and MOG one class classifiers. As regards the SOM classifier, it achieved recognition percentages higher than 80%. It was also indicated that *Ranunculus repens* identification was significantly lower, reaching an accuracy of 53.7% (Fig. 4.10). Regarding MOG classifier, the identification accuracies were higher than 80% in the case of the following weed species including *Cirsium arvense*, *Stellaria media*, *Urtica dioica* and *Medicago lupulina*, demonstrated with weed index 2, 4, 8, and 10 respectively (Fig. 4.11).

The SOM and MOG one class classifiers produced higher accuracies regarding the identification of different weed species. Simultaneously, the SVM and Autoencoder one-class classifiers produced non acceptable accuracies as it is demonstrated in Figs. 4.12 and 4.13.

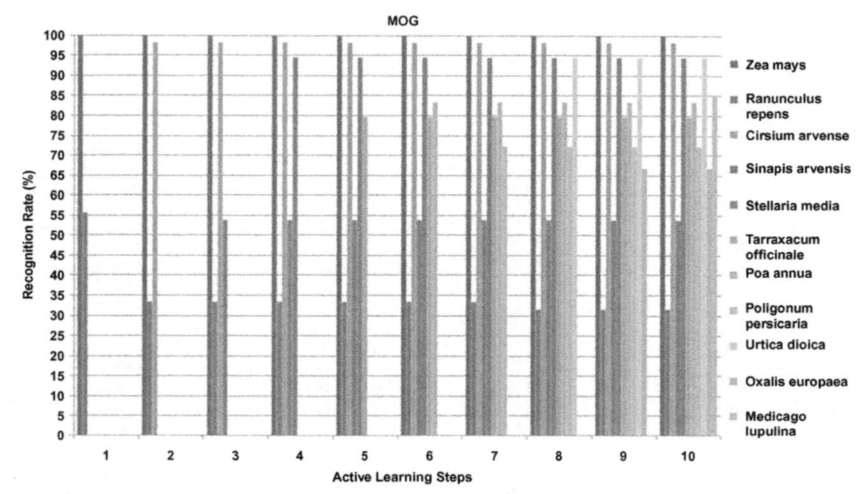

Fig. 4.10 Identification results for SOM based One class classifier with active learning approach achieved for each of the learning steps while perform active learning for the identification of new types of weeds (Pantazi, Moshou, & Bravo, 2016).

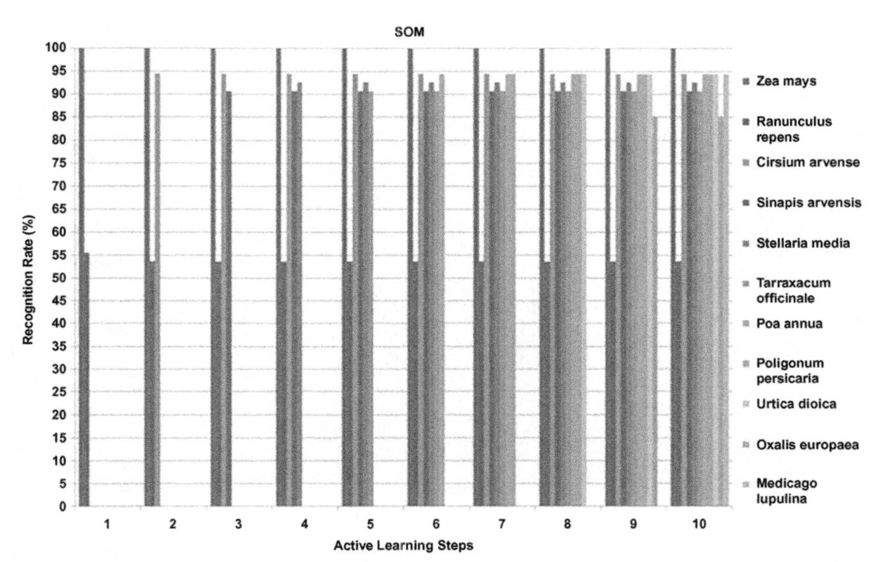

Fig. 4.11 Identification results for MOG based One class classifier with active learning approach achieved for each of the learning steps while perform active learning for the identification of new types of weeds (Pantazi et al., 2016).

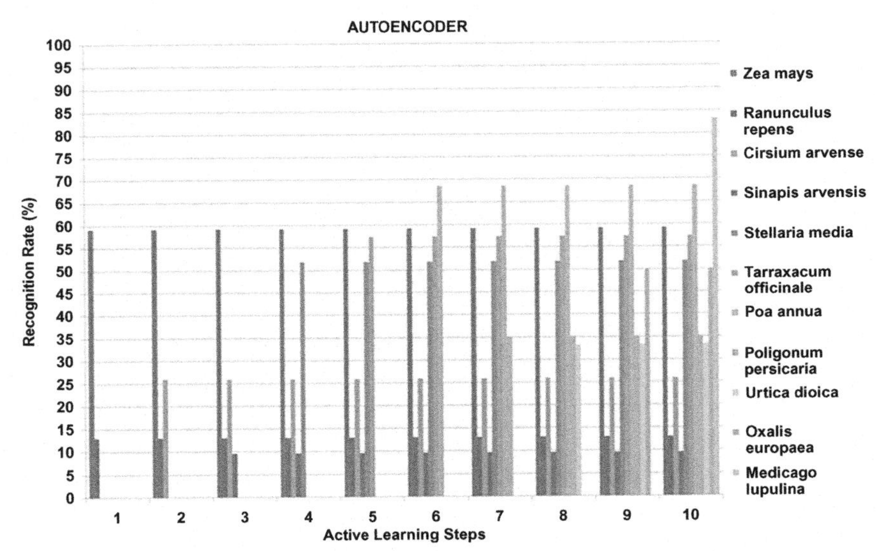

Fig. 4.12 Identification results for Autoencoder based One class classifier with active learning approach achieved for each of the learning steps while perform active learning for the identification of new types of weeds (Pantazi et al., 2016).

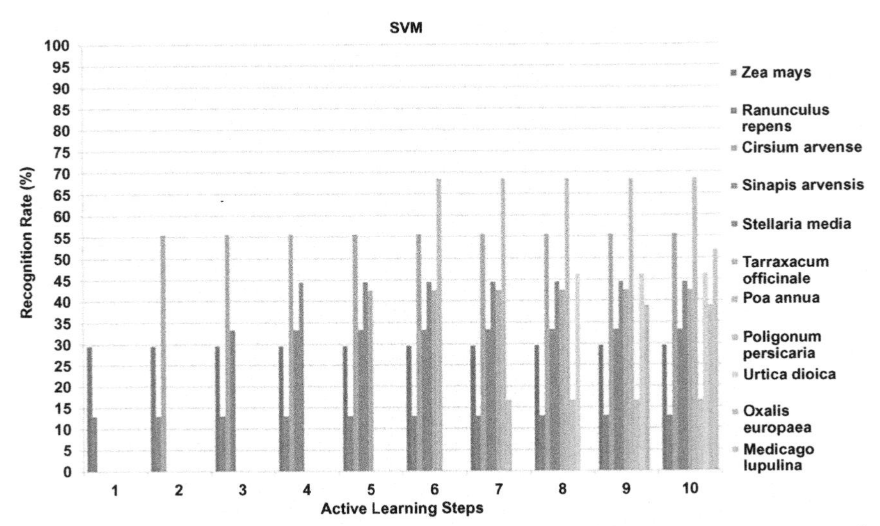

Fig. 4.13 Identification results for SVM one-class classifier with active learning approach achieved for each of the learning steps while perform active learning for the identification of new types of weeds (Pantazi et al., 2016).

The slight differentiation concerning the behaviour MOG and SOM is explained from the different probability density representation concerning weed recognition. MOG-represents probability density while the SOM classifier tries to reconstruct the data. The worse recognition rates can be partially explained by the similarity of the spectral signatures, which leads to confused classes. The augmentation of features for example related to the spectrum or texture might enhance the recognition accuracy.

For various weed types, the accuracy remains the same during the evolution of learning steps. This could be expected from the augmentation of new unseen weed species which can affect the correct identification of individual weeds. This can lead to mislabeling of other species when the weed classes increase in quantity. The proposed method involving Active Learning has enhances classification compared to regit classifier classifier structure because it offers flexibility of actively forming novel weed classes.

4.4.4 Discussion on SOM based models and active learning

In the current case study various Self Organizing Map classifiers employing supervised learning are applied aiming to take advantage of the reflectance data for recognizing different weed species and detecting *Zea mays* crop. More specifically, Hierarchical Self Organizing Map models like SKN, Xyf and CPANN have been trained so as to organize hyperspectral imaging data for recognizing whether hyperspectral signatures orginiate from either a crop (in the current case study the main crop is *Zea mays* or a weed species. In order to enhance classification the most effective wavebands were picked by waveband selection based on their effect on the classification accuracy.

It was indicated that both weed species recognition and *Zea mays* detection demonstrated an accuracy exceeding 90% for the majority of combinations. The spatial visualizations concerning both input and output weights is a predominant feature of supervised SOMs that facilitates the extraction of the most persistent correlations associating features and classes.

1. In the case of the active learning approach, the most performant results were obtained by the SOM and MOG classifiers. On the contrary, the Auto-encoder and SVM classifiers demonstrated lower performance accuracies. More specifically, the crop identification accuracy reached 100% for both MOG and SOM. In the case of MOG classifier, the identification accuracy between different weed species fluctuated from 31% to 98%, while the SOM classifier demonstrated a recognition capability from 53% to 94%.

References

AlSuwaidi, A., Grieve, B., & Yin, H. (2018). Combining spectral and texture features in hyperspectral image analysis for plant monitoring. *Measurement Science and Technology*, *29*(10)104001.

Barrero, O., & Perdomo, S. A. (2018). RGB and multispectral UAV image fusion for Gramineae weed detection in rice fields. *Precision Agriculture*, *19*(5), 809–822.

Barrett, S. H. (1983). Crop mimicry in weeds. *Economic Botany*, *37*(3), 255–282.

Bauriegel, E., Giebel, A., Geyer, M., Schmidt, U., & Herppich, W. B. (2011). Early detection of Fusarium infection in wheat using hyperspectral imaging. *Computers and Electronics in Agriculture*, *75*(2), 304–312.

Borregaard, T., Nielsen, H., Nørgaard, L., & Have, H. (2000). Crop–weed discrimination by line imaging spectroscopy. *Journal of Agricultural Engineering Research*, *75*(4), 389–400.

Clifton, L., Clifton, D. A., Zhang, Y., Watkinson, P., Tarassenko, P., & Yin, H. (2014). Probabilistic novelty detection with support vector machines. *IEEE Transactions on Reliability*, *63*(2), 455–467.

De Castro, A. I., Torres-Sánchez, J., Peña, J. M., Jiménez-Brenes, F. M., Csillik, O., & López-Granados, F. (2018). An automatic random forest-obia algorithm for early weed mapping between and within crop rows using UAV imagery. *Remote Sensing*, *10*(2), 285.

ElMasry, G., & Sun, D. W. (2010). Principles of hyperspectral imaging technology. In *Hyperspectral imaging for food quality analysis and control* (pp. 3–43). Academic Press.

Fernández-Quintanilla, C., Peña, J. M., Andújar, D., Dorado, J., Ribeiro, A., & López-Granados, F. (2018). Is the current state of the art of weed monitoring suitable for site-specific weed management in arable crops? *Weed Research*.

Goetz, A. F. (2009). Three decades of hyperspectral remote sensing of the earth: A personal view. *Remote Sensing of Environment*, *113*, S5–S16.

Goetz, A. F., Vane, G., Solomon, J. E., & Rock, B. N. (1985). Imaging spectrometry for earth remote sensing. *Science*, *228*(4704), 1147–1153.

Harker, K. N., & O'donovan, J. T. (2013). Recent weed control, weed management, and integrated weed management. *Weed Technology*, *27*(1), 1–11.

Herrala, E., Okkonen, J., Hyvarinen, T., Aikio, M., & Lammasniemi, J. (1994). Imaging spectrometer 20 for process industry applications. *SPIE*, *2248*, 33–40.

Jensen, T., Apan, A., Young, F., & Zeller, L. (2007). Detecting the attributes of a wheat crop using digital imagery acquired from a low-altitude platform. *Computers and Electronics in Agriculture*, *59*(1–2), 66–77.

Kohonen, T. (1988). *Self-organization and associative memory*. Berlin: Springer Verlag.

Kumar, N., Belhumeur, P. N., Biswas, A., Jacobs, D. W., Kress, W. J., Lopez, I. C., et al. (2012). Leafsnap: A computer vision system for automatic plant species identification. In *Computer Vision, ECCV 2012* (pp. 502–516): Springer.

Kuska, M., Wahabzada, M., Leucker, M., Dehne, H. -W., Kersting, K., Oerke, K. -C., et al. (2015). Hyperspectral phenotyping on the microscopic scale: Towards automated characterization of plant-pathogen interactions. *Plant Methods*, *11*, 28.

Lambert, J. P. T., Hicks, H. L., Childs, D. Z., & Freckleton, R. P. (2018). Evaluating the potential of unmanned aerial systems for mapping weeds at field scales: A case study with Alopecurus myosuroides. *Weed Research*, *58*(1), 35–45.

Laursen, M. S., Jørgensen, R. N., Midtiby, H. S., Mortensen, A. K., & Baby, S. (2017). Dicotyledon weed quantification algorithm for selective herbicide application in maize crops: Statistical evaluation of the potential herbicide savings. *World Academy of Science, Engineering and Technology, International Journal of Biological, Biomolecular, Agricultural, Food and Biotechnological Engineering*, *11*(4), 272–281.

Marini, F. (2009). Artificial neural networks in food analysis: Trends and perspectives. *Analytica Chimica Acta*, *635*, 121–131.

Mehl, P. M., Chen, Y. -R., Kim, M. S., & Chan, D. E. (2004). Development of hyperspectral imaging technique for the detection of apple surface defects and contaminations. *Journal of Food Engineering, 61*(1), 67–81.

Meier, U. (2001). Growth stages of mono- and dicotyledonous plants. *BBCH Monograph.* https://doi.org/10.5073/bbch0515.

Melssen, W., Wehrens, R., & Buydens, L. (2006). *Chemometrics and Intelligent Laboratory Systems, 83,* 99–113.

Moshou, D., Ramon, H., & De Baerdemaeker, J. (2002). A weed species spectral detector based on neural networks. *Precision Agriculture, 3*(3), 209–223.

Moshou, D., Vrindts, E., De Ketelaere, B., De Baerdemaeker, J., & Ramon, H. (2001). A neural network based plant classifier. *Computers and Electronics in Agriculture, 31*(1), 5–16.

Pantazi, X. E., Moshou, D., & Bravo, C. (2016). Active learning system for weed species recognition based on hyperspectral sensing. *Biosystems Engineering, 146,* 193–202.

Pimentel, M. A., Clifton, D. A., Clifton, L., & Tarassenko, L. (2014). A review of novelty detection. *Signal Processing, 99,* 215–249.

Rouse, J. W. J., Haas, R. H., Schell, J. A., & Deering, D. W. (1974). Monitoring vegetation systems in the Great Plains with ERTS. In *Third ERTS Symposium, NASA SP-351, Washington DC,* pp. 309–317.

Sánchez, M. T., & Pérez-Marín, D. (2011). Nondestructive measurement of fruit quality by NIR spectroscopy. In *Advances in post-harvest treatments and fruit quality and safety* (pp. 10–163). Hauppauge: Nova Science Publishers, Inc.

Søgaard, H. T. (2005). Weed classification by active shape models. *Biosystems Engineering, 91* (3), 271–281.

Tax, D. M. J. (2015). Data description toolbox. *DD tools: A Matlab toolbox for data description, outlier and novelty detection.*

Tellaeche, A., Pajares, G., Burgos-Artizzu, X. P., & Ribeiro, A. (2011). A computer vision approach for weeds identification through support vector machines. *Applied Soft Computing, 11*(1), 908–915.

Wang, X., Zhang, M., Zhu, J., & Geng, S. (2008). Spectral prediction of Phytophthora infestans infection on tomatoes using artificial neural network (ANN). *International Journal of Remote Sensing, 29*(6), 1693–1706.

Whelan, B., & Taylor, J. (2013). *Precision agriculture for grain production systems.* CSIRO Publishing.

Xie, C., Shao, Y., Li, X., & Hea, Y. (2015). Detection of early blight and late blight diseases on tomato leaves using hyperspectral imaging. *Scientific Reports, 5,* 16564.

Zhang, H., Paliwal, J., Jayas, D. S., & White, N. D. G. (2007). Classification of fungal infected wheat kernels using near-infrared reflectance hyperspectral imaging and support vector machine. *Transactions of the ASABE, 50*(5), 1779–1785.

Zupan, J., Novic, M., & Gasteiger, J. (1995). Neural networks with counter-propagation learning strategy used for modelling. *Chemometrics and Intelligent Laboratory Systems, 27,* 175–187.

CHAPTER 5

Tutorial II: Disease detection with fusion techniques

Contents

5.1 Introduction

According to FAO (2009), the increasing demand for cereals is predicted to exceed approximately 3 billion tonnes before the year 2050, in comparison with the current production which is estimated at about 2.1 billion tonnes (FAO, 2009). Taking into consideration that this global demand is significantly higher than the relative supply; Therefore, the embracement of new policies targeting on adaptation of novel technologies is crucial, to cover this need. An unpredictable enemy against this effort, several types of threats that poses significant challenge to the yield targets including climate change biotic stresses resulting from plant enemies (e.g. viruses, bacteria, fungi, pests, nutrient stress, lack of proper irrigation) responsible for the degradation of crop quality and possible yield losses. It is speculated that the absence of relevant prevention and treatment regarding the above mentioned plant enemies, would reach 18% for plant infections and 32% for weeds presence, independent of the applied crop management practices (Oerke & Dehne, 2004).

https://doi.org/10.1016/B978-0-12-814391-9.00005-4

Intensive cultivation practices after 1950 have induced an increased pathogen presence and susceptible crops. Apart from the spread of new diseases resulting from international commerce and favorable transmission conditions thanks to climate change, risk producing practices that trigger several types of fungal in crops concern using unsuitable crop genotypes targeting only on yield and improper fertilization not corresponding to the crop needs (Bebber, Holmes, & Gurr, 2014). Biotic stress factors like weeds and microorganisms (viruses, fungi, bacteria) decrease both yield and the market potential of agricultural production and imposes negative impact on the cost in agriculture because of phytosanitary chemical applications.

Accurate diagnosis and monitoring of the possible infection presence are fundamental towards the application of effective management practices. Crop diseases usually appear in dispersed locations in the crop growing areas and they exhibit a dynamic pattern in terms of spatio-temporal variability. Effective techniques focusing on effective crop infections detection are vital. For this reason, several scientific studies are carried out aiming at analyzing the behavior of crop infections in various settings (Al Masri, Hau, Dehne, Mahlein, & Oerke, 2017).

Precision agriculture is using information technology and sensor based monitoring solutions aiming to deal with timely effective identification and location of the invasion of crop diseases including the main foci and types of pathogen. It is characterized as crop cultivation practice that uses as parameters the spatio-temporal variations of crop and soil parameters for field management for providing efficient management decisions in real-time (Stafford, 2000).

Keeping into consideration, that the crop infection appearance relies on several ambient conditions and they exhibit a heterogeneous appearance pattern, optical sensing is regarded as an efficient tool for detecting main invasion spots and access the disease severity pattern in the field (Franke & Menz, 2007). It can be combined effectively with different machine learning approaches, in order to aid decision for the application of sustainable pest management program. The precise and sensor driven pesticides applications complying to current crop protection practices lead to possible decrease concerning in pesticide application which can subsequently diminish the economic loss and ecological consequences in agricultural domain (Gebbers & Adamchuk, 2010).

Currently phenotyping is related to non-contact imagery and sensor-driven analytics of crop characteristics in terms of anatomy, physiology, and biochemistry (Walter, Liebisch, & Hund, 2015). Novel methods using standard monitoring systems like optical sensors are employed serving as useful and efficient tools for nondestructive crop infection identification. There is an ample variety of imaging and noncontact sensors ready to support crop infection detection.

5.1.1 Optical sensing and their contribution to crop disease identification

Plants exhibit a variety of phenotypic characters regarding their optical features, which can be assessed by eye, while various become explicit after applying special sensors. These sensory systems are proven helpful for detecting these optical features by measuring several light properties. A quality based assessment of the genomic expression for a given crop activity triggered by the environmental conditions is performed in a visual manner by human experts. This is the so-called phenotyping process. Phenotyping is the crop visual depiction and appreciation in a holistic way during crop growth (Fiorani & Schurr, 2013). The assessment concerns the growth rate, structure, tissue coloring, health condition, and establishment of abiotic and biotic stress manifestations and their related severity symptoms. Conventional visual crop infection level assessment has the tendency to false estimations due to plant-pathogen interactions involving cellular and leaf changes and the subjective character of the assessment by a human expert.

In the initial infection period, fluorescence is the most effective sensing principle because it assesses the health condition since it estimates photosynthetic efficiency, which is related to the health state.

After the early metabolic changes, the fungus disperses in a radial pattern outwards from the infection focus. Subsequently, the initial focus area demonstrates necroses symptoms including loss of pigmentation, the destruction photosynthetic apparatus and the cell walls breakdown. At this point of time, we have the first manifestation of the infection patterns. The characteristics of the reflected light can reveal specific patterns that are correlated to the incumbent infection contributing to the detectability of infections after this stage. Pathogen propagules are detectable in the visual part of the spectrum (detectability depends on the specific pathogen); chlorophyll downgrade in the region of visual and the red-edge (550 nm; 650–720 nm); senility in the VIS and NIR regions (680–800 nm) appears as browning while dryness is evident in SWIR area (1400–1600 nm and 1900–2100 nm). Finally, alterations in the structures of the canopy and the collapse of leaf area is evident in the NIR.

During the period that disease spreads across the whole plant, this manifests as a widespread stress, appearing through a general pause of stomata functioning. Aiming to water loss reduction. The altered transpiration leads to heat containment that can be sensed with the help of thermography. A difficulty that arises in recording a consistent indicator of stress through thermography is the dependency of thermal response to ambient factors like air temperature, lighting conditions and wind. Hence, thermography is not reliable if is used on ground vehicles as proximal sensor.

Changes of leaf reflectance are due to leaf compositional alterations resulting from an infection. Diseases can influence the optical leaf characteristics at multiple wavebands. Therefore, disease-sensing tools can possibly rely on recording spectral responses in a variety of wavebands or a fusing information from several wavebands. Induct plants have a green appearance because the green light (ca. 550 nm) is reflected at a high degree in comparison to the other colors like blue, yellow and red. The infected leaf areas are manifested through distinct leaf lesions, related to necrosis or chlorosis appearing in these regions, resulting in elevating reflectance in the visual band and more specifically, where the chlorophyll is absorbed. More specific, reflectance alterations at wavelengths circa 670 nm, result in the red edge shift (the stipe slope that appears in the reflectance spectrum that starts from a weak visual signal and finishes at a very strong NIR signal and appears at 730 nm) to switch to lower wavelengths. Vice versa, the reduced biomass, correlated to senescence, signifies a lower growth and leaf loss resulting in a collapse of the canopy reflectance around the NIR region.

Optical sensing is predicted to be important factor in the forthcoming smart farming applications to enable sustainable farming thanks to efficient crop monitoring in precision agriculture. This operational model will become functional if multi-disciplinary expertise combining research, industrial and agronomic expertise collaborating to achieve a collective vision of functional methods (Mahlein, 2016).

5.1.2 Artificial intelligence approaches for crop disease monitoring

A shortcoming of optical sensors concerns the volume and the complex character of the data that is accumulated. In order to apply optical sensor data in more effective way, efficient data analytics and machine learning methods are regarded crucial. The data analysis has to comply with several objectives including the crops early stage detection the discrimination between different types of infection, the discrimination between biotic and abiotic stresses, and the estimation of the disease severity level. These objectives have to be achieved at a least at the same level obtained with a standard technique and a shorter processing time.

Approach based on data mining and machine-learning methods are capable of minimizing the effort and achieving early detection of infections through hyperspectral imaging (Behmann et al., 2015). Regarding machine learning techniques, unsupervised and supervised learning are employed for offering solutions to classification and clustering problems. On the other hand, unsupervised machine learning algorithms seek to discover data mining patterns automatically. To the contrary, for performing supervised learning there is a need for training data with labelled samples. Standard classifiers and clustering algorithms include k-means, SOMs, SVMs, and ANNs (Kersting et al., 2016; Pantazi, Moshou, Alexandridis, Whetton, & Mouazen, 2016).

It is currently known (CiresSan et al., 2010, Chollet, 2017) that feature selection can affect positively the classification, the processing time. When multiple data sources are available an effective fusion method is necessary for combining the obtained features for extracted from each source because the data from each source could be complementary (Hu, Chen, Ge, & Wang, 2018). Feature concatenation is a basic technique to fuse the obtained features; therefore, the accuracy is not assured since weighted fusion is required.

A variety of techniques is available for performing feature fusion from different data or information sources. As an illustration, a feature fusion regression kernel was employed to assimilate features extracted from satellite imagery (Hughes & Salathé, 2015). Spectral, textural, and pattern similarity produced by every object to create a feature vector and then use the combined vector to improve object classification and improved generalization. In Šulc and Matas (2017) a kernel PCA (KPCA) was applied to perform normalization from hyperspectral images, positioning spatial and terrain (LiDAR) information, hence thus decreasing their dimensionality and denoising effectively to perform final fusion.

The acquisition of specific crop hyperspectral patterns would be very useful, for wide spread information extraction related to crop condition assessment. Moreover, this technique of data acquisition could become part of operational field monitoring for example growth monitoring maps, meteorological data and prediction models. This could lead to a development of an effective tool for functional early disease warning and precision farming for crop protection.

5.1.3 Reference and advanced optical methods for plant disease detection

Standard methods targeting on the identification and the discrimination of several plant infections involve visual crop disease severity assessment by human experts, morphology based identification of microscopic pathogens features, and molecular, serological, and microbiological methods (Bock et al., 2010). These methods are incorporated a main features of plant protection services for both research and industry purposes. In the recent years, DNA-based and serological techniques have triggered the targeted characterization and crop disease severity assessment (Martinelli et al., 2014).

When employing molecular and serological techniques, different types of pathogen strains can be classified according to their virulence level or are invulnerable to specific fungicides. The diseases severity is not directly correlated to the infestation level of the pathogen, since the disease manifestation does not follow a linear correlation with the pathogen biomass (Nutter, 2001).

Traditionally, disease infection symptoms such as blight, tumors, rots, cankers, etc., or observable pathogen symptoms such as mycelium are utilized as indicators for detecting infections. Visual crop disease assessment has become capable of providing better and trustworthy estimations thanks to the meticulous protocols and standardization of the assessment process (Bock et al., 2010). However, the drawback concerning visual assessment lays on the temporal variability of the assessor experience, which brings often-significant credibility issues regarding the repeatability of the assessment (Bock et al., 2010). On the whole, these methods are characterized as time-consuming and also require experts with high levels of experience in the field of disease evaluation which leaves open the possibility of human bias.

Recently, novel sensory-based approaches have been introduced aiming to the detection, identification, and crop disease severity assessment (West et al., 2010). The current methods are capable of estimating the optical crop features from different spectral bands, by using spectral bands from the non-visible range. They take into account possible alterations affecting the crop status due to a possible infection that modifies the textural, structural and optical leaf properties negatively (West et al., 2010). The most efficient methods are considered sensory systems capable of measuring optical and thermal crop properties. Most of them were initially developed for military applications, earth remote sensing, satellite and aerial sensing and in certain manufacturing applications.

The early multispectral remote sensing systems were developed in 1964. Hyperspectral imagery appeared during the 1980s (Campbell, 2007). In agricultural engineering, remote sensing is a technique for attaining plant features in a non-destructive manner. The idea has been extended through proximal and limited-scale sensing concerning crop status (Oerke, Mahlein, & Steiner, 2014). The sensory systems have been incorporated to several platforms including agro-machinery and other aerial, space and ground vehicles and are listed as follows:

1. RGB-imaging

Visual imaging is used in the field of plant pathology for monitoring crop health status. Digital cameras produce images in RGB format (red, green, and blue) for crop monitoring and infection severity assessment. Regarding the technical features of these easy to use, portable devices, including the dynamic range of the imaging chip, spatial accuracy, or lens characteristics have evolved through the years. Nowadays, the wide spread use of high quality imaging sensors is common due to their availability in mobile devices so that phytopathology experts and farmers can utilize them when needed. There are also other common ways for crop monitoring through video sensors which are capable of obtaining frames from various plant parts, ranging from the lower parts such as roots to the higher ones like the clusters of flowers. The RGB cameras utilize the red, green, and blue channels on different resolutions for identifying possible crop biotic stresses during the growing period (Bock et al., 2010,). In addition, there are other morphological features and textural characteristics that can be correlated to the presence and identity of various plant infection symptoms (Neumann, Hallau, Klatt, Kersting, & Bauckhage, 2014).

The targeted selection of correlated features from preprocessing the RGB images tend to improve the classification performance (Behmann, Steinrücken, & Plümer, 2014). The digital image analysis is a commonly used technique for assessing different types of crop infections. ASSESS 2.0 and "Leaf Doctor," are the most known software based on digital image analysis. There is also the possibility of custom-made modules (Pethybridge & Nelson, 2015; http://www.plant-image-analysis.org/).

In ASSESS 2.0 software, the RGB characteristics of the images is visualized in histogram form, which are used to define threshold for further processing. The user through a user-friendly interface provides the parametrization needed to fine-tune the discrimination between healthy and diseased areas. Regarding the severity symptoms assessment, an extraction of the pixels classified as infection symptom pixels or they appear as an area after thresholding the image so that a mask is created that deletes the background around of the region of interest. Therefore, ASSESS 2.0 requires single leaves or well-defined background in order to isolate the leaf areas. Moreover, further constraints such as homogeneous focus, detail, and ambient lighting affect the overall performance and reliability for automatic image analysis. The image analysis seems to be affected by the several natural conditions including the acquisition angle, the pixel size as defined by the distance between the leaf and the camera. In the event of heterogeneous environmental settings and low image analysis the infection symptoms areas are hard to identify. The most critical condition for repeatability is a credible image acquisition protocol that produces consistent results.

2. Multi- and hyperspectral reflectance sensors

Spectral sensing devices are characterized from their spectral resolution, their spatial coverage, and the detection principle used by the sensor. Multispectral sensing systems appeared earlier than the other spectral approaches. These sensors usually gather the spectral response of targets in a few wide wavebands. The introduction of modern The hyperspectral sensory systems elevated the complexity of the recorded images by having a spectrum defined between 350 and 2500 nm and by some case carrying a very high spectral resolution that can be lower than 1 nm (Steiner, Bürling, & Oerke, 2008).

Unlike the non-imaging devices, which contain the average spectral response from specific area, hyperspectral imaging devices are capable of providing the spectral and spatial features of the target object. Hyperspectral information is appended by three-dimensional matrices (x, y, z). The spatial resolution affects strongly the detectability of the crop infections (West et al., 2003). Airborne devices are capable of identifying infected field patches by soil propagated pathogens (Hillnhütter, Schweizer, Kühnhold, & Sikora, 2010) or in later infected filed areas (Mahlein, Oerke, Steiner, & Dehne, 2012). Sensory of approximately 1 m spatial resolution are not able to discriminate isolated infection symptoms or infected leaves and crops. For this application, sensors that are mounted on proximal platforms are considered more appropriate (Oerke et al., 2014).

The reflection of the electromagnetic waves from plants is a complicated process, which emerges from a cascade of interactions, which are biophysical, and biochemical in nature. The visual region of the spectrum (from 400 to 700 nm) is correlated by the pigment concentration of the leaf. The near infrared region, from 700 to 1100 nm is highly related to the leaf architecture, structural interference with light scattering and dependence on leaf water concentration. The short-wave region from 1100 to 2500 nm is affected by the chemical components and water content (Jacquemoud & Ustin, 2001).

Different reflectance alterations caused by possible infections result from damages in the leaf assembly and its chemical components during infection that is characteristic. This alteration flow specific pattern like the evolution from chlorotic to necrotic symptoms or the occurrence of fungal colonies. In the case of biotrophic fungi an example of which is different types of powdery mildew and rust they influence leaf integrity and chlorophyll content in negligible at an early stage of infection. On the other hand, perthotrophic pathogens activity in crops often brings leaf degradation caused by the emitted toxins or enzymes to the infected crop area, which are characteristic for each pathogen, leading finally to necrotic lesions.

The aforementioned leaf infection manifestations facilitate the spectral detection of leaf infection. Mahlein et al. (2013) has presented the identification of leaf infections in sugar beet plants attained from leaf spectral signatures. In ground sensing, hyperspectral camera approaches are regarded an efficient and trustworthy method for the detection of fungi that emit mycotoxins in *Zea mays* plants (Del Fiore et al., 2010).

3. Thermal sensors

Infrared thermography (IRT) detects plant temperature as a parameter to differentiate the water content of plants (Jones et al., 2002), the crop microclimate (Lenthe, Oerke, & Dehne, 2007), and with transpiration related symptoms resulting from the establishment of an infection (Oerke, Steiner, Dehne, & Lindenthal, 2006). The infrared emissions in the spectral regions between 8 and 12 mm are detectable by thermal infrared sensors and is visualized in a raster format, where each pixel is assigned the level of the temperature reading from the observed target. In precision agriculture, thermal cameras are applied in various different settings ranging from aerial to microscopic thermal acquisition tasks. Nevertheless, the functionality of thermal cameras is compromised by ambient disturbances like variations in temperature, precipitation, sunlight, or wind interference. The temperature on the leaf is directly affected by the transpiration of the plant, which is resulting as an impact from the invasive activity of different pathogens (Jones et al., 2002). While several pathogens, like rust infection, cause localized foci, damage by root infection or systemic diseases like *Fusarium* spp. induces an impact on plant transpiration and affects the water circulation globally in the plant.

In the case of apple infection by fungus *V. inaequalis*, thermal imaging achieved a visual mapping of the fungus invasion area that was not possible with a visible light camera on apple crops, while fungal organs could be detected only with the help of a microscope (Oerke, Gerhards, Menz, & Sikora, 2010).

4. Fluorescence imaging

A variety of chlorophyll fluorescence factors have proven useful regarding the assessment of changes in the photosystems of plants due to biotic and abiotic stresses. Chlorophyll fluorescence cameras are usually sensors with a light excitation source for sensing the emission light from electron di-excitation which is interpreted as a signature of the specific stress factor (Bauriegel, Brabandt, & Gärber, U., and Herppich, W. B., 2014). Interpreting fluorescence imaging with signal processing algorithms has enabled the detection and severity assessment of fungal infections (Konanz, Kocs'anyi, & Buschmann, 2014).

A drawback of chlorophyll fluorescence cameras concerns the conditioning of the plants which has to comply to a specific protocol, which prevents its use in field circumstances. Hence, alternatives have been oriented towards the sunlight as an excitation source for obtaining the fluorescence factors in field conditions, thus enabling crop infection monitoring in ambient conditions (Rossini et al., 2014).

5.1.4 Combination of optical sensing with data mining algorithms

In precision farming, several applications have emerged that employ hyperspectral sensing for identifying the crop status. Lorente et al. (2013) have utilized hyperspectral camera in order to identify citrus decay infection by employing a wavelength pruning approach with a receiver operating characteristic curve (ROC) which was fed as input to a neural network algorithm. Ahmed, Al-Mamun, Bari, Hossain, and Kwan (2012) have used digital imaging for classifying crops and weeds by SVM algorithms. Liu, Zhang, Wang, and Wang (2013) combined effectively an SVM algorithm with a self-adaptive mutation particle swarm optimization (SAMPSO) algorithm aiming to label land use/cover.

Water stress in plants signifies the response of the plant to water scarcity. When this happens, the water deficiency starts affecting the plant's physiological activities. Water existence to plants is quantified as water potential. This potential is estimated using measurements of leaf water before dawn.

It has been estimated that the value of −0.8 MPa notifying a threshold for plant stress detection (Cleary, Zaerr, & Hamel, 1984).

Spectral indices were defined as indicators of plant water content as detectors of crop water stress. The correlation of the spectral indices to the plant water content is defined by the leaf structure. This explains the fact that only some specific spectral indices are more sensitive to the water stress appearance (Eitel et al., 2011). A couple of spectral indices that have been utilized are the normalized difference water index (Gao, 1995) while an additional one is the water band index (Penuelas, Baret, & Filella, 1995). The water band index results from the division of reflectance that appears at 900 nm to that of 970 nm (Penuelas et al., 1995). This index is affected by the plant water concentration in both the leaf and the canopy but it is more sensitive to variations in leaf water. This fact offers an advantage in the case of agriculture because drought affects the leaf water content at an earlier stage compared to the rest parts of the plant (Champagne, Staenz, Bannari, McNairn, & Deguise, 2003).

A crucial feature of plant growth concerns the accurate assessment of crop productivity with the use of non-destructive approaches. Important breakthroughs in the sensing devices used for chlorophyll fluorescence acquisition and advances in the comprehension of the relation between the plant fluorescence features and the physiological status have made the use of fluorescence very popular in crop physiology research. Rumpf et al. (2010) investigated the use of hyperspectral sensing for the automated detection of crop infections and severity assessment with SVM in which the chlorophyll concentration and other physiological factors were used as indicators for the infection severity level assessment.

Photosynthetic activity level is considered as a possible stress indicator and the plant adaptation is quantified by the chlorophyll fluorescence behavior (Strasser & Stirbet, 2001). Due to alterations in the chlorophyll fluorescence showing up at an early stage preceding other manifestations of tissue degradation, leading to the early detection of many stress types before any symptoms of destruction (Lichtenthaler, Babani, & Langsdorf, 2007). The division FV/FM, which results from the variable fluorescence (FV) after division with the maximal fluorescence (FM), defines the estimation of the photochemical efficiency of Photosystem II (PSII). Intact H plants demonstrate F_V/F_M ratio approaching 0.8 (Peterson, Oja, & Laisk, 2001). In studies that concern the practical use of chlorophyll fluorescence, the F_V/F_M ratio is associated with the presence of water stress (Cifre, Bota, Escalona, Medrano, & Flexas, 2005).

Performing fusion of data that emanate from sensors mounted on vehicle or they are installed on ground platforms facilitates a practical implementation of automated systems for recognizing different types of infections or nutrient efficiencies (Moshou et al., 2011). The PCA approach was effectively employed for observing the evolution of infection in wheat crops by *F. graminearum* (Bauriegel, Giebel, Geyer, Schmidt, & Herppich, 2011). Various studies have indicated that the spectral range between 350 and 2.500 nm did not manage to detect the infection from fungus under field conditions due to the reason that the narrow spectral wavelengths demonstrate a high correlation between them (Mewes, Franke, & Menz, 2011).

Mahlein et al. (2013) proposed a method for obtaining specific hyperspectral wavebands that were correlated to disease presence in the leaves of sugar beet. The specification provided by the hyperspectral-imaging sensor can form the basis of designing new low cost instruments, which will be based off the shelf optics like LEDs and C-MOS chips, resulting in devices that can be used by the wider public for crop monitoring (Grieve & Hammersley, 2015).

In the last decades, novel approaches utilizing machine learning architectures have been introduced. Wahabzada et al. (2015a) presented a data based and automatic approach using hyperspectral camera images depicting infected barley leaf areas. The spectral signatures were extracted in order to monitor the evolution of the symptoms during the infection development (Mahlein, 2016).

In the current case study, the fusion of features derived from multispectral sensors and hyperspectral cameras was employed to discriminate induct from infected wheat aiming to further utilize it as a detector for crop biotic stress.

Spectral reflection has been employed in order to discriminate water stress from symptoms of Septoria appearance by the use of a hyperspectral camera. At the same time, the crop health status was evaluated through fluorescence kinetics measurements with the help of a Plant Efficiency Analyser (PEA) fluorimeter (Hansatech Instruments, Norfolk, UK). A data fusion approach was combined with LSSVM approach aiming to discriminate water stress from infected wheat plants. The LSSVM is evaluated against various classifiers including the MLP and QDA for detecting and separating water stress plants from the diseased ones.

5.2 Experimental setup

Wheat cultivars sensitive to *Septoria tritici* have been picked for the current case study. The 'Hussar' variety was picked up due to its characteristic

Fig. 5.1 The experimental set up inside the greenhouse. The environmental conditions were kept at the following set points: temperature 9–11 °C; relative humidity: 50–60%; photoperiod: 16 h (Moshou, Pantazi, Kateris, & Gravalos, 2014).

sensitivity to *Septoria tririci* and its wide dispersion in Europe. The technique for crop material development inside the greenhouse simulated the growth conditions of winter wheat as they appear in the field in terms of soil, distance and ambient conditions (Fig. 5.1).

For the trials, four different health conditions of wheat crop were investigated including *Septoria* [s+, w-] infested, healthy [s-w-] water stressed [s-w+] or Septoria and water stress infected [s+w+]. Apart from the TDR data acquisition, pressure head data were obtained. The plants cultivated in the two trays were not provided sufficiently with the adequate amount of water until stress symptoms appear in the following four days. The threshold value of the pressure head that indicated the water stress onset was 10 m.

The field trials were performed in four neighboring boxes situated on trolleys, with a density around 300–350 plants m^{-2}. The data acquisition was performed in a manner that allowed the accurate simulation of field conditions. The depth of the trays that accommodated the water stressed crops was higher due the presence of time domain reflectometers (TDR) aiming to measure water content and did not affect negatively the plant growth.

For conducting the trials, some measurements were carried out at the same time on the same plant including:

- Transient fluorescence tests on wheat crops infected by *Septoria tritici*;
- hyperspectral signatures measured by a hyperspectral camera

5.3 Optical instrumentation

Hyperspectral signatures have been measured by a monochrome V10 Specim hyperspectral camera by Specim. The sensor creates a reflectance spectrum per single point from on a narrow line on the object that is targeted (Herrala et al., 1994). Three modules including a lens, a grated prism, and a Digital Video Camera (DVC) compose the hyperspectral imager. The spectral pixel was 7 nm in the spectral region between 460 and 900 nm.

Illumination was produced with the help of incandescent lamps within the greenhouse area. A couple of halogen lamps of 500 W each provided complementary lighting. The spectral signatures from wheat leaves were collected and preprocessed so that the spectral magnitudes were invariant to ambient lighting.

The spatial footprint of each pixel was equal to 0.65 mm and the acquisition was performed from an angle of 45°. There was no particular tension to the plant to the plant positioning with respect to ambient conditions and the averaged data values were considered for the whole scene.

The fluorescence measurement were carried out by using the PEA fluorimeter by Hansatech Instruments Ltd., UK. The excitation light for the fluorescence measurement was realized by ultra-bright light LEDs that had a peak waveband at 650 nm. Chlorophyll fluorescence signatures were obtained by a photocell after being filtered by a high pass filter of 50% transmission in the region of 720 nm. For achieving a recording of the transient fluorescence signal the acquisition period for the fluorescence signal was 1 s with a period of 10 μs during at the first 2 ms and then with a duration of 1 ms. The ambient light was blocked with the help of a leaf clip. Dark adaptation was imposed on the plants for a minimum time of 20 min. The clip was applied on the topmost side of the leaf in the central area of the leaf, for 1 s duration with the PEA. The Septoria infected leaves were measured on those leaves that carried early symptoms of the infection (only those turned yellow-brown).

5.4 Fusion of optical sensing data

Multi-sensing fusion systems augment signals emanating from various sensing platforms to produce decisions that would be infeasible to reach by a stand-alone sensor. Areas of deployment (target recognition, risks identification), earth observation applications (minerals mapping), human health status assessment, predictive maintenance of machines, navigation of robots

Table 5.1 Symbolism of the four investigated crop health status

s-w-	s+w-	s-w+	s+w+
Control treatment	inoculated with *Septoria tririci*	deficient water supply	inoculated treatment, deficient water supply

and multi-sensor control of industry machines. Data fusion is a simulation of the continuous functioning of human neural systems that try to combine signals from the human senses to produce decisions regarding the status of the environment that the human entity lives in.

For fusing various types of sensors, spectral signatures were augmented with transient fluorescence signal parameters. In Moshou et al. (2014) a quartet of differing treatments were applied aiming to create specific stress status that needed to be accessed. The applied treatments are presented in the Table 5.1 as follows:

Spectral signatures were composed of 21 signatures. Every signature corresponded to 21 nm wavebands throughout all the spectral bands of the spectrograph (which had region of 460–900 nm). The fluorescence signatures comprised of a couple of parameters given as follows: the F_0 denoting the fluorescence at 0.05 ms and the F_V/F_M ratio which corresponds to the efficiency ratio of the primary photochemical system while F_M denotes the peak of the fluorescence transient.

5.5 Results and discussion

5.5.1 Results and discussion for LSSVM and ARD

For calibrating the LSSVM algorithm, Matlab toolbox LS-SVMlab v1.8 (De Brabanter, De Brabanter, Suykens, & De Moor, 2010) was employed. A dataset of 846 samples comprising of both spectral and fluorescence parameters, were utilized. A dataset of 302 samples was retained for validation. Feature relevant determination was performed as integrated feature of the LSSVM algorithm with ARD (Van Gestel et al., 2002).

The ARD algorithm employs a diagonal matrix that applies weights on the parameters used as inputs in order to train the RBF-kernel. The inputs related to small weights are linked to small values relevance in the kernel function so that they can be neglected. In the presented application, the total amount of spectral signatures, connected with low value weights were detached and only six, were attained for training. Despite the fact that the

RBF kernel tends to show robustness when dealing with unrelated inputs, the generalizing to new inputs is optimized through the application of weights matrix.

The combination ARD with LSSVM algorithm produced six most relevant features from the initial 21 spectral features and more specifically the spectral bands related to indices 3, 5, 6, 8, 20 and 21 of the whole spectral region corresponding to the full capacity of the spectrograph. These spectral indices are associated with wavebands with a width of 21 nm each centered at 503, 545, 566, 608, 860 and 881 nm. The selection was driven by the hypothesis that the spectrum carries redundant information, which would not be needed towards a lower cost version that would run on a custom electronic platform. A supporting argument would be that the cost of the custom device that could measure the six wavebands would be similar with the cost of a handheld spectrometer for the same range. In the current case study, the retained bands were decided according to the presence of water stress while assuming that the *Septoria* infection is an unknown plant health condition. Considering the water stress scenario as a base case study to implement a binary classifier, the validation set included 155 samples, 75 of them infected by *Septoria* and 80 healthy and 147 samples without water stress symptoms, 70 infected by *Septoria* and 77 healthy.

An LSSVM was trained with the spectral and subsequently with the fluorescence features and results are depicted in Tables 5.2 and 5.3. Another LSSVM was calibrated by using the fused dataset deriving from spectral

Table 5.2 Classification results of an LSSVM vs QDA based on reflectance (the QDA performance is shown in parenthesis)

		Detected %	
		w+	w-
Real 100	**w+**	89 (76.3)	11 (23.7)
	w-	10.2 (13.4)	89.8 (86.6)

Table 5.3 Classification results of a LSSVM vs QDA based on fluorescence (the QDA performance is shown in parenthesis)

		Detected %	
		w+	w-
Real 100	**w+**	69 (69)	31 (31)
	w-	34.7 (34.7)	65.3 (65.3)

Table 5.4 Classification results of a LSSVM vs QDA (the QDA performance is shown in parenthesis)

		Detected %	
		w+	w-
Real 100	**w+**	Real 100	5.8 (7.9)
	w-	5.45 (6.3)	94.55 (93.7)

and fluorescence data which are demonstrated in Table 5.4. A multiclass LSSVM was implemented by using the "One vs. One" algorithm to be able to separate the four categories are associated with the combined classes that result from water stress and *Septoria* presence. Total incorrect classification levels were lower than 1% for water stress presence and *Septoria* infection (Table 5.5).

In Table 5.5, the chosen spectral bands that were the same with those selected by LSSVM ARD were utilized, since it was desirable to retain the same set of features for constructing future devices that could be used for operational scanning for water stress and disease identification. To be able to compare across two different methods for classification, a QDA classifier was employed. A higher error was produced from the QDA classifier compared to the LSSVM in the case of water stress identification based on spectral reflectance data (Table 5.1). This outcome can be attributed the fact that inclusion of six features crates a nonlinear problem which is more amenable to a treatment by a nonlinear classifier like SVM compared to the QDA which is a quadratic decision classifier. Fluorescence input data for the QDA classifier yielded the same results to the LSSVM regarding the detection of water stress (Table 5.2). This outcome is not unexpectable because fluorescence in based on two features which decreases the nonlinearity of the decision surface, leading to an nonlinear problem which can be handled by a

Table 5.5 Validation results by using the fusion of features with an LSSVM classifier vs MLP

		Detected%			
		s+w+	s-w+	s+w-	s-w-
Real 100	**s+w+**	100	0	0	0
	s-w+	0	98.75 (98.2)	0 (1.2)	1.25 (0.6)
	s+w-	0	0 (2.1)	100 (96.8)	0 (1.1)
	s-w-	0	1.3 (0.2)	0 (1.1)	98.7

quadratic decision function. The results of the fusion classifier (Table 5.3) for water stress detection demonstrated that the LSSVM showed improved results compared to the QDA. It is worth noting that the performance increase regarding water stress detection is evident when using fusion of fluorescence and reflectance data.

The same features that have been used for the LSSVM classifier for separating different stress combinations. An MLP classifier consisting of one hidden layer of 10 neurons was trained from a fusion data set combining spectral and fluorescence signatures. The selected number of hidden neurons was equal to ten after testing all the options between 5 and 25 using a step of 5. The criterion of selection was to keep the highest accuracy divided by the number of neurons. The overall false classification decreased the considerably for Septoria infection and water stress identification. However, the LSSVM approach demonstrated higher accuracy rates as it is depicted in Table 5.5, where the MLP performance results are annotated in parenthesis.

As it is demonstrated by the results, it is proven that the MLP model performance is almost equal to that of the LSSVM model. It is also clear that the feature fusion from a camera and a fluorometer can produce an accurate estimation by forming a software virtual sensory system that can give optimal results compared to isolated sensors. This modification improves the capability of detecting not only winter wheat plants infected solely by water stress but also discriminating those that are infected simultaneously by water stress and *Septoria*. The current approach is also capable of discriminating effectively healthy plants or plants that are infected either by *Septoria* or demonstrate water stress symptoms.

There is a multitude of advantages regarding the cost and the flexibility of presented of the current approach that lays mainly on the effective data fusion acquired from heterogeneous sensors. Practically, the fusion of information not only about sensor information but can include temporal information in the form of previous acquired observations. This outcome emanates from the capability of the fusion system to discriminate the simultaneous occurrence of abiotic and biotic stress, like water stress and *Septoria* infection.

5.5.2 Results and discussion for active learning

Regarding the active learning algorithm structure, it contained a four step procedure described as follows:

(1) The basic training set contained feature vectors, comprising of 23 elements consisting of 21 spectral wavelengths and 2 fluorescence signatures.

The feature vectors were attained from the variety of phenotypic responses that were caused by the onset of water stress, the Septoria symptoms or the simultaneous occurrence of both water stress and Septoria symptoms. A total of 894 feature vectors were assembled by combining all the four different crop health conditions. More specifically, 220 feature vectors were acquired from healthy while 235 from the water stressed and 228 from Septoria infection condition and 211 from the simultaneous occurrence of both water stress and Septoria infection condition. The trained one-class SOM was validated with 235 vectors corresponding to water stressed condition. The indicator of performance is the discrimination ability of identifying the unknown plant condition as an outlier compared to the target set (healthy condition).

(2) The primary baseline set was combined with outlier vectors that are derived from the identified outlier crop status and the produced set resulting set was ascribed as a new target set. The process outlier detection is iterated but the indicator of performance is to classification capability of new crop health status as outlier compared to the augmented target set that concerns the healthy condition and the unhealthy plant conditions (water stress, Septoria infected, both water stressed and Septoria infected).

(3) In the event that the new sample is a member of a class belonging to the baseline set, the outliers identification is performed per class through the baseline set and the sample is detected as belonging in one of the default subclasses of the augmented baseline set.

(4) Steps (2) and (3) are iterated and more precisely, the outliers identification and the augmentation are performed for unseen data that possibly correspond to the already known health status classes or might correspond to new non-labelled crop status classes. In the current case study, the iteration of the steps ended after the formulation of 4 distinct classes that correspond to different crop health status including the healthy condition and the 3 unhealthy plant conditions (water stress, Septoria infected, both water stressed and Septoria infected).

It is a significant advantage of the presented approach the procedure from steps 1 to 4 does not demand human supervision and, since it is based on outlier detection and augmentation actions that are executed in an automatic manner. The results of the Active Learning algorithm are annotated in Table 5.6.

Table 5.6 Detection rate of the active learning algorithm based on OC-SOM for four different crop health status

	s-w-	s-w+	s+w-	s+w+	outliers
s-w-	97.73%	0%	1.36%	0%	0.91%
s-w+	4.39%	85.53%	9.21%	0%	0.88%
s+w-	0%	0%	99.05%	0%	0.95%
s+w+	0%	0%	0%	99.15%	0.85%

The proposed Active Learning algorithm demonstrated comparable performance for Septoria and simultaneous occurrence of *Septoria* infection and water stress condition achieving 99%. In the case of water stress condition, the performance rate was found less high reaching a performance rate of 85% while in the case of healthy plants achieved a performance rate of 97%. The current performance behavior can be explained by the fact that the healthy conditions contains individual plants of late maturity that their health condition is not affected by the stresses occurred in the presented experiments. The low accuracy concerning the water stress symptoms can be ascribed from the symptoms correlation between *Septoria* and water stress, since they take place in the same optical band range. As a result, the proposed spectral and fluorescence techniques and their combination does not yield enough effective discrimination potential.

References

Ahmed, F., Al-Mamun, H. A., Bari, A. H., Hossain, E., & Kwan, P. (2012). Classification of crops and weeds from digital images: A support vector machine approach. *Crop Protection*, *40*, 98–104.

Al Masri, A., Hau, B., Dehne, H. -W., Mahlein, A. -K., & Oerke, E. -C. (2017). Impact of primary infection site of Fusarium species on head blight development in wheat ears evaluated by IR-thermography. *European Journal of Plant Pathology*, *147*, 855–868.

Bauriegel, E., Brabandt, H., Gärber, U., & Herppich, W. B. (2014). Chlorophyll fluorescence imaging to facilitate breeding of Bremia lactucae-resistant lettuce cultivars. *Computers and Electronics in Agriculture*, *105*, 74–82.

Bauriegel, E., Giebel, A., Geyer, M., Schmidt, U., & Herppich, W. B. (2011). Early detection of Fusarium infection in wheat using hyper-spectral imaging. *Computers and Electronics in Agriculture*, *75*(2), 304–312.

Bebber, D. P., Holmes, T., & Gurr, S. J. (2014). The global spread of crop pests and pathogens. *Global Ecology and Biogeography*, *23*, 1398–1407.

Behmann, J., Mahlein, A. -K., Paulus, S., Kuhlmann, H., Oerke, E. -C., & Plümer, L. (2015). Calibration of hyperspectral close-range pushbroom cameras for plant phenotyping. *ISPRS Journal of Photogrammetry and Remote Sensing*, *106*, 172–182.

Behmann, J., Steinrücken, J., & Plümer, L. (2014). Detection of early plant stress responses in hyperspectral images. *ISPRS Journal of Photogrammetry and Remote Sensing*, *93*, 98–111.

Bock, C., et al. (2010). Plant disease severity estimated visually, by digital photography and image analysis, and by hyperspectral imaging. *Critical Reviews in Plant Sciences*, *29*, 59–107.

Campbell, J. B. (2007). *Introduction to remote sensing* (4th ed.). New York: Guilford.

Champagne, C. M., Staenz, K., Bannari, A., McNairn, H., & Deguise, J. C. (2003). Validation of a hyperspectral curve-fitting model for the estimation of plant water content of agricultural canopies. *Remote Sensing of Environment*, *87*(2–3), 148–160.

Chollet, F. (2017). *Deep learning with python.* Manning Publications Company.

Cifre, J., Bota, J., Escalona, J. M., Medrano, H., & Flexas, J. (2005). Physiological tools for irrigation scheduling in grapevine (*Vitis vinifera* L.): An open gate to improve water-use efficiency? *Agriculture, Ecosystems & Environment*, *106*(2–3), 159–170.

CiresSan, D. C., et al. (2010). Deep, big, simple neural nets for handwritten digit recognition. *Neural Computation*, *22*, 3207–3220.

Cleary, B., Zaerr, J., & Hamel, J. (1984). Guidelines for measuring plant moisture stress with a pressure chamber. PMS Instrument Co, 2750.

De Brabanter, K., De Brabanter, J., Suykens, J. A., & De Moor, B. (2010). Optimized fixed-size kernel models for large data sets. *Computational Statistics & Data Analysis*, *54*(6), 1484–1504.

Del Fiore, A., Reverberri, M., Ricelli, A., Pinzari, F., Serranti, S., Fabbri, A. A., et al. (2010). Early detection of toxigenic fungi on maize by hyperspectral imaging analysis. *International Journal of Food Microbiology*, *144*, 64–71.

Eitel, J. U., Vierling, L. A., Litvak, M. E., Long, D. S., Schulthess, U., Ager, A. A., et al. (2011). Broadband, red-edge information from satellites improves early stress detection in a New Mexico conifer woodland. *Remote Sensing of Environment*, *115*(12), 3640–3646.

FAO (2009). http://www.fao.org/fileadmin/templates/wsfs/docs/Issues_papers/HLEF2050_Global_Agriculture.pdf.

Fiorani, F., & Schurr, U. (2013). Future scenarios for plant phenotyping. *Annual Review of Plant Biology*, *64*, 267–291.

Franke, J., & Menz, G. (2007). Multi-temporal wheat disease detection by multispectral remote sensing. *Precision Agriculture*, *8*, 161–172.

Gao, B. C. (1995). Normalized difference water index for remote sensing of vegetation liquid water from space. In 2480. *Imaging spectrometry* (pp. 225–237): International Society for Optics and Photonics.

Gebbers, R., & Adamchuk, V. I. (2010). Precision agriculture and food security. *Science*, *327*, 828–831.

Grieve, B., & Hammersley, S. (2015). Localized multispectral crop imaging sensors. In *IEEE Instrumentation and Measurement Society* (p. 6).

Herrala, J., Puolijoki, H., Impivaara, O., Liippo, K., Tala, E., & Nieminen, M. M. (1994). Bone mineral density in asthmatic women on high-dose inhaled beclomethasone dipropionate. *Bone*, *15*(6), 621–623.

Hillnhütter, C., Schweizer, A., Kühnhold, V., & Sikora, R. A. (2010). Remote sensing for the detection of soil-borne plant parasitic nematodes and fungal pathogens. In *Precision crop protection—The challenge and use of heterogeneity* (pp. 151–165).

Hu, Z., Chen, T., Ge, Q., & Wang, H. (2018). Observable degree analysis for multi-sensor fusion system. *Sensors*, *18*(12), 4197.

Hughes, D.P. and Salathé, M. (2015) An open access repository of images on plant health to enable the development of mobile disease diagnostics through machine learning and crowdsourcing. CoRR abs/1511.08060.

Jacquemoud, S., & Ustin, S. L. (2001). Leaf optical properties: a state of the art. In *Proceedings of the 8th international symposium physical measurements & signatures in remote sensing, 8–12 January 2001* pp. 223–232 CNES: Aussois, France.

Jones, H. G., Stoll, M., Santoa, T., de Sousa, C., Chaves, M. M., & Grant, O. M. (2002). Use of infrared thermography for monitoring stomatal closure in the field: Application to grapevine. *Journal of Experimental Botany, 53*, 2249–2260.

Kersting, K., Bauckhage, C., Wahabzada, M., Mahlein, A. -K., Steiner, U., Oerke, E. -C., et al. (2016). Feeding the world with big data: Uncovering spectral characteristics and dynamics of stressed plants. In J. Lässig, K. Kersting, & K. Morik (Eds.), *Computational sustainability* (pp. 99–120). Cham: Springer International Publishing.

Konanz, S., Kocs'anyi, L., & Buschmann, C. (2014). Advanced multi-color fluorescence imaging system for detection of biotic and abiotic stresses in leaves. *Agriculture, 4*, 79–95.

Lenthe, J. -H., Oerke, E. -C., & Dehne, H. -W. (2007). Digital infrared thermography for monitoring canopy health of wheat. *Precision Agriculture, 8*, 15–26.

Lichtenthaler, H. K., Babani, F., & Langsdorf, G. (2007). Chlorophyll fluorescence imaging of photosynthetic activity in sun and shade leaves of trees. *Photosynthesis Research, 93* (1–3), 235.

Liu, Y., Zhang, B., Wang, L. M., & Wang, N. (2013). A self-trained semisupervised SVM approach to the remote sensing land cover classification. *Computers & Geosciences, 59*, 98–107.

Lorente, D., Blasco, J., Serrano, A. J., Soria-Olivas, E., Aleixos, N., & Gómez-Sanchis, J. U. A. N. (2013). Comparison of ROC feature selection method for the detection of decay in citrus fruit using hyperspectral images. *Food and Bioprocess Technology, 6*(12), 3613–3619.

Mahlein, A. K. (2016). Plant disease detection by imaging sensors–parallels and specific demands for precision agriculture and plant phenotyping. *Plant Disease, 100*(2), 241–251.

Mahlein, A. -K., Oerke, E. -C., Steiner, U., & Dehne, H. -W. (2012). Recent advances in sensing plant diseases for precision crop protection. *European Journal of Plant Pathology, 133*, 197–209.

Mahlein, A. -K., Rumpf, T., Welke, P., Dehne, H. -W., Plümer, L., Steiner, U., et al. (2013). Development of spectral vegetation indices for detecting and identifying plant diseases. *Remote Sensing of Environment, 128*, 21–30.

Martinelli, F., Scalenghe, R., Davino, S., Panno, S., Scuderi, G., Ruisi, P., et al. (2014). Advanced methods for plant disease detection. A review. *Agronomy for Sustainable Development, 35*, 1–25.

Mewes, T., Franke, J., & Menz, G. (2011). Spectral requirements on airborne hyperspectral remote sensing data for wheat disease detection. *Precision Agriculture, 12*(6), 795.

Moshou, D., Bravo, C., Oberti, R., West, J. S., Ramon, H., Vougioukas, S., et al. (2011). Intelligent multi-sensor system for the detection and treatment of fungal diseases in arable crops. *Biosystems Engineering, 108*(4), 311–321.

Moshou, D., Pantazi, X. E., Kateris, D., & Gravalos, I. (2014). Water stress detection based on optical multisensor fusion with a least squares support vector machine classifier. *Biosystems Engineering, 117*, 15–22.

Neumann, M., Hallau, L., Klatt, B., Kersting, K., & Bauckhage, C. (2014). Erosion and features for cell phone image based plant disease classification. *Proceeding of the 22nd International Conference on Pattern Recognition (ICPR), Stockholm, Sweden, 24–28 August 2014*, (pp. 3315–3320).

Nutter, F. W., Jr (2001). Disease assessment terms and concepts. In O. C. Maloy & T. D. Murray (Eds.), *Encyclopedia of plant pathology* (pp.312–323). New York: John Wiley & Sons, Inc.

Oerke, E. C., & Dehne, H. W. (2004). Safeguarding production—losses in major crops and the role of crop protection. *Crop Protection, 23*(4), 275–285.

Oerke, E. -C., Gerhards, R., Menz, G., & Sikora, R. A. (Eds.), (2010). *Precision crop protection—The challenge and use of heterogeneity* (vol. 5). Dordrecht, Netherlands: Springer.

Oerke, E. -C., Mahlein, A. -K., & Steiner, U. (2014). Proximal sensing of plant diseases. In M. L. Gullino & P. J. M. Bonants (Eds.), *Detection and Diagnostic of Plant Pathogens, Plant Pathology in the 21st Century* (pp. 55–68). Dordrecht, the Netherlands: Springer Science and Business Media.

Oerke, E. -C., Steiner, U., Dehne, H. -W., & Lindenthal, M. (2006). Thermal imaging of cucumber leaves affected by downy mildew and environmental conditions. *Journal of Experimental Botany*, *57*, 2121–2132.

Pantazi, X. E., Moshou, D., Alexandridis, T., Whetton, R. L., & Mouazen, A. M. (2016). Wheat yield prediction using machine learning and advanced sensing techniques. *Computers and Electronics in Agriculture*, *121*, 57–65.

Penuelas, J., Baret, F., & Filella, I. (1995). Semi-empirical indices to assess carotenoids/chlorophyll a ratio from leaf spectral reflectance. *Photosynthetica*, *31*(2), 221–230.

Peterson, R. B., Oja, V., & Laisk, A. (2001). Chlorophyll fluorescence at 680 and 730 nm and leaf photosynthesis. *Photosynthesis Research*, *70*(2), 185–196.

Pethybridge, S. J., & Nelson, S. C. (2015). Leaf Doctor: A new portable application for quantifying plant disease severity. *Plant Disease*, *99*(10), 1310–1316.

Rossini, M., Alonso, L., Cogliati, S., Damm, A., Guanter, L., Julietta, T., et al. (2014). Measuring sun-induced chlorophyll fluorescence: an evaluation and synthesis of existing field data. pp. 1–5 In *5th International Workshop on Remote Sensing of Vegetation Fluorescence, 22–24 April 2014, Paris*.

Rumpf, T., Mahlein, A. K., Steiner, U., Oerke, E. C., Dehne, H. W., & Plümer, L. (2010). Early detection and classification of plant diseases with support vector machines based on hyperspectral reflectance. *Computers and Electronics in Agriculture*, *74*(1), 91–99.

Stafford, J. V. (2000). Implementing precision agriculture in the 21th century. *Journal of Agricultural Engineering Research*, *76*, 267–275.

Steiner, U., Bürling, K., & Oerke, E. -C. (2008). Sensorik für einen präzisierten Pflanzenschutz. *Gesunde Pflanz*, *60*, 131–141.

Strasser, R. J., & Stirbet, A. D. (2001). Estimation of the energetic connectivity of PS II centres in plants using the fluorescence rise O–J–I–P: Fitting of experimental data to three different PS II models. *Mathematics and Computers in Simulation*, *56*(4–5), 451–462.

Šulc, M., & Matas, J. (2017). Fine-grained recognition of plants from images. *Plant Methods*, *13*, 115.

Van Gestel, T. V., Suykens, J. A., Lanckriet, G., Lambrechts, A., Moor, B. D., & Vandewalle, J. (2002). Bayesian framework for least-squares support vector machine classifiers, Gaussian processes, and kernel Fisher discriminant analysis. *Neural Computation*, *14* (5), 1115–1147.

Walter, A., Liebisch, F., & Hund, A. (2015). Plant phenotyping: from bean weighing to image analysis. *Plant Methods*, *11*, 14.

West, J. S., Bravo, C., Oberti, R., Lemaire, D., Moshou, D., & McCartney, H. A. (2003). The potential of optical canopy measurement for targeted control of field crop diseases. *Annual Review of Phytopathology*, *41*(1), 593–614.

West, J. S., Bravo, C., Oberti, R., Moshou, D., Ramon, H., & McCartney, H. A. (2010). Detection of fungal diseases optically and pathogen inoculum by air sampling. In *Precision crop protection—The challenge and use of heterogeneity* (pp. 135–149).

Further reading

Nutter, F. W., Esker, P. D., & Netto, R. A. C. (2006). Disease assessment concepts and the advancements made in improving the accuracy and precision of plant disease data. *European Journal of Plant Pathology*, *115*, 95–103.

Wahabzada, M., Paulus, S., Kersting, K., & Mahlein, A. -K. (2015). Automated interpretation of 3D laserscanned point clouds for plant organ segmentation. *BMC Bioinformatics*, *16*, 248.

Xu, R., et al. (2018). Aerial images and convolutional neural network for cotton bloom detection. *Frontiers in Plant Science*, *8*, 2235.

CHAPTER 6

Tutorial III: Disease and nutrient stress detection

Contents

6.1 Introduction

Precision agriculture methodologies concerning the synergy of various tools including sensory systems, decision support and farm management software are capable of achieving sustainability targets regarding the minimization of environmental and financial impact (Gebbers & Adamchuk, 2010). It is estimated that 40% of the global yield potential is lost due to various pests presence (Oerke & Dehne, 2004). The occurrence of weed and disease establishment is based on the environmental factors and demonstrates a variable pattern within the field (Franke, Gebhardt, Menz, & Helfrich, 2009). Thus, the early detection of significant pest infected patches as well as defining management zones with variable infection severity is of high importance (Moshou, Bravo, et al., 2004; Moshou, Deprez, and Ramon, 2004). Pest monitoring and the decision procedure are fundamental for site-specific treatment of pests (Hillnhütter & Mahlein, 2008). To achieve a level of precision in crop management a prerequisite is a high resolution of spatio-temporal data availability.

For maintainable increase of crop production, fertilizers and pesticides have to be administered according to environmental compliance requirements. Pesticides are applied in a uniform way, while disease invasions form

https://doi.org/10.1016/B978-0-12-814391-9.00006-6

distinct foci. The patches occurrence can be decreased effectively by employing site-specific approaches avoiding spraying healthy areas and targeting solely on the infected ones aiming to minimize the environmental impact and maximize the financial gain. Regarding fertilizer dosage applications, these are customized according to the site-specific nutritional crop requirements. The intercorrelation of the spectral crop signatures corresponding to different types and levels of stresses, can function as the main components for an optical device that is capable of recognizing several disease foci and crop nutritional requirements. On the other hand, sensory systems including multispectral, hyperspectral imaging and chlorophyll fluorescence can provide detailed and high precision data concerning different crop health conditions (Mahlein, Steiner, Hillnhütter, Dehne, & Oerke, 2012).

Progress in sensor devices together with advances in information technology and earth observation systems introduces innovative tools for precision agriculture, by enabling the early detection of several types of pests (Rumpf et al., 2010).

The volume and the fitness of the acquired data by sensory systems has substantially increased, but since those systems are highly correlated to several biological variations, cannot lead to trustworthy conclusions. Considering also the inability of those systems quantifying crop physiological factors in a direct way, and their capability of measuring a spectrum of reflectance components associated to several crop signatures and the corresponding ambient conditions (Jensen, 2007), it is concluded that introduction of applications employing advanced data mining methods is crucial.

Lu, Ehsani, Shi, Castro, and Wang (2018) used a hand-held spectroradiometer to recognize various tomato leaf diseases by constructing a set with halogen lamps and black background in order extract the leaf area. The diseases that were investigated were late blight and bacterial spots, discriminated from healthy tomato leaves, yielding a high accuracy reaching 100%. The employed classification method was KNN combined with PCA.

He, Li, Qiao, and Jiang (2018) used a hyperspectral spectroradiometer applying a regression approach by extracting wavelet coefficients on hyperspectral data to recognize stripe rust occurrence on winter wheat leaves. The authors conclude that the detection of strip rust is possible by using hyperspectral sensing to correlate the disease severity with a decrease in chlorophyll content. The employed multivariate linear model that utilized wavelet transform features achieved an R^2 of 0.905.

6.1.1 Yellow rust disease detection

Wheat is regarded as significant grain crop worldwide and is associated with food security in many countries. A variety of wheat infections have imposed great threats affecting the sustainability of wheat yield in global basis. Yellow rust belongs to the family of fungal diseases and more specifically is caused by the pathogen called *Puccinia striiformis f.* sp. *tritici* (*Pst*). It is considered as one of the most devastating diseases in wheat (Wan et al., 2004; Zhang et al., 2011), mostly due to their frequent occurrence and their potential to spread as an epidemic, causing significant crop losses (Moshou, Bravo, et al., 2004; Moshou, Deprez, and Ramon, 2004) (exceeding 29.3% of the national total production in China) while at the same time decreasing the quality of the produced wheat (Wan et al., 2004). The current practice, involved observation of visual symptoms that are connected to the presence of the infection, but also damaging the environment due to the heavy use of fungicides (Huang et al., 2007; Zhang et al., 2011). Yellow rust has the propensity to appear sporadically in a location. Hence, it is needed to use a different way that is more applicable for assessing the locations in which the disease has established its presence.

It is commonly established that leaf water content, pigment concentrations, and the inner crop structure may deteriorate when an infection occurs, and their relevant physiological and biochemical alterations are depicted in its spectral response, for example different changing in spectral features and the deviation in reflectance value. A variety of research works have suggested the use of relevant wave bands or vegetation indexes (VIs) that are capable of detecting and observing crop infections at leaf and canopy level. a photochemical reflectance index (PRI) was suggested by Huang et al. (2007, 2014) for yellow rust identification at crop and also at field level ($R^2 = 0.91$), and introduced a vegetation index called YRI (yellow rust index), to discriminate yellow rust from asymptomatic wheat and simultaneously detect powdery mildew and the presence of aphids, respectively. Zhang et al. (2012) discovered that the physiological reflectance index (PhRI) was correlated to the presence of yellow rust disease throughout to all growth stages. Devadas, Lamb, Simpfendorfer, and Backhouse (2009) concluded a single spectral index was not sufficient of detecting different types of wheat rust while instead, the anthocyanin reflectance index (ARI) is capable to tell apart yellow rust disease from asymptomatic wheat.

Plant disease and stress mapping could be accomplished with the help of air-borne systems. Unfortunately, the currently used satellite sensor cannot provide practical disease recognition (even if the gathered data belong to relevant wavebands) due to inappropriate spatial resolution. In the best case, satellite imagery are useful tools for aiding the spotting of extended areas of disease or stress occurrence in crop as an early warning that can be examined by the farmer. Moreover, in frequent observations and variable visibility due to intermittent cloud appearance, lead to the conclusion that satellites are not reliable source of information when required. To the contrary, airborne sensing platforms overcome these constraints because their access to the acquisition object can be controlled by manipulating the positioning of the platform.

Nitrogen (N) is regarded globally as one of the most important nutrients closely linked to plant growth due to its crucial role to the photosynthetic procedure (Andrews, Raven, & Lea, 2013). Simultaneously, N widely known for its impact to both environment and economy. Consequently, the optimal administration of N based fertilizer application in different crops has been the case study of many spectroscopy studies (Cao et al., 2017; Goron, Nederend, Stewart, Deen, & Raizada, 2017).

Additionally, to the nitrogen content, the distribution of color alterations in canopy might be attributed to certain interconnected factors, including nutrients, dehydration and infections. This lead to the need for collecting information about crop which can be used to aid the recognition of the type of stress situations that the crop is facing and lead to actionable insights and associated managements inductions.

The determination of the plant N content can be categorized into two categories: destructive and non-destructive. The most frequently used technique of destructive evaluation is a laboratory analysis based on the Kjeldahl technique requires a lot of effort, is time consuming and of high cost (Vigneau, Ecarnot, Rabatel, & Roumet, 2011). Regarding the non-destructive approach, it is mainly dependent on canopy reflection in the visible–NIR spectral region (400–900 nm) aiming to offer an optical remote sensing technique for assessing the of the N crop status. The acquisition is carried out locally, decreasing the amount of field samples that are needed and thus shortening the time and the expense of field acquisition, preprocessing and lab evaluation. A large amount of work research has been allocated to non-destructive evaluation of the N condition, determination plants by remote sensing technologies (Tremblay, Wang, & Cerovic, 2012) and spectral indices closely associated to the N crop content have been extracted from hyperspectral signatures (Tian et al., 2011).

Crop yield prediction and N content determination are handled simultaneously due to the direct connection with fertilizer administration strategies. Crop yield targets are commonly utilized for estimating the crops N needs, both at the pre-season and in-season period. For deriving precision fertilization schedules for N fertilization, especially during season, an evaluation of both is required.

6.1.2 Machine learning in crop status recognition

The main objective of PA is to aid decision making though acquisition of information that can lead to improved decisions concerning crop interventions in a spatial and temporal dimension (Whelan & Taylor, 2013). More specific, changes in crop condition, have to be followed by the discrimination between biotic and abiotic stress in a way that can lead to enable the appropriate decision regarding treatment.

In the last decade, the crop protection technology has started adopting different tool that are based on various Machine Learning architectures (Ahmed, Al-Mamun, Bari, Hossain, & Kwan, 2012). The efficiency of these architectures relies on their ability of finding associations and intercorrelations between data for aiding the detection of reason that leads to crop stress.

Mehra, Cowger, Gross, and Ojiambo (2016) applied ML methods including ANNs, categorical and regression trees and Random Forests (RFs) for pre-planting risk prediction of *Stagonospora nodorum* blotch that affects winter wheat crop. The proposed models for risk assessment have been proven efficient of applying suitable disease treatments before planting the winter wheat crop.

Machine learning methods combined with hyperspectral imagery are capable of exploring the physiological and structural crop behavior and sense possible alterations in the crop physiology due to different ambient conditions (Wahabzada et al., 2016). Goldstein et al. (2017) has proven that field parameters including as soil moisture content, ambient conditions, irrigation practices, and the expected yield are fused successfully by employing ML models aiming to form an automated irrigation schedule. Gutiérrez, Diago, Fernández-Novales, and Tardaguila (2018) have applied thermal imaging by employing two ML models including Rotation Forests and Decision Trees, in order to assess water status in wine crops for irrigation purposes.

Spectral vegetation indices (VIs) are special designed to capture the functional links between the crop features and observations from spectral sensors (Wiegand, Richardson, & Kanemasu, 1979). Apart from the Normalized Difference Vegetation Index (NDVI) (Johnson, 2014). a various vegetation indices have been introduced, such as the two-band Enhanced Vegetation

Index (EVI2) and Normalized Difference Water Index (NDWI) (Satir & Berberoglu, 2016).

Yuan, Lin, and Wang (2015) extracted the most appropriate bands aiming to lower the data dimension without any loss of information. They introduced an unsupervised selection band selection framework, which progressively eliminated the redundant bands. Trials on hyperspectral classification and color visualization have been carried out, demonstrating that the applied technique is more accurate compared to traditional pointwise art selection techniques.

The use of VIs compromises valuable information, which is existent in the raw spectra. To the contrary, retaining the full spectral information, provides the advantage of keeping important information that can reveal relevant patterns in huge datasets might be important if there is noise in the data. Römer et al. (2012) proved by using the information of the reflectance spectrum transformed through polynomial approximations and machine learning can enable the detection of early stage of biotic stress with high performance. To improve the comprehension of plant stress dynamics, it is necessary to obtain data with a high resolution both and in and temporal domain They aid the construction of spatial dispersion maps of symptoms which can provide a better comprehension of the plant reaction to the biotic or abiotic stressor and can be proven helpful for optimal sampling for further laboratory analysis. As a result, it is evident that Machine Learning methods can improve the speed and accuracy of data analytics, in comparison to statistical discrimination algorithms. Moshou, Bravo, et al. (2004) demonstrated one of the early machine learning based approaches for to detecting yellow rust in winter wheat by using reflectance responses, reaching a classification accuracy between 95% and 99% (Moshou, Deprez, and Ramon, 2004).

The need for active learning in order to recognize disease is dictated by the drawbacks of the usual approach which is based on one crop measurement which is not sufficient for identifying the factor that causes the crop stress. The current case study aims to realize a functional pest/nutrient crop protection system. The investigated system proposes a One Class Classifier that is capable of identifying type of stress that occurs, relying on unforeseen spectra by having been trained with spectra from asymptomatic plants. The realized Active Learning classifier is able to learn iteratively in order to identify various malignant conditions by appending new classes of new stresses by using an One Class Classifier against the previously identified target class. This is accomplished through a repeatable way, One Class Classification class augmentation of the already identified class.

Moreover, for the current case study, a second approach has been employed based on different Self Organizing Maps. This technique utilized a supervised learning method for classifying spectral signatures aiming to estimate the health condition in winter wheat. Three different crop health conditions corresponding asymptomatic, nitrogen stressed and yellow rust infected plants were validated. Cross-validation has been applied for assessing the validity of the results in a relation to the capacity of the networks and their relevant generalization ability. Finally, the ability of the employed self-organizing models in visualizing the spectral features topology, was evaluated.

6.2 Materials and methods

Yellow rust patches were formed in 6 areas of winter wheat, and surrounded by guradrows with a 3m width. The control of non-target diseases was achieved by using fungicides in those areas that was regarded as necessary. Spectral signatures corresponding to asymptomatic, infected and nutrient deficient winter wheat crops were attained through hyperspectral imagery, consisting of a V10 Spectrograph (Specim, Oulou, Finland). The plant spectra were acquired at a range from 400 to 1000nm. In the nutrient deficient crops, there was no fertilization applied for approximately one week.

Optical sensor measurements were carried out in the most distinct three patches of Yellow Rust diseased plants. Then, the observations of these measurements were contrasted with those acquired from three induct plots and a substantial area that was not received any type of fertilization. The spectral measurements were attained at six plots of the field, randomly chosen. For calibrating the employed classifiers, feature vectors were formed by using the mean values of the 21 spectral bands of the spectrograph. Light intensity normalization was performed to eliminate the high spectral variability due to canopy structure and various illumination levels. Discrimination was enhanced by applying a spatial average window, which was wide as one plant. Using five wavebands with a width of 20nm, the precise detection of the different crop stress conditions and of the asymptomatic crops was performed. Three wavebands, including 725, 680, and 475nm +/− 10nm achieved high performance when using normalized features and a window of 300 pixels for the calculation of averages. Another two extra wavebands, corresponding to 750 and 630nm +/− 10nm were added due to their necessity for obtaining the NDVI. This ultimate step had as an objective to achieve a leaf-soil separation.

6.2.1 Active learning scheme

The employed active learning scheme is given as follows:

(1) The initial training set comprised of 5 features (derived from relevant spectral bands) emanating from various crop status related to nitrogen deficiency, or to yellow rust disease. A total number of 9058 vectors corresponding to 3 different crop status were utilized, obtained from 3846 asymptomatic, 1343 nitrogen deficient and 3869 with Yellow Rust infected crops. Then, the calibrated one-class SOM was tested with 1312 spectra obtained from asymptomatic, 442 spectra obtained from nitrogen deficient and 1308 spectra from infected crops. The achievement indicator was the successful classification of a novel plant condition recognized as an outlier compared with asymptomatic plants, constituted the target set.

(2) The primal baseline set was combined with outlier samples that originated from the detected outlier crop status. The subsequent set replaced the target set as the new baseline set. The described procedure was then iterated, having as indicator the capacity to classify novel crop conditions as an outlier compared with a combined target set which comprises of most of the asymptomatic crops but also the just included plant conditions.

(3) In the occasion that a novel sample originates from a class which already belonging to the baseline set, then the discovery of outliers carries on inside each class in the target set that has to internally per class in the baseline set and then it gets classified into one of the classes that constitute the baseline set.

(4) Steps (2) and (3) were iterated, and more precisely the discovery of outliers and combination were performed for novel data which could be classified as an existing crop condition class or it would reveal unclassified types of crop status. In the current study, the application of the repetition of steps was finished after the formation of 3 classes that were linked to asymptomatic crops and the 2 health status classes that correspond to nitrogen deficiency and yellow rust disease.

It important to be noted that the algorithm steps, (1) to (4) does not rely on any human decision but relies exclusively on outlier discovery and the data set combination step which are performed iteratively.

6.2.2 Results and discussion for active learning

Yellow Rust diseased crops were successfully discriminated from the nitrogen deficient and the asymptomatic ones, employing One Class SOM

Table 6.1 One Class SOM Recognition performance (%) for asymptomatic (A), nitrogen deficient (ND) and yellow rust diseased (YR) crops. The last column denotes the falsely recognized crops as outliers

	Detected by the OCSOM (%)			
From 100%	**Asymptomatic**	**N deficient**	**Yellow rust infected**	**Outliers**
Real asymptomatic	93.51	0.00	1.95	4.55
Real N deficient	0.00	95.15	0.00	4.85
Real yellow rust infected	0.65	0.00	94.44	4.91

Table 6.2 One Class SVM Recognition performance (%) for asymptomatic (A), nitrogen deficient (ND) and yellow rust diseased (YR) crops. The last column denotes the falsely recognized crops as outliers

	Detected by the OCSOM (%)			
From 100%	**Asymptomatic**	**N deficient**	**Yellow rust infected**	**Outliers**
Real asymptomatic	87.40	0.00	10.00	2.60
Real N deficient	0.00	98.88	0.00	1.12
Real yellow rust infected	7.75	0.00	89.02	3.23

(Table 6.1) and One Class SVM (Table 6.2). A recognition accuracy exceeding 94% was attained by One Class SOM with 10 × 10 neurons for nitrogen deficient and Yellow Rust infected crops with 1.95% recognized as infected and 4.55% as outliers. Regarding the One Class SVM approach (with Gaussian kernel spread equal to 1.5), there were more misclassifications that could been accepted, more specifically, some asymptomatic crops falsely recognized as stressed (10% infected and 2.6% as outliers).

Compared to similar classification approaches based on supervised learning where the number of the investigated categories is already known, for the current active learning approach it is not mandatory to receive external supervisory information. On the contrary, the proposed technique utilizes only outlier detection and repetitive knowledge extraction by augmenting a new class.

6.3 Results and discussion for hierarchical self organizing classifiers

The second approach was based on Hierarchical Self-Organizing Maps, employing the models CPANN, SKN and XYF which formed square shaped networks comprising of 30 × 30 neurons. The three models were trained to discriminate asymptomatic from nitrogen deficient and yellow rust infected crops from the spectral characteristics. In order to estimate cross validation generalization performance, 25% of the data was defined as the testing set while the rest 75% was used for training. The recognition performances for each of the three employed models are demonstrated in Tables 6.3, 6.4, and 6.5, it is evident that the XYF network demonstrated the most accurate classification results. Fig. 6.1 depicts the component layers of the training features and output categorical features related to classes and their color-coded layers across the 30 × 30 grid to illustrate the associated relations.

The proposed hierarchical training algorithms enable a more trustable functioning in the control of affecting factors and a more direct visualization of correlation and information mining connecting crop status and sensorial information. The visual depiction of the layers of the trained SOM display the associations between the crop health condition and the training features of the neural network are associated with the spectral signatures of the infected plants.

Table 6.3 SKN true class recognition performance

Asymptomatic	Nitrogen deficient	Yellow Rust diseased
97.79%	0%	2.2%
0%	100%	0%
4.78%	0%	95.22%

Table 6.4 XYF f true class recognition performance

Asymptomatic	Nitrogen deficient	Yellow Rust diseased
97.27%	0%	2.72%
0.37%	99.63%	0%
5.17%	0%	99.83%

Table 6.5 CPANN true class recognition performance

Asymptomatic	Nitrogen deficient	Yellow Rust diseased
96.36	0%	3.64%
0.37%	99.63%	0%
5.68%	0%	94.32%

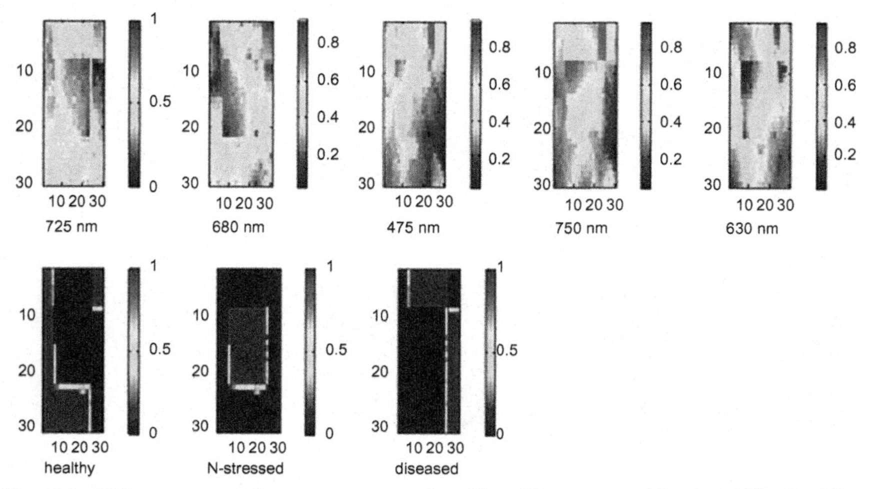

Fig. 6.1 XYF component layers corresponding 30 × 30 neuron grid, where The first line shows the spectral features while the second denotes the output class symbolized per SOM neuron (healthy, N-stressed, yellow rust diseased) (Pantazi et al., 2017).

The relation connecting input and output weights of the Self Organizing Hierarchical models describes the function that connects the spectral signatures with a binary class, which is represented by the output weights. The output weights are taking two discrete values zero or one, which are symbolized as two different color values in the component layers. The binary values are shown as red color for the unity and blue color for zero. The polygons that are linked to the output layer, are complementary with all the other component layers meaning that the output maps correspond to different regions in the weight space. This outcome can be explained by the training regime of the network that during training, builds the association between input and output values. Each output vector is regarded as a triplet comprising of solely one unity value, meanwhile all other elements are set to zero. By observing Fig. 6.1, it is obvious, that the 725 and 630 nm wavelengths are associated with infection occurrence while the nitrogen stress is not directly associated with any specific wavelength. The impact of infection is thoroughly illustrated by the inverse correlation between the weights of the SOM units that are linked to the 725 nm waveband and the output weights that are depicted in the third output weight layer.

The global effect is emerging from the fusion of the spectral information, which is assigning factors of importance during calibration so that they become optimal weights that allow the networks to learn and recreate the

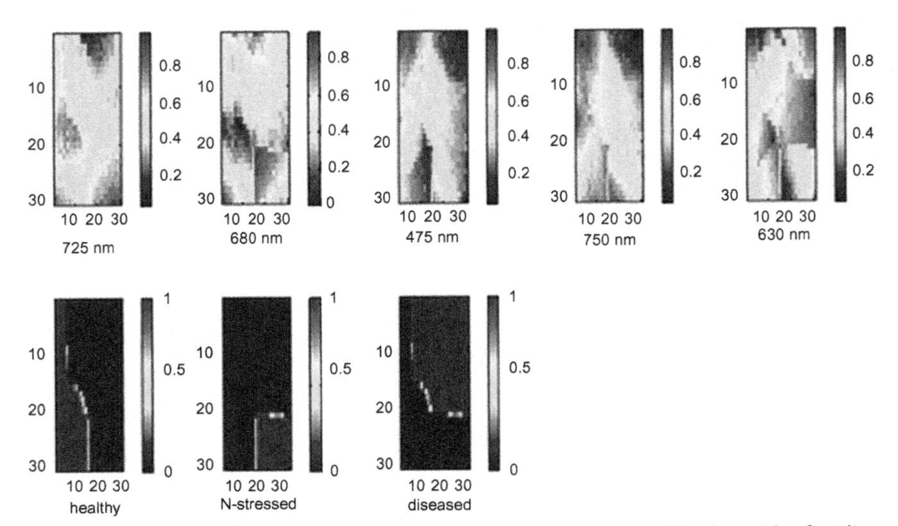

Fig. 6.2 SKN component layers corresponding 30 × 30 neuron grid, where The first line shows the spectral features while the second denotes the output class symbolized per SOM neuron (healthy, N-stressed, yellow rust diseased) (Pantazi et al., 2017).

associative mapping between spectral signatures and health condition. In Fig. 6.2, the impact of infection is thoroughly illustrated by the negative correlation of the SOM neuron weights that are connected to the 725 nm and to the 750 nm wavebands, corresponding to the yellow rust infection output map. Simultaneously, the 630 nm band appears as close associated to the output weight of the infected crop by yellow rust. In Fig. 6.3, the 725 nm and the 630 nm band show high correlation to the nitrogen stress layer and infected condition respectively.

As it is illustrated in Figs. 6.4 and 6.5 it assumed the mean weights in both SKN and XYF models demonstrate similar performance as in the CPANN.

The only deviation characterizes the 630 nm band for the yellow rust infected condition which demonstrates high separation, reaching the conclusion that the associated cluster is evidently separated with respect to the other two clusters yielding improved recognition accuracy regarding the yellow rust infected spectra. On the other hand, this recognition capability is not demonstrated in the discrimination performance (Tables 6.3, 6.4, 6.5), indicating a higher correlation from the 725 nm band, achieving a high accuracy to the XYF model.

By observing Fig. 6.6, it can be concluded that the infected crops show reduced reflection at 725 and 750 nm, while at the same time this high reflectance increased at the orange wavelength. This phenomenon can be

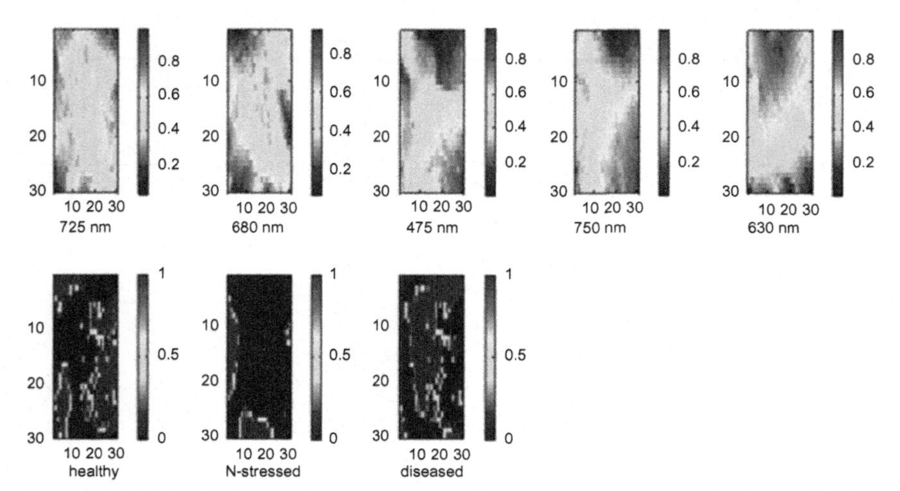

Fig. 6.3 CPANN component layers corresponding 30 × 30 neuron grid, where The first line shows the spectral features while the second denotes the output class symbolized per SOM neuron (healthy, stressed, yellow rust diseased) (Pantazi et al., 2017).

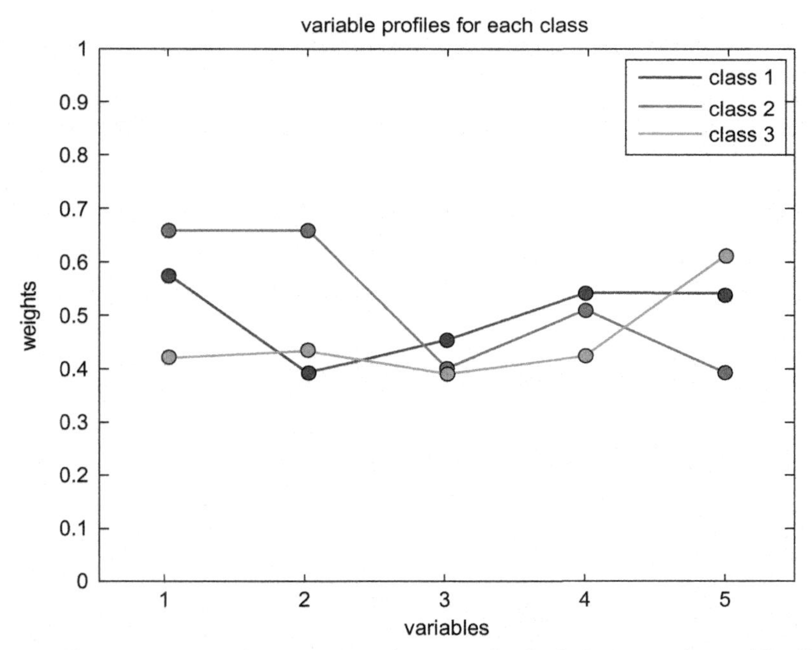

Fig. 6.4 The average weights corresponding to individual classes performed by SKN model. The asymptomatic class is denoted with blue color (class 1), while nitrogen stress and yellow rust infected class are denoted with red (class 2) and green (class 3) respectively. The variables from 1 to 5 correspond to 725 nm, 680 nm, 475 nm, 750 nm, and 630 nm respectively.

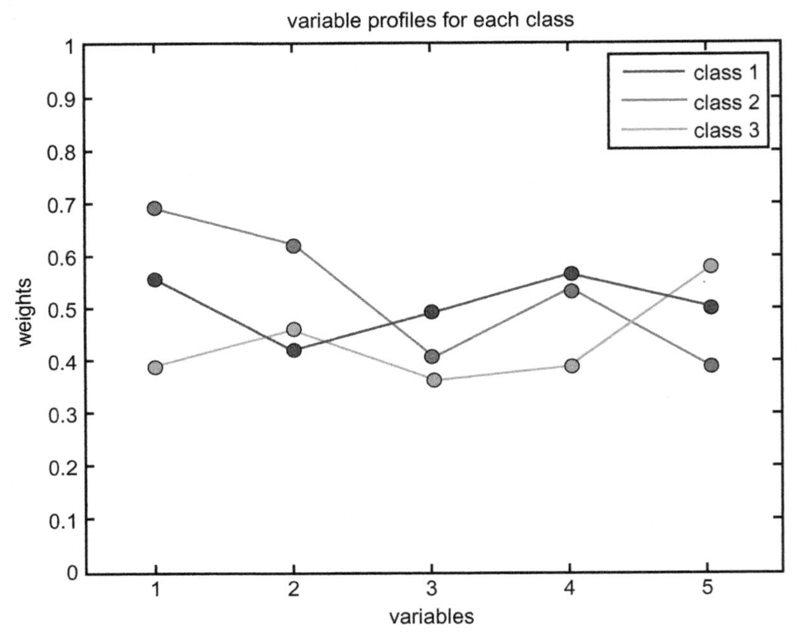

Fig. 6.5 The average weights corresponding to individual classes performed by XYF model. The asymptomatic class is denoted with blue color|(class 1), while nitrogen stress and yellow rust infected class are denoted with red (class 2) and green (class 3) respectively. The variables from 1 to 5 correspond to 725 nm, 680 nm, 475 nm, 750 nm, and 630 nm respectively.

easily explained due to the cell wall collapse, which leads to the reflectance reduction at the near infrared. The nitrogen stressed crops exhibit the inverse behavior at the region of 725 and 680 nm while they show a similar behavior at the blue region.

The visual depiction of the cluster averages (Figs. 6.1, 6.2, and 6.3) demonstrate similar trends compared to the hierarchical SOM layer maps (Figs. 6.4, 6.5, and 6.6). The benefit of hierarchical SOM layer maps lays on their ability to display precisely the inner classes associations, depicted in the output weights, and their relation to the input feature behavior which are visualized in the component layers. The sensitive separation of the infected crop condition of the component XYF layers into two discrete subclusters, is cannot be obtained just by plotting the cluster averages. This lead to the conclusion that hierarchical SOM component layer visualization are able to tackle an arbitrary cluster distribution and improve the visualization potential of subtle effects and correlation which might not be possible by improving statistical properties of clusters.

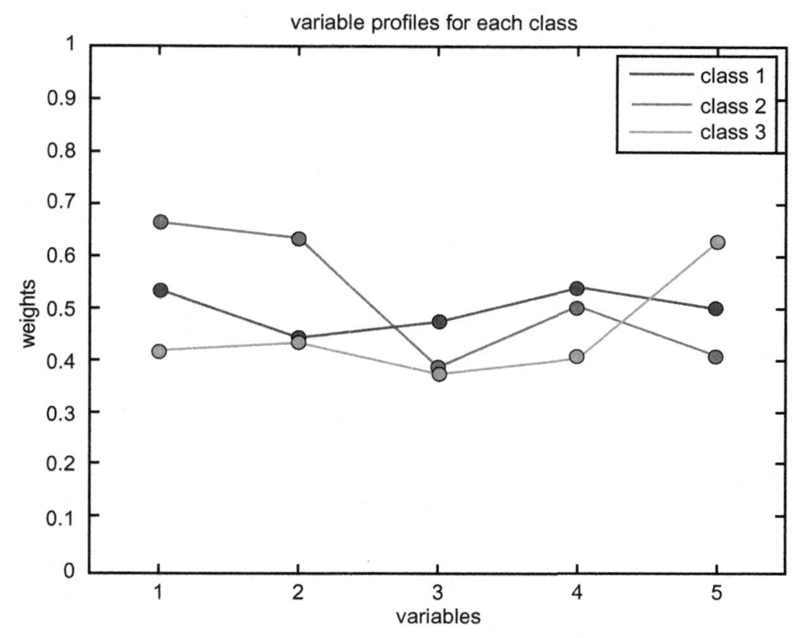

Fig. 6.6 The average weights corresponding to individual classes performed by CPANN model. The asymptomatic class is denoted with blue color (class 1), while nitrogen stress and yellow rust infected class are denoted with red (class 2) and green (class 3) respectively. The variables from 1 to 5 correspond to 725 nm, 680 nm, 475 nm, 750 nm, and 630 nm respectively.

References

Ahmed, F., Al-Mamun, H. A., Bari, A. H., Hossain, E., & Kwan, P. (2012). Classification of crops and weeds from digital images: A support vector machine approach. *Crop Protection*, *40*, 98–104.

Andrews, M., Raven, J. A., & Lea, P. J. (2013). Do plants need nitrate? The mechanisms by which nitrogen form affects plants. *The Annals of Applied Biology*, *163*, 174–199.

Cao, Q., Miao, Y., Li, F., Gao, X., Liu, B., Lu, D., et al. (2017). Developing a new crop circle active canopy sensor-based precision nitrogen management strategy for winter wheat in North China plain. *Precision Agriculture*, *18*, 2–18.

Devadas, R., Lamb, D. W., Simpfendorfer, S., & Backhouse, D. (2009). Evaluating ten spectral vegetation indices for identifying rust infection in individual wheat leaves. *Precision Agriculture*, *10*, 459–470.

Franke, J., Gebhardt, S., Menz, G., & Helfrich, H. -P. (2009). Geostatistical analysis of the spatiotemporal dynamics of powdery mildew and leaf rust in wheat. *Phytopathology*, *99* (8), 974–984.

Gebbers, R., & Adamchuk, V. I. (2010). Precision agriculture and food security. *Science*, *327*, 828–831.

Goldstein, A., Fink, L., Meitin, A., Bohadana, S., Lutenberg, O., & Ravid, G. (2017). Applying machine learning on sensor data for irrigation recommendations: Revealing the agronomist's tacit knowledge. *Precision Agriculture*, *19*(3), 421–444.

Goron, T., Nederend, J., Stewart, G., Deen, B., & Raizada, M. (2017). Mid-season leaf glutamine predicts end-season maize grain yield and nitrogen content in response to nitrogen fertilization under field conditions. *Agronomy*, *7*, 41.

Gutiérrez, S., Diago, M. P., Fernández-Novales, J., & Tardaguila, J. (2018). Vineyard water status assessment using on-the-go thermal imaging and machine learning. *PLoS One*, *13*, e0192037.

He, R., Li, H., Qiao, X., & Jiang, J. (2018). Using wavelet analysis of hyperspectral remote-sensing data to estimate canopy chlorophyll content of winter wheat under stripe rust stress. *International Journal of Remote Sensing*, *39*(12), 4059–4076.

Hillnhütter, C., & Mahlein, A. K. (2008). Early detection and localization of sugar beet diseases: New approaches. *Gesunde Pflanzen*, *60*(4), 143–149.

Huang, W., Davidw, L., Zheng, N., Zhang, Y., Liu, L., & Wang, J. (2007). Identification of yellow rust in wheat using in-situ spectral reflectance measurements and airborne hyperspectral imaging. *Precision Agriculture*, *8*, 187–197.

Huang, W., Guan, Q., Luo, J., Zhang, J., Zhao, J., Liang, D., et al. (2014). New optimized spectral indices for identifying and monitoring winter wheat diseases. *IEEE Journal of Selected Topics in Applied Earth Observations and Remote Sensing*, *7*, 2516–2524.

Jensen, J. R. (2007). In D. Kaveney (Ed.), *Remote sensing of the environment: an earth resource perspective* (2nd ed.). Upper Saddle River, NJ: Prentice Hall.

Johnson, D. M. (2014). An assessment of pre- and within-season remotely sensed variables for forecasting corn and soybean yields in the United States. *Remote Sensing of Environment*, *141*, 116–128.

Lu, J., Ehsani, R., Shi, Y., Castro, A. I., & Wang, S. (2018). Detection of multi-tomato leaf diseases (late blight, target and bacterial spots) in different stages by using a spectral-based sensor. *Scientific Reports*, *8*(1), 2793.

Mahlein, A. K., Steiner, U., Hillnhütter, C., Dehne, H. W., & Oerke, E. C. (2012). Hyperspectral imaging for small-scale analysis of symptoms caused by different sugar beet diseases. *Plant Methods*, *8*(1), 3.

Mehra, L. K., Cowger, C., Gross, K., & Ojiambo, P. S. (2016). Predicting pre-planting risk of Stagonospora nodorum blotch in winter wheat using machine learning models. *Frontiers in Plant Science*, *7*, 390.

Moshou, D., Bravo, C., West, J., Wahlen, S., Mccartney, A., & Ramona, H. (2004). Automatic detection of 'yellow rust in wheat using reflectance measurements and neural networks. *Computers and Electronics in Agriculture*, *44*, 173–188.

Moshou, D., Deprez, K., & Ramon, H. (2004). Prediction of spreading processes using a supervised self-organizing map. *Mathematics and Computers in Simulation*, *65*(1–2), 77–85.

Oerke, E. -C., & Dehne, H. -W. (2004). Safeguarding production—losses in major crops and the role of crop protection. *Crop Protection*, *23*, 275–285.

Pantazi, X. E., Moshou, D., Oberti, R., West, J., Mouazen, A. M., & Bochtis, D. (2017). Detection of biotic and abiotic stresses in crops by using hierarchical self organizing classifiers. *Precision Agriculture*, *18*(3), 383–393.

Römer, C., Wahabzada, M., Ballvora, A., Pinto, F., Rossini, M., Panigada, C., et al. (2012). Early drought stress detection in cereals: Simplex volume maximisation for hyperspectral image analysis. *Functional Plant Biology*, *39*(11), 878–890.

Rumpf, T., Mahlein, A. K., Steiner, U., Oerke, E. C., Dehne, H. W., & Plümer, L. (2010). Early detection and classification of plant diseases with support vector machines based on hyperspectral reflectance. *Computers and Electronics in Agriculture*, *74*(1), 91–99.

Satir, O., & Berberoglu, S. (2016). Crop yield prediction under soil salinity using satellite derived vegetation indices. *Field Crops Research*, *192*, 134–143.

Tian, Y. C., Yao, X., Yang, J., Cao, W. X., Hannaway, D. B., & Zhu, Y. (2011). Assessing newly developed and published vegetation indices for estimating rice leaf nitrogen concentration with ground- and space-based hyperspectral reflectance. *Field Crops Research*, *120*, 299–310.

Tremblay, N., Wang, Z., & Cerovic, Z. G. (2012). Sensing crop nitrogen status with fluorescence indicators. A review. *Agronomy for Sustainable Development*, *32*, 451–464.
Vigneau, N., Ecarnot, M., Rabatel, G., & Roumet, P. (2011). Potential of field hyperspectral imaging as a non destructive method to assess leaf nitrogen content in wheat. *Field Crops Research*, *122*, 25–31.
Wahabzada, M., Mahlein, A., Bauckhage, C., Steiner, U., Oerke, E., & Kersting, K. (2016). Plant phenotyping using probabilistic topic models: Uncovering the hyperspectral, language of plants. *Scientific Reports*, *6*, 22482.
Wan, A., Zhao, Z., Chen, X., He, Z., Jin, S., Jia, Q., et al. (2004). Wheat stripe rust epidemic and virulence of Puccinia striiformis f. Sp. tritici in China in 2002. *Plant Disease*, *88*, 896–904.
Whelan, B. M., & Taylor, J. A. (2013). *Precision agriculture for grain production systems*. Australia: CSIRO Publishing.
Wiegand, C. L., Richardson, A. J., & Kanemasu, E. T. (1979). Leaf area index estimates for wheat from LANDSAT and their implications for evapotranspiration and crop modeling. *Agronomy Journal*, *71*, 336–342.
Yuan, Y., Lin, J., & Wang, Q. (2015). Hyperspectral image classification via multitask joint sparse representation and stepwise MRF optimization. *IEEE Transactions on Cybernetics*, *46*(12), 2966–2977.
Zhang, J.; Huang, W.; Li, J.; Yang, G.; Luo, J.; Gu, X; Wang, J. Development, evaluation and application of a spectral knowledge base to detect yellow rust in winter wheat. Precision Agriculture 2011, 12, 716–731. [CrossRef].
Zhang, J., Pu, R., Huang, W., Yuan, L., Luo, J., & Wang, J. (2012). Using in-situ hyperspectral data for detecting and discriminating yellow rust disease from nutrient stresses. *Field Crops Research*, *134*, 165–174.

Further reading

Zadoks, J. C. (1961). Yellow rust on wheat studies in epidemiology and physiologic specialization. *Tijdschrift over Plantenziekten*, *67*, 69–256.

CHAPTER 7

Tutorial IV: Leaf disease recognition

Contents

7.1 Introduction

Crop infection detection is critical for crop growth monitoring and safeguarding the quality of production. For this purpose, optical recognition can be affectivity utilized by specialists for monitoring possible changes in crop leaves, therefore a highly experienced and specialized background. New techniques handling data mining by AI is able to enhance trustworthy the credibility of a possible infection recognition and thus can be featured into actionable insights to enable interventions.

Methods using AI combined with image feature analysis are capable of providing a more precise and effective tool for crop disease detection. Crop infection recognition using visual information is regarded challenging, more specifically, on its automated form. This task is often hard to be tackled, because of the special characteristics in terms of morphology and but also because of lighting conditions. Crop infection evaluation is a critical step for adopting accurate and efficient crop management practices encompassing predictive algorithms and various means of treatment.

Intelligent Data Mining and Fusion Systems in Agriculture
https://doi.org/10.1016/B978-0-12-814391-9.00007-8

Fungal diseases are regarded responsible for severe yield degradation, yield low productivity and financial losses, ranging between 5 and 80% depending on the severity of infection, ambient factors related to climate and terrain, and lack of resistance due to susceptible genotype (Šrobárová & Kakalíková, 2007). It is crucial to focus on the relation that exists between various unexpected deviations in the characteristics of plant leaf morphology and the variety of the environmental factors that would act as are possible stress factors to a specific crop that demonstrates stress symptoms (Gaunt, 1995).

7.2 State of the art

Various techniques are used in the present targeting on identifying plant infections by applying digital image processing. Crop disease identification was achieved in sugarcane leaves by threshold segmentation to isolate the infected leaf location and triangle thresholding to obtain the lesioning area, reaching an average accuracy of 98.60% (Patil & Bodhe, 2011).

Texture feature extraction has also been used for plant disease detection. Patil and Kumar (2017) introduced plant disease detection approach base on texture feature extraction including morphological features, uniformity, and relative associations derived by estimating the co-occurrence matrix of the gray image of corn leaves combined with color extraction, giving a trustworthy training feature based set for image enhancement and improved detection.

Rothe and Kshirsagar (2015) employed a Back propagation approach for classifying diseased leaf images of cotton. During the training procedure, seven invariant moments were extracted from three leaves images depicting infections from three different diseases respectively. The utilized snake segmentation algorithm achieved a classification of 85.52%, providing an efficient solution for set apart the area of interest (the infected one). However, the proposed method is characterized as time consuming. Zhang, Wu, You, and Zhang (2017) proposed a leaf disease detection procedure in cucumber, by setting apart the infected leaf area through an efficient combination of k-means clustering and color and shape feature extraction, obtaining a score of 85.7% correct recognition. A similar approach has been presented by Guo, Liu, and Li (2014). In this approach texture and color features were combined by a Bayesian approach. The investigated diseases where downy mildew, anthracnose, powdery mildew, and gray mold. The utilized model achieved accuracy rates of 94.0%, 86.7%, 88.8%, and 84.4% respectively. Vianna, Oliveira, and Cunha (2017) proposed a pattern simulation approach for recognizing the globally threatening of late blight in tomato leaves. In this paper, 20 networks were tested, from which the best network produced a prediction of 97.99%. Fiel and Sablatnig (2013) have diversified leaf infection detection by applying Bag of Words with SIFT descriptors in 5 different tree species, attaining an accuracy of 93.6%.

Meunkaewjinda, Kumsawat, Attakitmongcol, and Srikaew (2008) presented an automated crop infection diagnosis detection system which utilized various artificial intelligent algorithms in infected vine leaves. Self-organizing feature map and back propagation neural network has been applied in order to detect the colors of vine leaves. Additionally, an altered modified SOM has been applied for segmenting the image and support vector machine is used as a classifier, demonstrating a performance of 86.03%. Self-organizing feature map is also used to detect disease of cotton leaves (Gulhane & Gurjar, 2011).

Kebapci, Yanikoglu, and Unal (2011) developed a plant retrieval system taking into consideration the color, shape and texture characteristics by employing the color histogram, the color co-occurrence matrix and a modified Gabor method approach. The experiment has been performed for crop type recognition in domestic plants reaching 73% for successful recognition.

Pethybridge and Nelson (2015) have presented an IOS application called 'Leaf Doctor", capable of discriminating the possibly infected from the healthy plant tissue areas. In this approach, eight colors denoting the healthy tissue areas form the threshold assisting the recognition of the healthy ones. With respect to the six corresponding infections, the algorithm's sensitivity was accessed according to targeted range of 10 color corrected images attaining a $R^2 > 0.79$. Nevertheless, some of the main drawbacks of this approach is that for the image processing, a black background is required. The black background has proven to be most of the times responsible for the majority of misclassifications since it can be falsely mixed up with alike colored pixels linked to infection symptoms. Moreover, considering the fact that the samples were destructively taken, makes the method less credible. The employed algorithm was difficult to come to a trustworthy conclusion in variable ambient field conditions and alterations, as a result more stable and trustworthy algorithms are required.

The current case study demonstrates a novel application of recognizing four different health status including healthy, downy mildew, powdery mildew and black rot in different tissue samples through One Class Classification. The presented application is able to assess images obtained in real conditions without additional processing and can decide with certainty on the occurrence of infection even at its early stage appearance. The current approach is comprised of a learning procedure with training data relate to each one of the four different crop status. A previously unseen and unlabeled feature vector is used as input to a one-class classifiers committee, triggering activations. In the occasion that multiple activations take place, then a conflict occurs since more than one classifiers are regarded as contenders of the new feature vector ownership. The current case study demonstrates high generalization potential regarding the leaf plant disease identification in several crops, where the training set was comprised solely of vine leaf samples.

7.3 Materials and methods

The Local Binary Pattern (LBP) is a texture analysis approach, that able to label pixels by defining their proximity limits, forming a binary outcome. The LBP approach takes advantage of its capability of maintaining a stable behavior to the upcoming gray scale level deviations that occur, such as illumination and image processing under complex real-time conditions, proving its computational potential to be implemented in several commercial applications.

The LBP operator has been introduced by Li, Wu, Wang, and Zhang (2008) and Llado, Oliver, Freixenet, Marti, and Marti (2009) considering texture features including pattern including the relative pattern's strength applied with a 3×3 grid, where the center value denotes a threshold. The LBP code is formulated as presented: the thresholded values are scaled by the corresponding weights of the pixels. Since the neighborhood is defined by eight pixels, a set of $2^8 = 256$ discrete labels are produced, corresponding to the relative threshold's gray levels and remaining pixels of the grid. The contrast indicator (C) is defined as the subtraction of the mean of the pixel levels that lie below the threshold defined from the mean of pixel levels larger or equal to the threshold. In the case that the thresholded pixel levels are equal to zero or one, the contrast value is equal to zero. Fig. 7.1 depicts an illustration of LBP operator function.

example

6	5	2
7	6	1
9	8	7

thresholded

1	0	0
1		0
1	1	1

weights

1	2	4
128		8
64	32	16

Pattern = 11110001

LBP = 1 + 16 + 32 + 64 + 128 = 241

C = (6+7+9+8+7)/5 – (5+2+1)/3 = 4.7

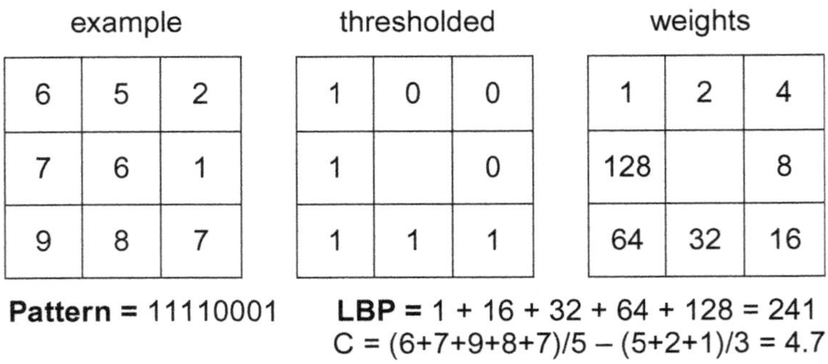

Fig. 7.1 Illustration of LBP operator function.

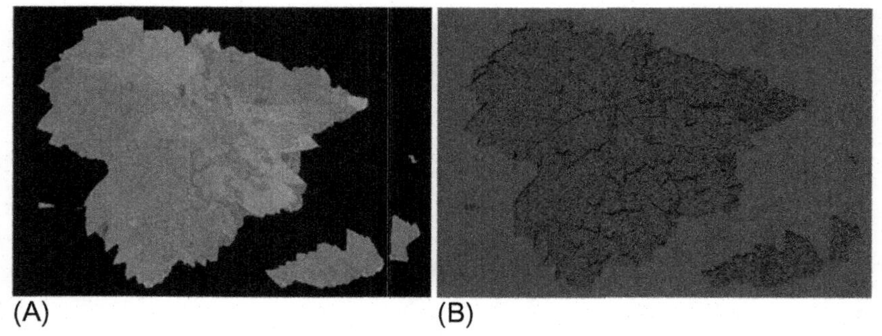

Fig. 7.2 (A) Hue channel illustration of a vine leaf (Pantazi, Moshou, & Tamouridou, 2019). (B) LBP leaf image of vine crop (Pantazi et al., 2019).

The LBP operator is able to transform an image into a matrix of integer levels that define the detailed local structure of the image (Fig. 7.1). These levels or their derived statistical operators like histogram, are the parameters used for producing image analytics. The LBP operator is mainly applicable to binary images as well as multi-channel images, video streams and multi-dimensional data. The hue channel is derived from image of the leaf as depicted in Fig. 7.2A. The LBP transform application on the Hue channel derived of the segmented image produces an LBP leaf image, as illustrated in Fig. 7.2B.

7.4 Application of LBP in disease recognition of infected plants

7.4.1 Image segmentation

As a first step, an image of the diseased leaf that manifests visual symptoms is obtained by a smartphone or tablet (Fig. 7.3A). Afterwards, image segmentation takes places aiming to obtain the leaf area and exclude the image background. Then, a Hue Saturation Value (HSV) transform is imposed to the segmented area (Fig. 7.3B). The GrabCut algorithm operation is summarized in the following steps:

1. The central image points correspond to foreground while the surrounding regions correspond to background. Then, a rectangle is formed including the target region while the internal pixels are recognized as not known and the external ones are denoted as known.
2. An image segmentation is performed by applying Gaussian Mixture Models (GMMs) to the foreground and background utilizing the Orchard-Bouman algorithm to create the clusters.

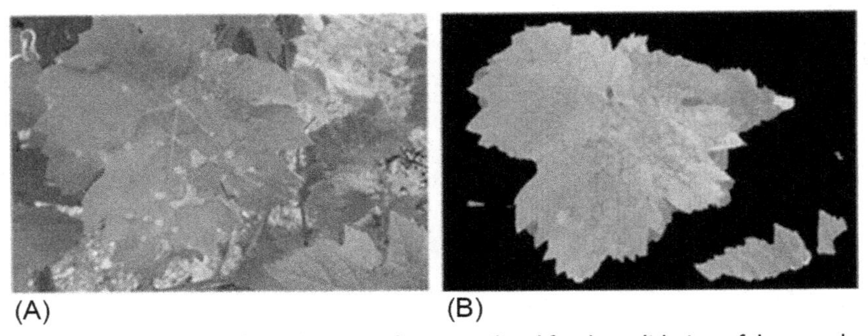

Fig. 7.3 (A) Randomly selected image of a vine utilized for the validation of the powdery mildew One Class Classifier (Pantazi et al., 2019). (B) Depiction of the segmented area of interest of the vine leaf illustrated in Fig. 7.1 after HSV transform and the GrabCut algorithm's operation (Pantazi et al., 2019).

3. All pixels are classified as foreground or background through the allocation of Gaussian component corresponding to the foreground or the background GMMs.
4. The arising pixel sets yield the learning adaptation, targeting on the creation of new GMMs.
5. A graph is constructed, activating the GrabCut algorithm for assigning a new foreground class.
6. The former steps (4–6) are iterated until all the pixel sets are classified. The GrabCut algorithm creates K components of multivariate GMM corresponding to the background. A similar procedure is followed for K components corresponding to the foreground. The GMM components are labeled according to the color statistics corresponding to each cluster.

7.4.2 Creation of LBP histogram

The GrabCut algorithm is applied individually on various single channels of the initial image because of the LBP operator is applied in a single channel. On the initial RGB image two separate transforms options were tested: the first transform obtained in HSV format while the second transform obtained gray scale image. From the HSV transform it has become apparent that the LBP histogram applied on the Hue channel, yielded the highest performance in comparison to the gray scale transform in relation to the detection of disease crop symptoms.

The LBP operator manages the labelling of the pixels corresponding to the obtained images based on their mutual distance in the Hue channel, and consequently takes in to account the textural characteristics concerning contrast and Hue variation as illustrated in Fig. 7.3B. The LBP histogram depicts the appearance rate of Hue levels from 0 to 255. For avoiding noisy values in the histogram, the 256 bins has been reduced to 32. An illustration of the reduced histogram is given in Fig. 7.4, where the x and y axes, denote the amount of bins and the allocation of pixels per bin (%) respectively.

The inhomogeneity of symptoms in symptomatic leaves induces deviations in texture of a local nature which are evident which appears in the LBP histogram as higher pixel counts in specific bins (Fig. 7.4).

As it is shown in Fig. 7.4, the healthy leaf status is associated with subtle alterations in the local texture structure. The pixel count in certain bins in the LBP histograms corresponds to other local deviations that are associated to the collapse of leaf structure produced by the pathogen invasion. Equally, the black rot occurrence provokes high pixel count in the first bin and an additional peak in the sixth bin. In the case of powdery mildew occurrence, lower pixel count is demonstrated in the first and a peak in the ninth bin. This is explained by the contribution of the pixel count to the leaf structure collapse that finally lead to the discrimination of black rot from powdery mildew symptoms.

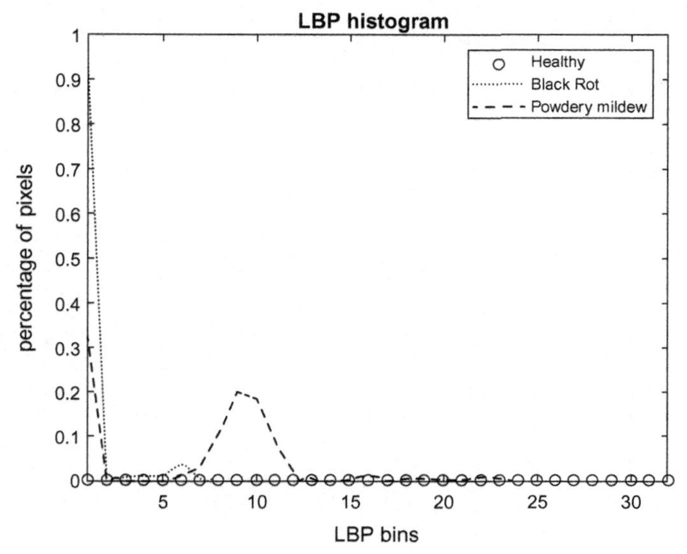

Fig. 7.4 LBP histogram depicting different health conditions (healthy, black rot and powdery mildew infected) in vine crop (Pantazi et al., 2019).

7.4.3 One class classification

The LBP histograms are used for calibrating One Class Classifiers. More exactly, One Class SVMs (OC-SVMs) were employed to classify leaf images according to their health status. For developing these classifiers eight leaf images from vine plants corresponding to each crop health status (healthy, powdery mildew, downy mildew, black rot) constituted sufficient training dataset (Figs. 7.5–7.8).

One Class SVMs were assessed by cross validation through application to samples corresponding to health crop status (healthy, powdery mildew, downy mildew, black rot), demonstrating an accurate calibration performance when tested with random images.

Fig. 7.5 The dataset utilized for calibrating the One Class SVM model for the healthy condition (Pantazi et al., 2019).

Fig. 7.6 The dataset utilized for calibrating the One Class SVM model for the powdery mildew infection (Pantazi et al., 2019).

Fig. 7.7 The dataset utilized for calibrating the One Class SVM model for the black rot infection (Pantazi et al., 2019).

Fig. 7.8 The dataset utilized for calibrating the One Class SVM model for the downy mildew infection (Pantazi et al., 2019).

7.4.3.1 One class support vector machines (OCSVMs)

Classification is performed to discriminate different sample types, whilst regression intends to produce a specific output for available data samples. Unlike the previously mentioned models, One Class Classifiers take advantage of solely target data to express a binary classification argument based on its membership or not to the target dataset. The One class Classification can be also characterized by the term domain description, denoting a set of objects, capable of recognizing those that are similar or dissimilar (labeled as outliers) to the calibration set assisted by novelty detection.

There is a plethora of data domain definition or outlier detection techniques in the literature. Tax and Duin (2004) proposed a type of One Class SVM Classification, called Support Vector Data Description (SVDD).

The SVDD is capable of classifying an unknown object into either target or outlier, based on the hypothesis that target data are bounded in the feature domain. A sphere, with a center a and minimum radius R containing most of N samples $\{x_i; i=1;::; N\}$. Slack variables ξ_i are defined, following the constraints below:

$$(x_i-\alpha)(x_i-a)^T \leq R^2+\xi_\iota \quad \xi_\iota \geq 0 \tag{7.1}$$

The R and the ξ_i are expressed as an equality by using a weight to slack variables:

$$F(R, a, \xi_i)=R^2+C\sum_i \xi_\iota \tag{7.2}$$

where C, symbolizes sphere's volume and it is directly relevant to number of data point exceptions. Consequently, the result is expressed in LaGrange multiplier equation as follows:

$$L(R, a, a_i, \xi_i)=R^2+C\sum_i \xi_i-\sum_i a_i\left\{R^2+\xi_i-\left(x_i^2+2\alpha x_i+\alpha^2\right)\right\} -\sum_i \gamma_i \xi_i \tag{7.3}$$

where $\alpha_i \geq 0$ and $\gamma_i \geq 0$ are LaGrange multipliers. By defining the partial derivatives to zero values, three conditions are formed as follows:

$$\sum_\iota \alpha_i=1 \quad a=\frac{\sum_i a_i x_i}{\sum_i a_i}=\sum_i a_i x_i \quad 0 \leq a_i \leq C \tag{7.4}$$

After optimization with respect to α_ι while satisfying the restrictions expressed in Eq. (7.4) the LaGranzian is expressed as below:

$$L=\sum_i a_i(x_i \cdot x_i)-\sum_{i,j} a_i a_j\left(x_i \cdot x_j\right) \tag{7.5}$$

A data sample z belongs to the sphere if its distance from the center is smaller compared to the R (Eq. 7.6):

$$(z-a)(z-a)^T=(\mathrm{z} \cdot \mathrm{z})-2\sum_i a_i(z \cdot x_i)+\sum_{i,j} a_i a_j\left(\mathrm{x}_i \cdot x_j\right) \leq R^2 \tag{7.6}$$

For a reduced size dataset, the calibration dataset is defined by the formulation expressed through the Eq.(7.6), in which support vectors are denoted by nonzero α_ι. In the special case where $C < 1$, the support vectors equate C. Those data samples originate from the external sphere's space and consequently they are considered as outliers. The above presented methodology is defined as the Support Vector Domain Description (SVDD). The OCSVM constructs a model by training the classifier using data samples provided by the SVDD. Consequently, the test data are classified by utilizing the variance of the calibration samples (Schölkopf, Platt, Shawe-Taylor, Smola, & Williamson, 2001).

7.4.3.2 One-class support-vector-learning for multi class problems

One-Class Classifiers are usually employed to resolve to multi-decision problems. Test data are sampled in random corresponding to various classes. For every one of the individual health condition classes, a classifier is employed and the sphere is constructed aiming to reach a conclusion about for every point emanating from the tested dataset. In the occasion of a newly presented sample, two possibilities arise:

1. In the situation arising when one classifier is assigned a new data sample, it is characterized by the relevant class.
2. Conflict resolution is activated in the case that a couple or more classifiers are assigned for the same data sample. In the occasion that a new sample is presented and not assigned by the classifier, it is considered an outlier.

The conflict is handled by employing the Nearest-Support-Vector Strategy.

7.4.3.3 Nearest-support-vector strategy

The distance of a new data sample (z) from the support vectors that correspond to each classifier takes its label (Eq. 7.7). By considering that:

$$SV(t) = \left\{\hat{x}_1^t, \ldots, \hat{x}_{1t'}^t\right\} \tag{7.7}$$

The SV denotes the set of support vectors, and i, t their quantity with the classifier t.

Aiming to classify z, the decision function $f(z)$ calculate the distances of the data sample to the support vectors in order find the most proximal individually discover in which classifiers it is assigned (Eq. 7.8).

$$f(z) = \text{argmin}_{l \in C} \|\hat{x}_i^l - z\| \tag{7.8}$$

where C symbolizes each one of the competing classifiers.

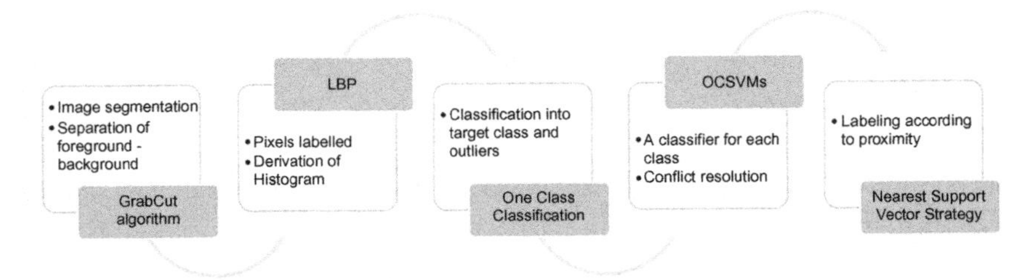

Fig. 7.9 Illustration of the followed steps (Pantazi et al., 2019).

All the steps followed in the current case study are presented schematically in Fig. 7.9.

7.4.4 Generalization of the vines powdery mildew model to different crop diseases

7.4.4.1 Validation of the developed disease classification framework

A set of eight vine leaf images corresponding to each of the investigated health status were used, including healthy condition, powdery mildew, black rot and downy mildew infected as calibration dataset to the One Class SVM. The calibrated OC-SVM with vine leaves was further employed for disease detection in eighteen different plant species achieved an excellent generalization for all individual cases. Conflict resolution has been proven to be a critical factor, for the health status assessment exceeding 50% of the cases, achieving a 100% in most of the cases.

The classifiers assessing the healthy plant status and the three investigated infections (powdery mildew, black rot, downy mildew) were calibrated using solely vine leaf images but the generalization potential is far greater since it can recognize the afore mentioned health conditions in variety of crops.

For the detection of powdery mildew occurrence ten samples of each crop were utilized and has proven to be effective in crop health status recognition by One Class Classifier or a multitude of One Class Classifiers, giving rise to a conflict situation. The examined crops were: *Vitis vinifera*, *Brassica oleraceae var capitata*, *Capsicum annuum*, *Cucurbita maxima*, *Cucurbita pepo*, *Cumis melo*, *Cucumis sativus*, *Citrullus lanatus*, *Lagenaria siceratia*, *Cucurbita moschata*, *Fragaria ananassa*, *Pisum sativum*, *Solanum melongena*, and *Malus pumila*. *Solanum lycopersicum* was *an* exception since it was the only examined crop that has been tested by the proposed method, but wasn't successfully identified.

In the occasion of black rot identification, the classification algorithm was tested on ten samples corresponding to each crop and reached high performance either with or without conflict resolution for the following crops: *Vitis vinifera, Brassica oleraceae var italica, Fragaria ananassa, Brassica oleracea var capitate, Malus pumila* and *Pyrus communis*.

Regarding downey mildew recognition, ten samples of each of the following crops were tested either with or without the help of conflict resolution: *Vitis vinifera, Citrullus lanatus, Cucumis sativus, Lactuca sativa, Brassica oleracea var botrytis*. Table 7.1 summarizes the recognition potential of the presented approach regarding the health condition assessment for the investigated crop species. The marked cells represent the all the tested combinations whether it is successful or not.

As it is depicted in Table 7.1, the healthy status is correctly identified for all investigated plants. The majority of the investigated plants for powdery

Table 7.1 Catalogue of the crops investigated for health condition assessment

	Healthy status	Powdery mildew	Black rot	Downy mildew
Vitis vinifera	x	x	x	x
Brassica oleraceae var. capitata	x	x	x	Fail
Brassica oleraceae var. botrytis	x			x
Brassica oleraceae var. italica	x		x	
Cucurbita pepo	x	x		
Lagernaria siceratia	x	x		x
Lactuca sativa	x			x
Cucumis sativus	x	x		
Cucumis melo	x	x		
Citrullus lanatus	x	x		x
Cucurbita maxima	x	x		
Solanum melongena	x	x		
Cucurbita moschata	x	x		
Capsicum annuum	x	x		
Fragaria ananassa	x	x	x	
Pisum sativum	x	x		
Pyrus communnis	x		x	
Malus pumila	x	x	x	
Solanum lycopersicum	x	Fail		

mildew infection, have been correctly recognized except the *Solanum lycopersicum* which was falsely recognized as healthy (annotated as Fail in Table 7.1). *Brassica oleraceae var* capitate was also the second crop that the model didn't manage to recognize regarding the downy mildew infection.

By observing Table 7.1, 46 combinations were tested, from which the 44 achieved correct classification yielding an overall identification accuracy of 95%. In each discrete condition, the correct classification rate was 100% concerning healthy and black rot infested leaf samples while, 93% and 83.3% regarding powdery mildew and downy mildew infested leaf samples respectively. In the occasion of powdery mildew recognition, there have been fourteen samples correctly classified out of fifteen tested, while for downy mildew, five samples were successfully recognized out of six tested.

For estimating the model's generalization capability, the testing samples were selected from a pool of images that did not belong to the calibration dataset.

7.4.4.2 The generalization potential of one class SVM from vines powdery mildew to other plant diseases

The One Class SVM classifier annotates vine leaves as infected by powdery mildew assigns to them the label 'powderyHSV' and stores the classifier's parameters as XML. Fig. 7.10A illustrates the evaluation of the classifiers performance on *Cucumis sativus* leaves. The generalization ability powderyHSV classifier was confirmed by the correct recognition of the infestation

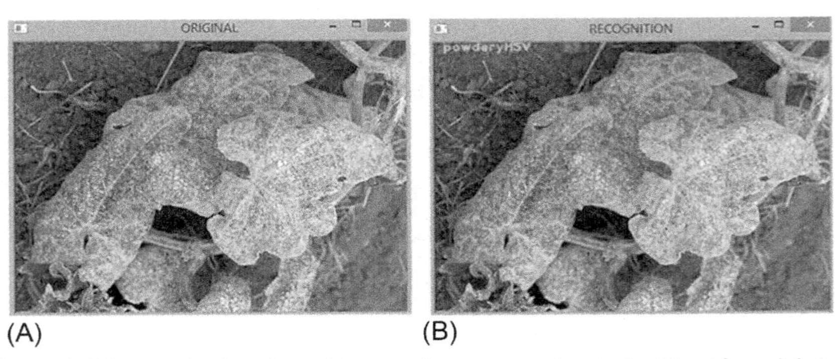

Fig. 7.10 (A) A randomly selected image of *Cucumis sativus* leaf utilized for validating the one class classifier corresponding to powdery mildew (Pantazi et al., 2019). (B) Annotation of *Cucumis sativus* as powdery mildew infected by the One Class Classifier (Pantazi et al., 2019).

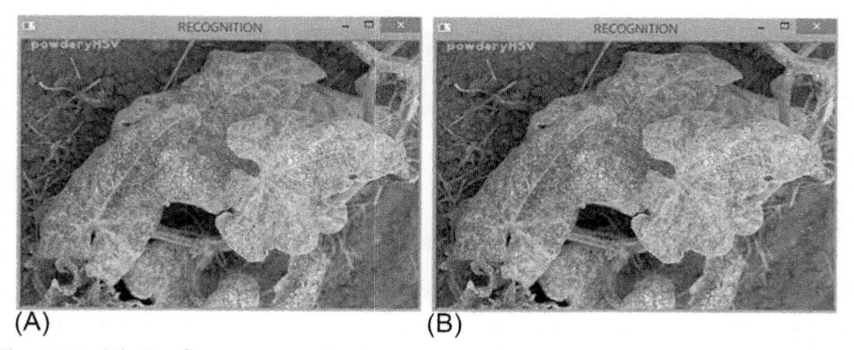

Fig. 7.11 (A) Conflict emerges between One Class SVM models (powderyHSV Vs BlackrotHSV) in a *Cucumis sativus* leaf (Pantazi et al., 2019). (B) Conflict resolution between One Class SVM models (powderyHSV Vs BlackrotHSV) reaches to assess correctly the powdery mildew infected leaf. (Pantazi et al., 2019).

on several plant species apart from the one that was used for training (Fig. 7.10B).

The PowderyHSV model can recognize powdery mildew infection in various crops achieving accurate generalization in nine species such as *Vitis vinifera*, *Brassica oleracea*, *Lagenaria siceraria*, *Cucumis sativus*, *Cucurbita pepo*, *Citrullus lanatus*, *Cucumis melo*, *Cucurbita moschata* and *Cucurbita maxima*.

Regarding the testing of the classifiers, ten samples were selected from each of the species. For guaranteeing the correct classification of the investigated sample images, another model called 'blackrotHSV' is activated. The two models are activated both by the observes symptoms simultaneously as shown in Fig. 7.11A. For coming to conclusion with respect to the leaf health status, conflict resolution takes place taking into consideration the distances between the samples and the support vectors. The classifier denoting the sample's label is successfully assigned as powderyHSV (Fig. 7.11B). In the same way, conflict resolution was consistently achieved for different combinations of the employed OC-SVMs classifiers for powdery, downy mildew and blackrot identification and has consistently attained accurate decisions concerning the leaf health condition of the investigated crops.

7.4.4.3 Generalization of the vines black rot model to different crops diseases

The One Class SVM classifier named 'blackrotHSV' recognizes correctly black rot infected leaves following the calibration with a dataset consisted

Fig. 7.12 (A) A randomly selected image of *Fragaria ananassa* leaf utilized for validating the one class classifier corresponding to black rot (Pantazi et al., 2019). (B) The selected area of interest in *Fragaria ananassa* image is extracted by the GrabCut algorithm (Pantazi et al., 2019). (C) The output of the OC-SVM classifiers is displayed at the top left of the screen (Pantazi et al., 2019).

of vine leaf images. The parameters of the classifers are stored in XML format for increased portability. Fig. 7.12A–C demonstrate the classifier's identification performance on *Fragaria ananassa* leaf images.

7.5 Segmentation technique

The current method is able to achieving successfully generalize disease ditection by using for the training of OC-SVMs a small amount of vine leaves samples, (8) arbitrarily chosen, with respect to the four different health conditions. The originality of the of the presented study lays on the GrabCut's algorithm ability to isolate the leaf area from that belonging to the foreground. In the occasion of multiple leaves present to the image even from other crop species, the GrabCut has proven capable of isolating efficiently the individual sample that is validated.

In the current case study, no special conditions were necessary for the image acquisition, because the algorithm can isolate the region of the leaf from any natural background, in field conditions. A significant advantage of the current method is that it is nondestructive, which works efficiently without the need of detaching the leaf from the crop. These characteristics constitue to proof that the presented method is robust and operational, because the functionality of the method is not affected by variations in shading, leaf orientation and ambient lighting. Regarding calibration, no camera settings were needed like focus and shutter settings. The afore mentioned features prove that the methodology is user friendly, highly functional and adjustable to a variety of alternative scenarios in different case studies.

The operation of the GrabCut algorithmis robust against variable RBG values due to the use of Gaussian Mixtures.

7.6 Classification process and the features extraction process

The current case study achieves an accurate recognition of specific leaf diseases by means of novelty detection. In some cases, the phenology of some diseases appears to be similar in terms of occurrence, meaning that for the same crop there are also detected as outliers. The close distance between the feature vector and the one class classifier spheres explains the operation of the algorithm. The characterization of both infected and healthy crops expressed in HSV features and more specifically the Hue channel of healthy crop images is the cornerstone is that defines the structure the OC-SVM for vine leaves. This behavior is closely related to the feature that has been utilized in the LBP histogram.

The LBP is responsible for the robustness against variations to ambient factors including scaling, translation and rotations or illumination fluctuations. The crucial information that is associated to the crop health status is exploited by the LBP histogram as variations in textural characteristics that emanate from the symptom and the visual distribution of the symptoms. The variation of the background is overcome due to ability of the GrabCut to remain unaffected by the possible intrinsic factors that could possibly affect the segmentation process arising from the image background and acquisition factors. The acquisition factors related to orientation and lighting are confronted by the LBP histograms thanks to their variability to these factors. However, various diseases or nutrient stresses can cause similar symptoms that cannot be easily identified by the validated OC-SVMs. Even in this case the presented conflict resolution can act in an independent manner, by exploiting the distance of the feature vector to the support distance so as to determine the class.

The proposed technique unites the AI's assets with progressive image analysis aiming to extract image characteristics that identify the textural alterations correlated to crop health status. The proposed approach exploits the synergistic effect between OC-SVMs reaching high precision in assessing crop health status through conflict resolution arising from the between opposing classifiers. The proposed technique can be exploited towards the determination of crop health status in order to decide the best management option in the framework of precision crop protection.

References

Fiel, S., & Sablatnig, R. (2013, August). Writer identification and writer retrieval using the fisher vector on visual vocabularies. In *2013 12th international conference on document analysis and recognition* (pp. 545–549): IEEE.

Gaunt, R. E. (1995). New technologies in disease measurement and yield loss appraisal. *Canadian Journal of Plant Pathology*, *17*(2), 185–189.

Gulhane, V. A., & Gurjar, A. A. (2011). Detection of diseases on cotton leaves and its possible diagnosis. *International Journal of Image Processing (IJIP)*, *5*(5), 590–598.

Guo, P., Liu, T., & Li, N. (2014). Design of automatic recognition of cucumber disease image. *Information Technology Journal*, *13*(13), 2129–2136.

Kebapci, H., Yanikoglu, B., & Unal, G. (2011). Plant image retrieval using color, shape and texture features. *The Computer Journal*, *54*(9), 1475–1490.

Li, J., Wu, W., Wang, T., & Zhang, Y. (2008). One step beyond histograms: Image representation using Markov stationary features. In *Proceedings of IEEE conference on computer vision and pattern recognition* (pp. 1–8): .

Llado, X., Oliver, A., Freixenet, J., Marti, R., & Marti, J. (2009). A textural approach for mass false positive reduction in mammography. *Computerized Medical Imaging and Graphics*, *33*, 415–422.

Meunkaewjinda, A., Kumsawat, P., Attakitmongcol, K., & Srikaew, A. (2008, May). Grape leaf disease detection from color imagery using hybrid intelligent system. In *Vol. 1. 2008 5th international conference on electrical engineering/electronics, computer, telecommunications and information technology* (pp. 513–516): IEEE.

Pantazi, X. E., Moshou, D., & Tamouridou, A. A. (2019). Automated leaf disease detection in different crop species through image features analysis and one class classifiers. *Computers and Electronics in Agriculture*, *156*, 96–104.

Patil, S. B., & Bodhe, S. K. (2011). Leaf disease severity measurement using image processing. *International Journal of Engineering and Technology*, *3*(5), 297–301.

Patil, J. K., & Kumar, R. (2017). Analysis of content based image retrieval for plant leaf diseases using color, shape and texture features. *Engineering in Agriculture, Environment and Food*, *10*(2), 69–78.

Pethybridge, S. J., & Nelson, S. C. (2015). Leaf doctor: A new portable application for quantifying plant disease severity. *Plant Disease*, *99*(10), 1310–1316.

Rothe, P. R., & Kshirsagar, R. V. (2015, January). Cotton leaf disease identification using pattern recognition techniques. In *2015 international conference on pervasive computing (ICPC)* (pp. 1–6): IEEE.

Schölkopf, B., Platt, J. C., Shawe-Taylor, J., Smola, A. J., & Williamson, R. C. (2001). Estimating the support of a high-dimensional distribution. *Neural Computation*, *13*(7), 1443–1471.

Šrobárová, A., & Kakalíková, L. (2007). Fungal disease of grapevines. *European Journal of Plant Science and Biotechnology*, *1*, 84–90.

Tax, D., & Duin, R. (2004). Support vector data description. *Machine Learning*, *54*(1), 45–66.

Vianna, G. K., Oliveira, G. S., & Cunha, G. V. (2017). A neuro-automata decision support system for the control of late blight in tomato crops. *World Academy of Science, Engineering and Technology, International Journal of Computer and Information Engineering*, *11*(4), 455–462 [Mar 2].

Zhang, S., Wu, X., You, Z., & Zhang, L. (2017). Leaf image based cucumber disease recognition using sparse representation classification. *Computers and Electronics in Agriculture*, *134*, 135–141.

CHAPTER 8
Tutorial V: Yield prediction

Contents

8.1 Introduction

Artificial neural networks (ANNs) have gained popularity an effective tool for offering solutions to a wide variety of different case studies of biological and agricultural background. Their effectiveness emanates from their ability to model complex relationships between observation data from sensors and predicted variables without relying on assumptions about the model structure hence they can predict the real nature of the nonlinear relation between input and output data. This allows the definition of arbitrary nonlinear relations that arise in real world problems like PA and are associated with crop status and other environmental factors (Uno et al., 2005).

Yield prediction is a major challenge in precision agriculture, closely associated to the adoption of best management practices, crop pricing and security. The yield prediction allows assessing the variability and the reasons that evoked this variability so that the parameters affecting yield including proper irrigation, fertilization, crop protection and field interventions to be applied site-specifically customized to the crops requirements. Various techniques and methodologies have been developed to predict crop yield in agriculture.

Intelligent Data Mining and Fusion Systems in Agriculture
https://doi.org/10.1016/B978-0-12-814391-9.00008-X

Drummond, Sudduth, Joshi, Birrell, and Kitchen (2003) employed more diverse methodology for calibrating ANNs with the same training set for predicting the yield. They concentrate on training performance, generalizing on new data, and exclusion on outlier data in order to decide which is the best methodology. The authors experimented with different variants of backpropagation for the training procedure, proving that all the forms of trainable neural networks performed significantly better than the linear ones, and that rprop was more effective than simple backpropagation.

Yu et al. (2010) proposed an group of feedforward ANNs to perform fertilization predictions. They indicate that two problems arise when a feedforward network is utilized in this way. Initially, they discovered that associating a maximal output yield before processing results in substantial regarding the rates of the fertilizer application. It has ben also concluded that the utilization of a single ANN, results in low forecast performance and generalization potential. The authors suggested an ANN where the input are the nutrient concentration and rate of fertilization while the prediction target is the yield. They calibrated various neural networks by using backpropagation through a bagging process and clustered the networks by using the k means algorithm. Following that, they picked up an ANN, representing each cluster to construct a group of ANNs. The combining weights were determined by using Lagrange multipliers with a set of constraints that force the sum of the weights to become equal to unity. Following that, a nonlinear objective function was constructed from the group output and applied nonlinear programming to find the optimal rate.

Kuwata and Shibasaki (2015) demonstrated a deep neural network (DNN) approach by using as input several parameters including, the canopy surface temperature, Absorbed Photosynthetically Active Radiation, water stress index and NDVI, to be fed to a DNN and a Support Vector Machine Regression. By employing the DNN, the RSME is equal to eight for predicting the yield in corn. You, Li, Low, Lobell, and Ermon (2017) applied a CNN combined with a Long-Short Term Memory (LSTM) network aiming to classify histograms emanating from multispectral remote sensing images. This CNN has attained the best RMSE values, equal to seven.

While considering different ANNs architectures that are found in literature, SOMs are the most applicable and effective machine learning tools, capable of offering solutions to arising nonlinear data analytics case studies (Marini, 2009). Furthermore, they are able to provide insight by functioning in a unsupervised way, similar to data clustering.

Data mining is an independent field of study that usually applies machine learning techniques. Pantazi et al. (2016) developed Supervised Self Organizing Maps methodology, seeking to perform an analysis on sensor data and construct an updated knowledge content. The presented approach assigned input nodes that were associated to the main parameters of wheat crop production cultivation. The SOM used that data to predict the wheat yield and productivity.

Some of the non-linear prediction models utilize traditional soil sampling (e.g. 1 sample per ha) and destructive laboratory methods which are laborious, time demanding and of high cost. An alternative approach proposed by Mouazen (2006) relies on high sampling resolution with a nondestructive, spectral soil sensor. This sensor has proven to estimate key soil parameters relevant to crop yield with variable accuracy, including soil organic matter content (SOMC), cation exchange capacity (CEC), moisture content (MC), total nitrogen (TN), pH, calcium (Ca), clay (CC), magnesium (Mg), organic carbon (OC), phosphorous (P), and plasticity index (PI) (Mouazen, 2014). There have been several approaches, focusing on data fusion of characteristic soil properties highly associated to crop growth indicators like NDVI to forecast yield in arable crops. An approach that would be interesting to follow concerns the visualization of the yield affecting parameters in order to discover correlations between them and the yield trend.

The current case study focuses on crop yield prediction by adopting a novel approach, relying on a hierarchical SOMs with enhanced supervised capability. The presented machine learning algorithms enable the fusion of high resolution multi-layer data on soil and on crop, by constructing a sensor fusion neural network model which learns to estimate the geo-spatial arrangement of wheat yield, with high performance in comparison with current techniques. These algorithms facilitate an enhanced visualization potential of soil, crop properties and yield productivity.

8.2 Materials and methods

In the current case study, three Self Organizing Map architectures including CPANN, SKN and XYF by employing Supervised Learning to find possible relations between precision farming observations and yield productivity levels. For the application of this methodology, physicochemical soil properties were acquired with the use of an on-line vis-NIR sensor which was assimilated with crop growth parameters by implementing a sensor fusion algorithm.

8.3 Experimental setup

The experimental field called Horn End (22 ha), was situated at Duck End Farm, in Wilstead, Bedfordshire, U.K. (Fig. 8.1). The field utilizes crop rotation for winter crops including wheat, barley and oil-seed rape. The soil type was determined as 'Haplic Luvisols'. Clay, clay loam, sandy clay loam and loam presence (USDA classification) characterized the textures of investigated soil samples. The terrain of the investigated area was mostly flat with an elevation range of 30–38 m, estimating by using the differential global positioning system (DGPS). The presented survey was carried out during 2013 cropping season.

8.3.1 Crop parameters affecting yield

The crop growth parameter NDVI, has been employed for the assessment of the related crop growth properties of the investigated field area (Rouse, Haas, Schell, & Deering, 1974). NDVI was estimated with the help of earth observation data provided by the UK-DMC-2 Disaster Monitoring Constellation for International Imaging, (DMCii) satellite on two different dates

Fig. 8.1 Horns End field as presented though a web mapping service (Pantazi, Moshou, Alexandridis, Whetton, & Mouazen, 2016).

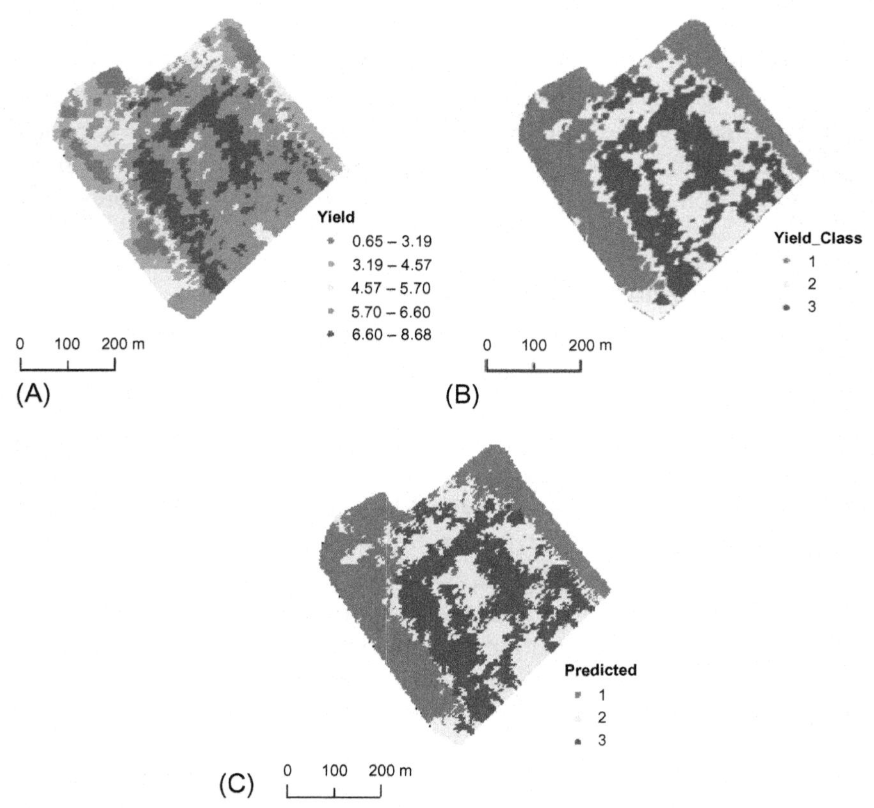

Fig. 8.2 Actual yield map of winter wheat in 2013 (A) and the predicted yield map which is formulated by categorizing the actual yield data into three isofrequency classes (B), in comparison to the predicted yield map, as results from the Kohonen network (SKN) (C) (Pantazi et al., 2016).

(2.05–3.06.2013) as illustrated in Fig. 8.2. The UK-DMC-2 was utilized for attaining multispectral images on three different bands including green, red and NIR area. The spatial resolution was estimated as 22 m, with a dynamic range of 14 bit.

The image calibration involved ortho-rectification, reflectance calibration of each band, and NDVI estimation, according to the provided equation:

$$\text{NDVI} = (\text{NIR} - \text{R})/(\text{NIR} + \text{R}) \quad (9.1)$$

where NIR denote the reflectance in the near-infrared and R the reflectance in the red region. The NDVI was recalculated to a 5×5 m grid, yielding a total of 8798 values.

Field data was acquired with the help of h a New Holland CX8070 combine harvester, carrying a yield sensor. The data collection acquisition took place during the 2013 summer period. For ensuring trustworthy yield estimations in the harvested area, special attention has been paid to the harvester so as to keep a full header during harvesting procedure. The yield was interpolated to match the same grid as the 5 × 5 m NDVI, yielding 8798 values.

8.3.2 Soil parameters affecting yield

By considering the launch of visible and near infrared (vis-NIR) spectrophotometers entering the market together with chemometrics software, the employment of vis–NIR spectroscopy has been widely adopted for assisting soil analysis. Further developments enabled the relative to vis-NIR spectroscopy applications to measure more accurately and effectively several soil parameters like the Moisture Content (MC) pH, Soil Organic Matter (SOMC), Total Nitrogen (TN) and Organic Carbon (OC) (Mouazen, Maleki, De Baerdemaeker, & Ramon, 2007).

A variety of calibration procedures have made possible the parallel acquisition of soil parameters estimations. Shibusawa, Made Anom, Sato, & Sasao, 2001 has presented an on-line vis-NIR (400–1700 nm) sensor for the estimation of pH, MC, NO_3-N, SOMC. Mouazen (2006) has proposed a simpler design compared to the one presented by Shibusawa et al. (2001) which was characterized by the lack of sapphire window optical configuration. The sensor system was capable of providing accurate estimations regarding the MC, TC, pH, TN, Ca, Mg, available P and cation exchange capacity (CEC) for several soil types across the European area (Kuang & Mouazen, 2013).

On-line soil estimations took place in the Horn's End field during crop harvesting period of summer 2013. Data acquisition was performed in parallel transects with mean velocity circa 1.5–2 km/h. A mobile AgroSpec fiber type, mobile, vis–NIR spectrophotometer (Tec5 Technology for Spectroscopy, Germany) with a spectral range between 305 and 2200 nm was used for soil diffuse spectra acquisition (Kuang & Mouazen, 2013).

The source soil spectra data has been registered and retained for temporal analysis. While performing the online data acquisition, soil samples were gathered for the assessment of sampling correctness of the relevant soil parameters. A collection of 60 soil samples were acquired and subjected to laboratory testing following standard protocols. Partial Least Squares Regression (PLSR) has been employed for the estimation of soil properties, specifically Ca, Mg, total N, CEC, OC, MC and pH. Further details are provided by Kuang and Mouazen (2013). The predicted values were obtained after interpolation on 5 m × 5 m grid, matching the sampling locations of 8798 points that correspond to the NDVI and crop yield.

8.3.3 Prediction of crop yield

For the prediction of crop yield three hierarchical SOM models including SKN, XYF and CPANN were employed by utilizing Matlab (MathWorks, Natick, Massachusetts). The detailed scientific background of the aforementioned models is given in Chapter 2, Section 2.9.

The dataset corresponding to the investigated soil parameters were acquired with the help of an online soil sensor. Then, they were combined in the same vector with the NDVI values obtained by the satellite imagery and historic yield data, resulting to 8798 feature vectors. The feature vectors were formed through the fusion of both soil and crop properties.

To prevent bias occurring during the training procedure, the data vectors that were attained from fusion, were transformed in a way that their mean variant was removed and the standard deviation was equal to one. The steps followed included the subtraction of the mean vector and the division with the help of the standard deviation operating on each one of the training dataset samples. To perform the yield prediction, the fusion vectors were supplied to the three models. The values of the yield were grouped in three classes each one containing 2933 samples in ascending mode, belonging to low, medium and high yield class.

8.4 Results and discussion

8.4.1 Accuracy of yield prediction with supervised models

The three hierarchical SOM models were calibrated with the 8798 data vectors as input and the isofrequency yield classes as targets. To evaluate whether the applied models generalize properly or not, cross validation procedure was followed by retaining a random selection of 25% of all samples for testing and training on the rest of the samples. Table 8.2 depicts the cross validation performance of the three employed models. For obtaining an operational prediction model the procedure of independent validation was followed by retaining 1000 samples for testing and the rest for calibration with the remaining 7798 samples. The outcomes of the independent validation are demonstrated in Table 8.1.

Table 8.1 Statistics of the data samples of the soil parameters corresponding to the training and independent validation datasets (Pantazi et al., 2014)

	Calibration set (7798 vectors)				Validation set (1000 vectors)			
Parameter vectors	**Min**	**Max**	**Mean**	**SD**	**Min**	**Max**	**Mean**	**SD**
NDVI	0.413	0.721	0.620	0.0702	0.423	0.720	0.619	0.072
Ca	6.757	57.854	26.317	10.106	7.690	55.282	26.324	10.286
CEC	9.514	20.893	13.702	2.229	9.497	20.642	13.716	2.307
MC	11.669	24.611	16.617	2.269	11.695	24.556	16.615	2.285
Mg	0.195	2.047	1.062	0.270	0.242	2.078	1.059	0.279
OC	1.250	2.284	1.753	0.170	1.323	2.214	1.750	0.169
P	1.170	2.860	1.951	0.207	1.341	2.729	1.954	0.207
pH	4.304	8.159	6.031	0.567	4.604	7.939	6.027	0.539
TN	0.158	0.364	0.2450	0.034	0.168	0.344	0.248	0.034
Y2011	3.611	7.650	5.437	0.746	3.748	7.562	5.425	0.725
Y2012	0.224	9.410	0.4.346	1.084	0.565	8.626	4.353	1.113

Table 8.2 Cross validation performances of the applied hierarchical models (SKN, XYF, and CPANN), consisting of 30 × 30 neurons for wheat yield prediction, using as predictors the normalized NDVI, sensor obtained soil parameters and yield data from previous years

	Network prediction (%)		
Actual yield isofrequency class	**Low**	**Medium**	**High**
SKN			
Low	91.3	7.23	1.4
Medium	7.84	70.54	21.61
High	1.26	15.62	83.12
CP-ANN			
Low	91.48	7.43	1.09
Medium	10.19	68.56	21.24
High	1.87	23.26	74.86
XY-F			
Low	92.15	7.09	0.75
Medium	8.9	72.48	18.62
High	1.29	20.56	78.14

The optimal results for wheat yield prediction regarding cross validation and independent validation were achieved by the SKN model for predicting the low yield class. The performance of the prediction was 91.3% both for the case of cross validation and independent validation as summarized in Tables 8.2 and 8.3. For the case of medium yield class, the performance was decreased compared to that of the low yield class, ranging from 70.54%, achieved by SKN during cross validation procedure as high as 85.15% achieved by XY-F during independent validation. The yield prediction for the high yield class was achieved with SKN networks, reaching a prediction level of 83.12% during cross validation procedure and 81.48% during independent validation.

Being compared to all cases, the optimal prediction performance is achieved for the low yield class, where intervention is required such as high fertilization application, while in the cases of satisfactory yield, no intervention is needed. Halcro, Corstanje, and Mouazen (2013) proposed a fertilizer recommendation that was inversely proportional to the yield, assigning an increased nitrogen fertilizer dose to the poorer fertility field areas. However, the same authors concluded that such suggestion was not for the most cases effective.

Table 8.3 Independent validation performances of the applied hierarchical models (SKN, XYF and CPANN), consisting of 30 × 30 neurons for wheat yield prediction, using as predictors the normalized NDVI, sensor obtained soil parameters and yield data from previous years

	Network prediction (%)		
Actual yield Isofrequency class	**Low**	**Medium**	**High**
SKN			
Low	91.3	6.96	1.74
Medium	10.87	64.35	24.78
High	1.54	16.98	81.48
CP-ANN			
Low	90.09	9.29	0.62
Medium	9.57	69.86	20.58
High	2.11	24.40	73.49
XY-F			
Low	87.91	11.21	0.89
Medium	5.76	85.15	9.09
High	2.11	38.67	59.21

The assumption that poor field areas are associated to lowest yield productivity has proven incorrect, due to the fact that as soil physical factors, more specific high water capacity and poor soil drainage have a critical impact in crop growth and yield productivity. However, the supplementation with extra nutrient in a location with a water logging problems highly unlikely to result in an enhancement of crop growth and restoring the yield to normal levels.

8.4.2 Yield maps

An illustration of both measured and predicted yield maps is given in Fig. 8.2. The predicted yield is allocated in three levels denoted with red, yellow and blue, for high, medium and low yield. The patterns corresponding to the actual and the predicted yield demonstrate high resemblance between them as it is illustrated in Fig. 8.2B and C. The kappa value was utilized in an attempt to quantify the similarity between the actual and the predicted yield maps (Congalton, 1991). The estimated kappa has reached 0.8386, proving the high performance of predicting the actual yield classes depicted in map 2c.

The calibration data for the hierarchical SOM form a specific topology, that takes advantage of the vectors' proximity for clusters formulation. The most prominent patterns in the calibration dataset force a specific topology leading to emergence of visual relationships connecting the components of the calibration data. In the specific application of yield prediction, the topological patterns allow the visualization of causal relationships connecting parameters affecting yield and their inter-correlations.

The SOM lattice facilitates visual assessment of the SOM parameters in a color pattern array in which the relative magnitudes are displayed in an ascending order in a color range between blue and red. Figs. 8.3, 8.4, and 8.5 demonstrate the three employed SOM models following the training vectors. The first subplot depicts the NDVI SOM parameter, while the rest of the vector parameters are associated to Ca, Mg, P, MC, TN, OC, and pH obtained with the help of the soil spectral sensor utilized by the Mouazen, 2006. The last two subplots illustrated in Fig. 8.3 represent an overview

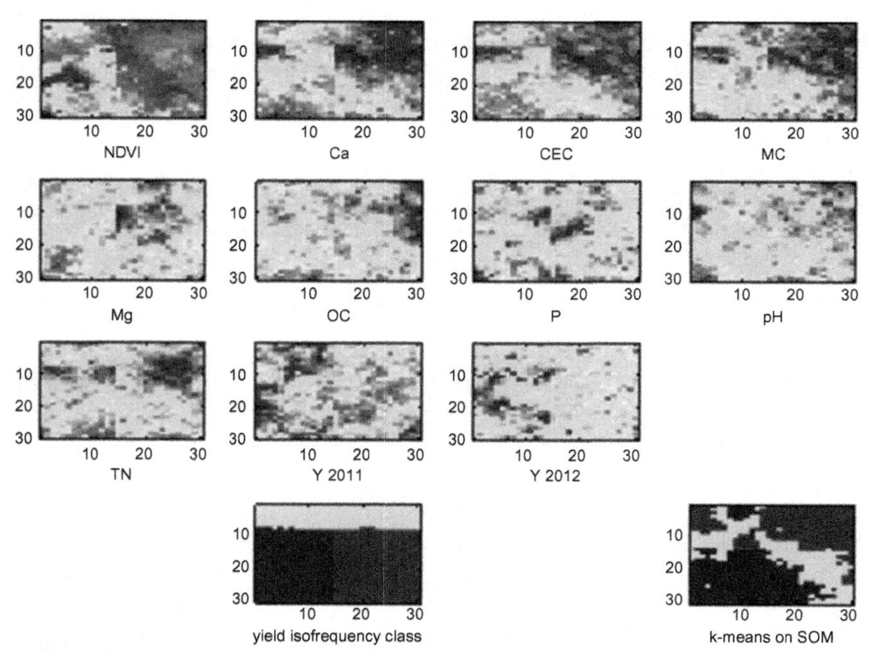

Fig. 8.3 SKN predicted maps of NDVI, soil properties, historic yields during 2011–2012 period, actual yield classes and SOM predicted yield classes. The relative magnitudes are displayed in an ascending order in a color range between blue (in monochrome it is given in black) and red (in monochrome it is given in deep gray) (Pantazi et al., 2016).

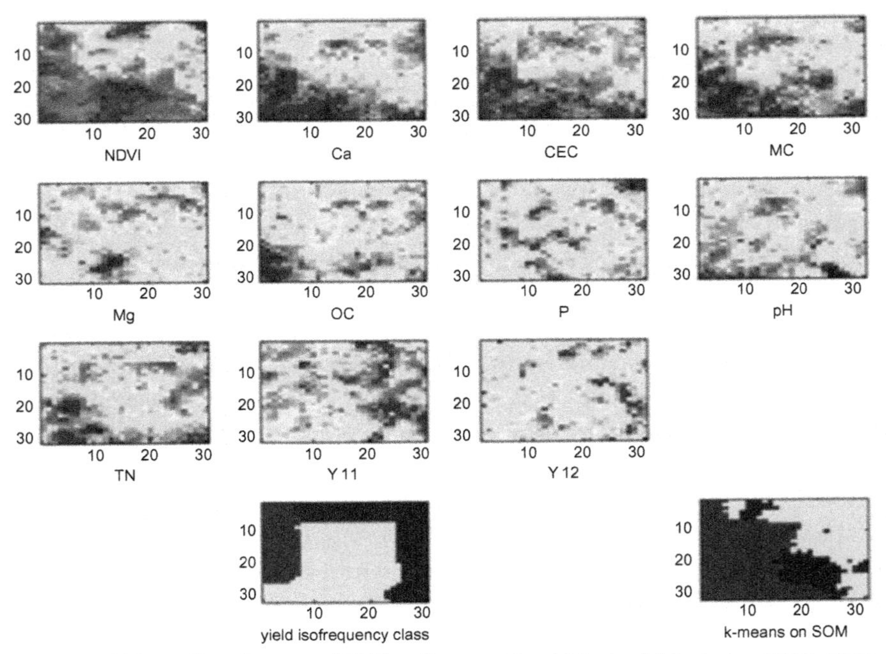

Fig. 8.4 XY-F predicted maps of NDVI, soil properties, historic yields during 2011–2012 period, actual yield classes and SOM predicted yield classes. The relative magnitudes are displayed in an ascending order in a color range between blue (in monochrome it is given black) and red (in monochrome it is given in deep gray) (Pantazi et al., 2016).

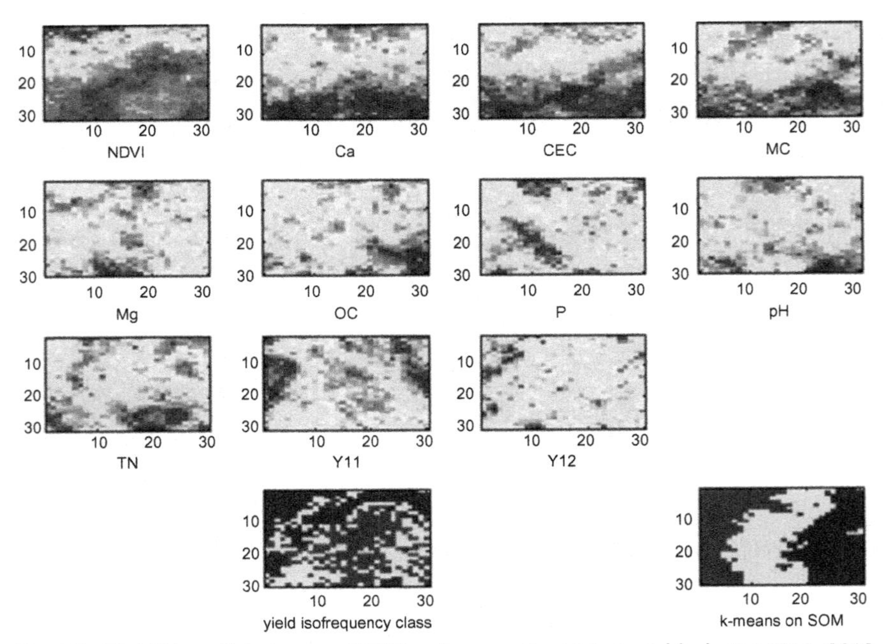

Fig. 8.5 CP-ANN predicted maps of NDVI, soil properties, historic yields during 2011–2012 period, actual yield classes and SOM predicted yield classes. The relative magnitudes are displayed in an ascending order in a color range between blue (in monochrome it is given in black) and red (in monochrome it is given in deep gray) (Pantazi et al., 2016).

historic yield during the 2011–2012 period, corresponding to classes of the actual yield and to classes of the SOM predicted yield and allocated in three groups by employing the k-means algorithm (MacQueen, 1967). Regarding the SKN model, all the training data including the class component, are used to train the network, resulting in a co-development of the input and the output parameters leading to the emergence a vertical line in the SOM parameters layers.

By accessing Figs. 8.3, 8.4, and 8.5, the NDVI pattern demonstrates significant correspondence to the predicted yield The majority of the soil parameters show a similar pattern to the lower yield class, while high values of some soil parameters and soil moisture content are observable in the low yield regions. After a conversation about this issue with the farmer, following field surveying, it has been found that a water logging problem affected negatively the low yield areas that appeared to be abundant in nutrients.

Comparing the three models that have been applied for the presented case study, each network results in a different topological structure, which can be explained from the initialization and the learning dynamics. By observing the pattern in different SOM parameters, the tendencies appear to be similar. However, taking into account that each vector is associated to different yield structure, the SOM maps manage to depict effectively that spatially dispersed point in the field can belong to the same cluster. Consequently, the added value of the SOM maps concerns the provision of knowledge relevant to the parameters that have an impact on yield productivity in an accurate manner due to the topology preservation property, as it has been explained above.

The conclusions derived from the current application can be characterized as innovative compared to relevant studies that have been published previously. In literature, there have been employed several yield prediction techniques relying on solely on limited soil properties (Papageorgiou, Aggelopoulou, Gemtos, & Nanos, 2013), attained with common soil sampling as compared to the online soil spectral scanner which is a nondestructive approach. Crop parameters have been used by Uno et al. (2005) expressed as vegetation indices for yield prediction. The synergy of soil and crop features as they show strong correlation with the yield historical data leads to a more explainable crop yield prediction. On the other hand, it must be noticed that there is no need to make the hypothesis regions of high fertility are directly associated to increased yield and high quality productivity. Hence, a more integrated approach has to be followed for accurate predictions, where field sensing and practical insight obtained from farmers has to be acquired prior to formulation of treatment actions concerning fertilizers application.

The introduced hierarchical map architectures appear as reliable techniques for offering solutions when classification problems arise thanks to their ability to tackle non-linear classification problems. Significant asset of these neural network classifiers is connected to their ability to perform local learning, contrary to the global updates necessary for training common ANN architectures, like the MLP.

References

Congalton, R. G. (1991). A review of assessing the accuracy of classifications of remotely sensed data. *Remote Sensing of Environment*, *37*(1), 35–46.

Drummond, S. T., Sudduth, K. A., Joshi, A., Birrell, S. J., & Kitchen, N. R. (2003). Statistical and neural methods for site-specific yield prediction. *Transactions of ASAE*, *46*(1), 5.

Halcro, G., Corstanje, R., & Mouazen, A. M. (2013). Site-specific land management of cereal crops based on management zone delineation by proximal soil sensing. In J. Stafford (Ed.), *Precision Agriculture 2013, Proceedings of the 10th European Conference on Precision Agriculture* (pp. 475–481). Wageningen Academic Publishers: Wageningen.

Kuang, B., & Mouazen, A. M. (2013). Effect of spiking strategy and ratio on calibration of on-line visible and near infrared soil sensor for measurement in European farms. *Soil and Tillage Research*, *128*, 125–136.

Kuwata, K., & Shibasaki, R. (2015). Estimating crop yields with deep learning and remotely sensed data. In *2015 IEEE international geoscience and remote sensing symposium (IGARSS)* (pp. 858–861): IEEE.

MacQueen, J. B. (1967). Some methods for classification and analysis of multivariate observations. *Proceedings of 5th Berkeley Symposium on Mathematical Statistics and Probability* (pp. 281–297) University of California Press. MR 0214227. Zbl 0214.46201. Retrieved 2009-04-07.

Marini, F. (2009). Artificial neural networks in foodstuff analyses: Trends and perspectives—A review. *Analytica Chimica Acta*, *635*(2), 121–131.

Mouazen, A. M. (2006). *Soil survey device*. International publication published under the patent cooperation treaty (PCT) World Intellectual Property Organization, International Bureau. International Publication Number: WO2006/015463; PCT/BE2005/000129; IPC: G01N21/00; G01N21/00.

Mouazen, A. (2014). Active learning system for autonomous combined biotic and abiotic crop stress detection. In *RHEA Conference, 21–23 May, Madrid, Spain* (pp. 167–176).

Mouazen, A. M., Maleki, M. R., De Baerdemaeker, J., & Ramon, H. (2007). On-line measurement of some selected soil properties using a VIS-NIR sensor. *Soil and Tillage Research*, *93*, 13–27.

Pantazi, X. E., Moshou, D., Alexandridis, T., Whetton, R. L., & Mouazen, A. M. (2016). Wheat yield prediction using machine learning and advanced sensing techniques. *Computers and Electronics in Agriculture*, *121*, 57–65.

Papageorgiou, E. I., Aggelopoulou, K. D., Gemtos, T. A., & Nanos, G. D. (2013). Yield prediction in apples using Fuzzy Cognitive Map learning approach. *Computers and Electronics in Agriculture*, *91*, 19–29.

Rouse, J., Jr., Haas, R. H., Schell, J. A., & Deering, D. W. (1974). Monitoring vegetation systems in the Great Plains with ERTS. *NASA*.
Shibusawa, S., Made Anom, S. W., Sato, H. P., & Sasao, A. (2001). Soil mapping using the real-time soil spectrometer.In G. Gerenier & S. Blackmore (Eds.), *agro Montpellier: Vol. 2*, ECPA 2001 (pp. 485–490). Montpellier, France.
Uno, Y., Prasher, S. O., Lacroix, R., Goel, P. K., Karimi, Y., Viau, A., et al. (2005). Artificial neural networks to predict corn yield from compact airborne spectrographic imager data. *Computers and Electronics in Agriculture*, *47*(2), 149–161.
You, J., Li, X., Low, M., Lobell, D. B., & Ermon, S. (2017). Deep gaussian process for crop yield prediction based on remote sensing data. In *Proceedings of the thirty-first AAAI conference on artificial intelligence* (pp. 4559–4565).
Yu, H., Liu, D., Chen, G., Wan, B., Wang, S., & Yang, B. (2010). A neural network ensemble method for precision fertilization modeling. *Mathematical and Computer Modelling*, *51*, 1375–1382.

CHAPTER 9

Tutorial VI: Postharvest phenotyping

Contents

9.1 Introduction

A critical factor concerning the production of vegetables is the stage that are harvested because it is closely related to their nutrients concentration which is affected by age, because of senescence. At present, the absence of reliable methods which are capable of assessing lettuce plants senescence which consequently lead to lack of a standardized protocol for labeling their different growth stages. Among different lettuce hybrids the head size differs a lot, as a result it cannot be considered as a robust indicator of the growth stage. Due to this fact, a more trustworthy indicator is needed to characterize the growth stage according to the senescence stage.

A commonly used approach for assessing senescence in lettuce plants is proximal sensing in the form of spectroscopy or fluorescence sensors. Chlorophyll fluorescence is a standard method for assessing the status of plant photosynthesis in a nondestructive way. Frequent fluorescence based applications concern crop monitoring for evaluating possible resistance to ambient stresses and for optimizing glasshouse vegetable quality and postharvest vegetable management. It is common knowledge that leaves' dark-adaption leads to a fluorescence transient response when exposed to light. The abrupt variations in fluorescence response that follow light excitation induce to a peak and are characteristic as an indicator of photosynthesis differences that

Intelligent Data Mining and Fusion Systems in Agriculture
https://doi.org/10.1016/B978-0-12-814391-9.00009-1

are representative of each plant. The light radiation taken by plants is transformed into chemical energy, heat and fluorescence.

SOMs applications have grown immensely in the last ten years in different sectors and currently they belong to the most important and efficient AI tools offering solutions in an unsupervised fashion, without requiring any target data (Marini, 2009). To increase their capacity for handling supervised problems, specific extension have been applied to them. To make this possible, architectures like CP-ANNs, which are SOM variance, (Zupan, Novic, & Gasteiger, 1995), have been introduced by appending an output layer on the original SOM.

For classification, CP-ANNs are regarded effective methods for solving non-linear classification problems. Different variations of CP-ANNs have led to the formation of new supervised models including associated learning training procedures like SKNs and XY-Fs (Melssen, Wehrens, & Buydens, 2006). The current case study presents a supervised learning approach with the help of SOMs aiming to build a classifier for fluorescence kinetics data aiming to assess different harvesting stages in lettuce plants. To accomplish this, two distinct fluorescence kinetics curves expressions were introduced; one associated to fluorometer variables while the other was the outcome of fitting polynomials on the Kautsky Curves.

9.2 Experimental setup

Chlorophyll fluorescence is regarded an important indicator for improving plant breeding and assess fast plant tolerance and resistance in crop breeding programs. Particularly, regarding the case of lettuce crop, the quality indicators and their and nutrient content varies considerably according to the growth stage. For the presented case study, different hybrids of lettuce plants including Atoll, Mastamar, and Starfighter of Batavia type as well as hybrids of Picos FM, Picos CLX and Bacio, of Romana type as illustrated in Fig. 9.1A were grown under heated glass greenhouse conditions from 15/10 to 27/12/2012. The harvesting stages were at baby, immature and mature at the 46th, 60th, and 70th day of growth, respectively.

The chlorophyll fluorescence parameters were extracted for a couple of the plant middle leaves corresponding to each hybrid, belonging to a distinct harvesting stage. The fluorescence kinetics parameters exhibited characteristic profiles that can enable the discrimination of the harvesting stages. FluorPen FP 100-MAX-LM of SCI (Fig. 9.1B), was the used sensor, capable

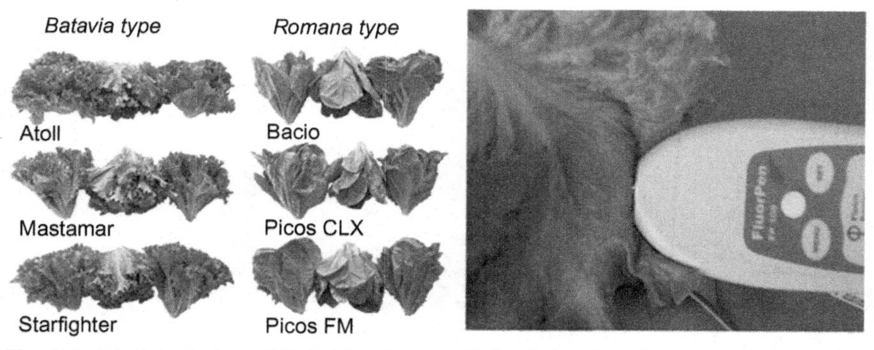

Fig. 9.1 (A) Batavia type (Atoll, Mastamar and Starfighter) and Romana type (Bacio, Picos CLX, Picos FM) lettuce hybrids. (B) Fluorometer FluorPen FP 100-MAX-LM.

of acquiring fluorescence kinetics with the help of OJIP protocol This protocol produces parameters from the fluorescence transient (Strasser et al., 2006).

The fluorescence parameters were used as calibration data for constructing the supervised Self Organizing Maps (SOMs) models targeting on the different harvesting stage assessment. Fluorescence kinetics have been applied previously by Moshou, Wahlen, Strasser, Schenk, and Ramon (2003) for investigating the mealiness occurrence in apples with the help of SOMs aiming to differentiate the mealiness levels.

9.3 Fluorescence parameters

A variety of fluorescence kinetics parameters have been already utilized in the past. The abrupt transients in fluorescence emissions, appearing when inducing photosynthesis following the exposition of a dark-adapted leaf to light, can serve as signatures that can accurately identify the specific condition of the photosynthetic health status (Fig. 9.2).

With the help of fluorometer excitation source, the fluorescence is induced by ultra-bright Light Emitting Diodes (LED) exhibit a peak response at 650 nm. Chlorophyll fluorescent transients were acquired by employing a photocell being filtered by a high-pass filter at 720 nm. The acquisition duration was set equal to 1 s with a sampling step of 10 μs for the period between 0 ms and 2 ms and following that, the sampling step was set at 1 ms, yielding 1200 values for the whole duration. The fluorescence transients, known as Kautsky curves are depicted in Fig. 9.2.

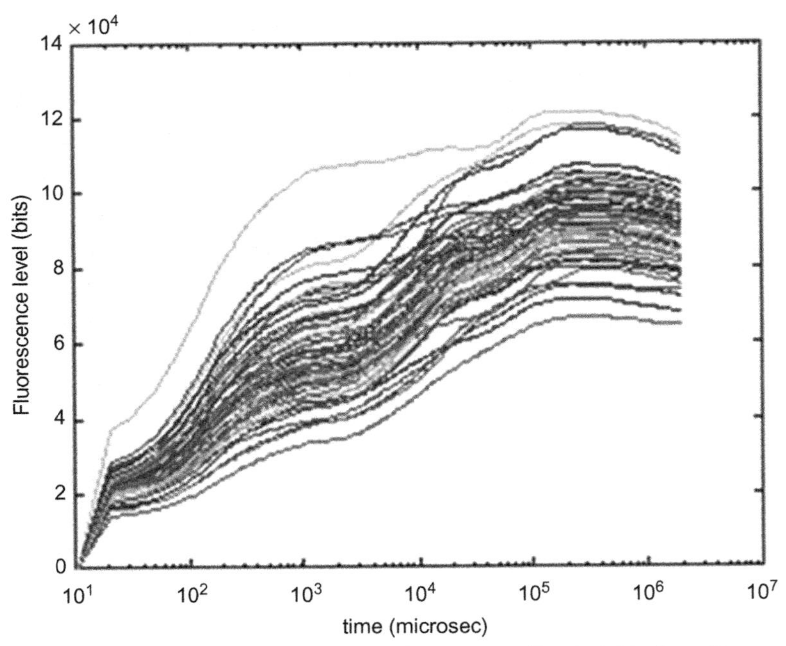

Fig. 9.2 Kautsky Curves resulting from the fluorometer (Pantazi, Moshou, Kasampalis, Tsouvaltzis, & Kateris, 2013).

9.4 Data analysis

9.4.1 FluorPen 100 fluorescence parameters

The fluorometer extracts in an automatic manner individual Kautsky curves characteristics. The attained parameters for the current case study are provided in Table 9.1. The fluorescence transients followed a normalization process through dividing the fluorescence value F_0 obtained after 50 μs. This value is associated to the initial fluorescence F_0, and is defined as the fluorescence response associated with the oxidized form of the electron acceptors.

The performance of normalization, alleviates random variations between the measurements caused by ambient factors, which do not characterize of the inner plant properties. Typical examples of these disturbing factors concern the perturbation of the excitation light level due to the battery charge depletion, and variability of the excited leaf area induced by the leaf curvature and textural inhomogeneities, for example lenticels, were minimized. Fluorescence transient response was sampled with high sampling frequency, yielding a large dataset from each acquisition session. Due to the amount of

Table 9.1 Fluorescence transient parameters produced by fluorometer

Formula abbreviation	Formula explanation
1. Bckg	Background
2. F_0	$F_{50}\mu s$, fluorescence intensity at 50 μs
3. F_J	Fluorescence intensity at J-step (at 2 μs)
4. F_i	Fluorescence intensity at i-step (at 60 μs)
5. F_M	Maximal fluorescence intensity
6. F_V	F_M - F_0 (maximal variable fluorescence)
7. V_J	$(F_J - F_0)/(F_M - F_0)$
8. V_i	$(Fi - F_0)/(F_M - F_0)$
9. F_M/F_0	
10. F_v/F_0	
11. F_v/F_M	
12. M_0 or $(dV/dt)_0$	$TR_0/RC - ET_0/RC = 4\ (F_{300} - F_0)/(F_M - F_0)$
13. Area	Area between fluorescence curve and F_M (background subtracted)
14. Fix Area	Area below the fluorescence curve between $F_{40}\mu s$ and $F_1 s$ (background subtracted)
15. S_m	Area/(F_M - F_0) (multiple turn-over)
16. S_s	the smallest SM turn-over (single turn-over)
17. N	S_M, M_0. ($1/V_J$) turn-over number Q_A
18. φ_{Po} or TR_0/ABS	$(1 - F_0)/F_M$ (or F_V/F_M)
19. ψ_0 or ET_0/TR_0	$1 - V_J$
20. φ_{Eo} or ET_0/ABS	$(1 - (F_0/F_M)).\ \psi_0$
21. φ_{DIo}	$1 - \boldsymbol{\varphi_{Po}} - (F_0/F_M)$ where $DI_0 = ABS- TR_0$
22. φ_{Pav}	$\boldsymbol{\varphi_{Po}}$ (S_M/t_{FM}), t_{FM} = time to reach F_M (in μs)
23. PI_{ABS}	(RC/ABS). $(\varphi_{Po}/(1- \varphi_{Po}))$. $(\psi_o/(1- \psi_o))$.
24. ABS/RC	M_0. $(1/V_J)$. $(1/\varphi_{Po})$
25. TR_0/RC	M_0. $(1/V_J)$
26. ET_0/RC	M_0. $(1/V_J)$. ψ_o
27. DI_0/RC	(ABS/RC) – (TR_0/RC)

data, it would be beneficial to retain a reduced subset of parameters, while at the same time, reserving the same level of information concerning discrimination as that of the original transient responses. The fluorescent transients F expressed in bits were modeled by a by a 10th order polynomial against log (time).

$$F = \left(\beta_0 + \beta_1 x + \beta_{10} x^{10}\right) \text{ with } x = \log(\text{time}) \tag{9.1}$$

The regression coefficients from β_0 to β_{10} formed an 11-dimensional vector that was calibrated to produce the classification model. The rank of the polynomial was selected based on the fitting performance.

9.5 Results and discussion

The CPANN, SKN and XYF models were calibrated with the fluorescence kinetics parameters produced by the fluorometer the polynomial Kautsky Curves modeling so as identify the 3 different harvesting stages of lettuce plants. The data was formed by 210 lettuce plants, consisting of 60 first harvesting stage plants, 60 s harvesting stage plants and 90 third harvesting stag plants. Cross validation was employed in order to access how the networks can cope with unforeseen data by generalizing their learned associations to new data when they are used operationally. This was achieved by producing four groups in random and the iterating the training and testing procedure with all the possible combinations consisting of three groups for training and the remaining group for validation. The outcomes of testing different architectures concerning the size of the SOM are given in Tables 9.2 and 9.3. A variety of map sizes were validated including 3×3, 5×5, 8×8, 10×10 and increasing to 30×30 with a step of 5. The grid of SOMs was square. A profound estimation concerns the improvement of results proportionally to the size of the network. An additional estimation is relevant to the utilization of the polynomial features that lead to improvement of generalization performance of the small size networks. The above estimation confirms the fact the polynomial features are able to capture information from the Kautsky Curves that were not modeled from the fluorometer parameters. To be able to evaluate the specific impact of each of the

Table 9.2 Recognition results comparison concerning the actual recognition of the three growth stages produced by the three network architectures when tested with the fluorescence parameters from Table 9.1

Type of used network	Successful recognition of actual growth at stage 1	Successful recognition of actual growth at stage 2	Successful recognition of actual growth at stage 3
Xyf 30×30	100%	95%	100%
Cpann 30×30	100%	95%	100%
Skn 30×30	100%	96.7%	100%

Table 9.3 Recognition results comparison concerning the actual recognition of the three growth stages produced by the three network architectures extracted from the parameters of the polynomials that approximated the Kautsky curves

Type of used network	Successful recognition of actual growth at stage 1	Successful recognition of actual growth at stage 2	Successful recognition of actual growth at stage 3
Xyf 30×30	100%	98.3%	97.8%
Cpann 30×30	100%	96.7%	96.7%
Skn 30×30	98.3%	98.3%	98.9%

fluorescence based features used by the Self-Organizing Map, the average of the weight values of the SOM neurons that are assigned to individual classes have been estimated. The parameters of high contribution will demonstrate a trend to acquire larger mean weight values due to contributing positively to acquisition of specific class in combination with specific parameter. The mean weights belonging to each class are depicted in Fig. 9.3.

While observing the Tables 9.2 and 9.3, both classifiers have specific advantages with respect to classification of individual classes. More analytically, the classifier that uses fluorometer features is particularly effective for

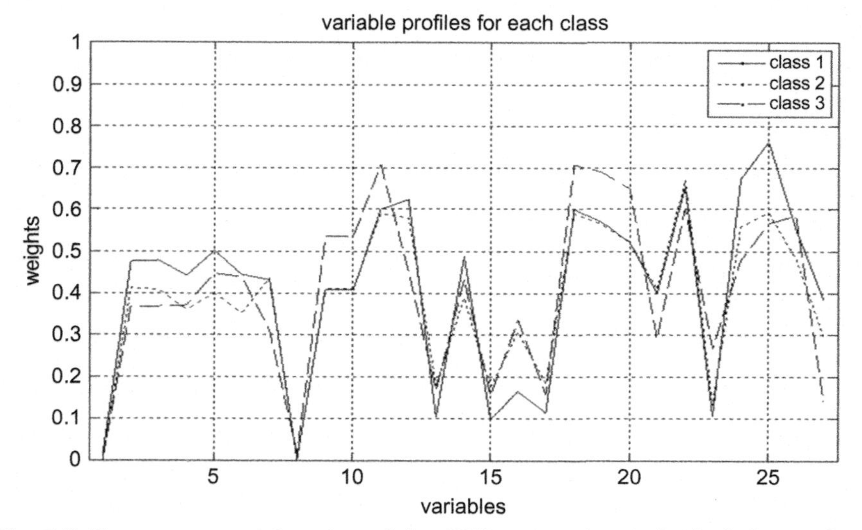

Fig. 9.3 The average weight value of the SOM assigned to individual classes that correspond to the three discrete harvesting stages (Pantazi et al., 2013).

predicting classes 1 and 3 and the opposite for class 2. Regarding the network architectures with dimensions 10×10 and 20×20, the accuracy of both classifiers derived from fluorescence polynomial features performed worse.

From Fig. 9.3 we can deduce that the variables which are more relevant for the separation in three harvesting stages of lettuce plants are the 25th variable (ET_0/RC) corresponding to the first harvesting stage, the 7th variable (V_J) corresponding to the second harvesting stage and 9th, 10th, and 18th variable (F_M/F_O), (F_v/F_O), (TR_0/ABS)) corresponding to the third harvesting stage.

To perform a visual assessment of the correlation of the fluorescence variables to the class separation, the first principal components (capturing the 95.65% of the global information) extracted from the SOM weights are depicted in Fig. 9.4. The associated variables show a clear classes separability since they are located in separate plane. The accurate classification outcome provided by Supervised Kohonen Network comprising of 30 neurons reached a 100–96%, 67–100% for classes 1-2-3 respectively. A comparison of the mentioned results with Tables 9.2 and 9.3 leads to the conclusion that

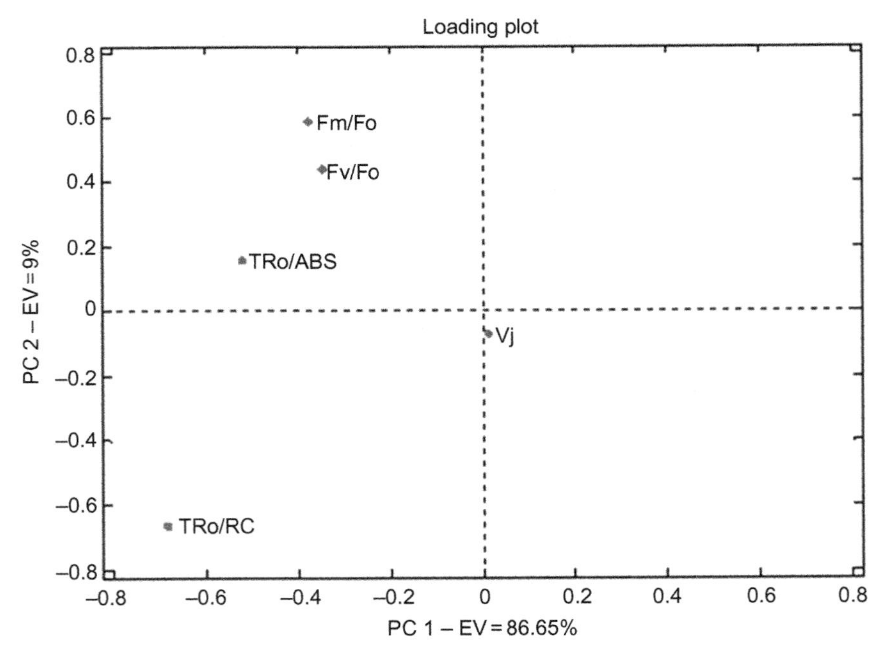

Fig. 9.4 Loading plot corresponding to the first two principal components estimated from the SOM weights (Pantazi et al., 2013).

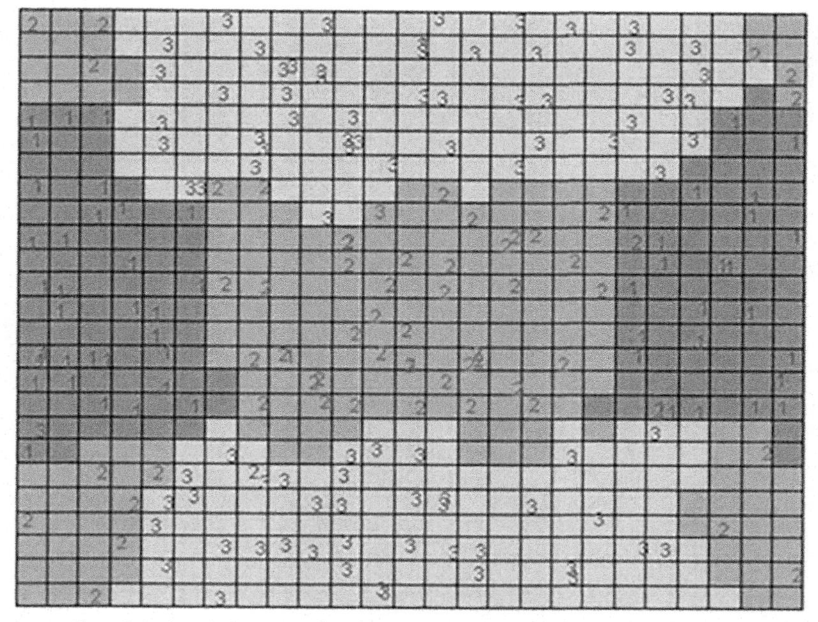

Fig. 9.5 Class labels of the SOM according to color shown together with the actual testing dataset super-imposed (Pantazi et al., 2013).

the chosen variables provide a promising result without the help of the fluorescence variables.

In Fig. 9.5, the color labeling denotes the class label determined from the training set data for the first, second and third harvesting stage. The appended digits show the actual sizes of the testing data. It is evident that the majority of the estimated classes from the testing dataset are estimated accurately with a minor percentage of misclassified denoted with different color.

The 2nd, 3rd, 4th, 5th, and 6th parameters show high level similarity to the degree that they appear identical. This explains the fact that the means corresponding to the three harvesting stages of the lettuce show not significant phenotypic difference concerning physiological mechanisms activity expressed by the fluorescence parameters, given in Table 9.1. Parameters 11, 18, 19, 20, and 26 demonstrate the same behavior and the condition appears similar between the 9th and 10th parameter as well. The 26th and 27th parameters are regarded as complementary compared to the second harvesting stage, explaining their high contribution to the correct classification outcome of this class (Fig. 9.6).

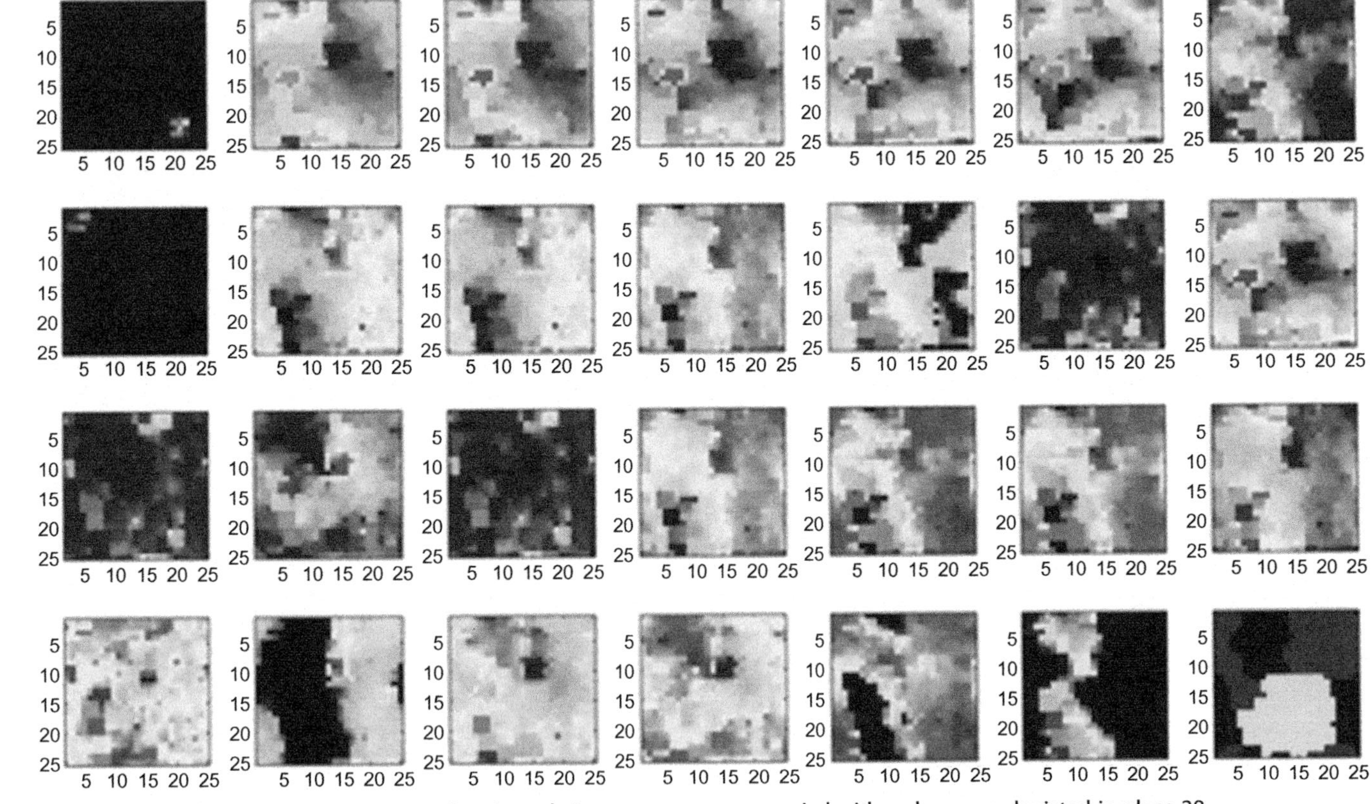

Fig. 9.6 The 27 fluorescence parameters showing relative response accompanied with a class map depicted in place 28.

References

Marini, F. (2009). Artificial neural networks in food analysis: Trends and perspectives. *Analytica Chimica Acta, 635*, 121–131.

Melssen, W., Wehrens, R., & Buydens, L. (2006). Supervised Kohonen networks for classification problems. *Chemometrics and Intelligent Laboratory Systems, 83*, 99–113.

Moshou, D., Wahlen, S., Strasser, R., Schenk, A., & Ramon, H. (2003). Apple Mealiness. Detection using fluorescence and self- organizing maps. *Computers and Electronics in Agriculture, 40*, 103–114.

Pantazi, X. E., Moshou, D., Kasampalis, D., Tsouvaltzis, P., & Kateris, D. (2013). Automatic detection of different harvesting stages in lettuce plants by using chlorophyll fluorescence kinetics and supervised self organizing maps (SOMs). In *International conference on engineering applications of neural networks* (pp. 360–369). Berlin, Heidelberg: Springer.

Strasser, F., Luftner, D., Possinger, K., Ernst, G., Ruhstaller, T., Meissner, W., et al. (2006). Comparison of orally administered cannabis extract and delta-9-tetrahydrocannabinol in treating patients with cancer-related anorexia-cachexia syndrome: A multicenter, phase III, randomized, double-blind, placebo-controlled clinical trial from the Cannabis-In-Cachexia-Study-Group. *Journal of Clinical Oncology, 24*(21), 3394–3400.

Zupan, J., Novic, M., & Gasteiger, J. (1995). Neural networks with counter-propagation learning strategy used for modelling. *Chemometrics and Intelligent Laboratory Systems, 27*, 175–187.

CHAPTER 10

General overview of the proposed data mining and fusion techniques in agriculture

Contents

10.1 Practical benefits for agriculture field

The current work proposes advanced remote and proximal soil and crop sensing, fusion and information management combined with high efficiency application methods (e.g., row fertilizer application, and precision spraying). By adopting the proposed technologies, the future farms will have to integrate and embrace not only the proposed but also to develop new technologies for several reasons:

(i) to ensure food security
(ii) to enhance intensive but sustainable crop production
(iii) to contribute to an increase to farm productivity to meet the demand of an increasing population under the anticipated climate change

Intelligent Data Mining and Fusion Systems in Agriculture
https://doi.org/10.1016/B978-0-12-814391-9.00010-8

(iv) to introduce ICT to agriculture and obtain the benefits and attract young people,

(v) to reduce environmental impacts of agriculture related to pesticides and nutrients

(vi) to fully integrate public goods produced by agriculture into local and regional contexts via integrated information systems and decision making of stakeholders at the regional level to optimize their natural, cultural and economic assets; and

(vii) to ensure with informed stakeholders and objective decision making based on information and facts that global agriculture continues to reflect globally the ecological and cultural diversity.

The presented approaches have large impact on sustainable intensive crop production, capable of progressing the development of the new paradigm for agriculture in the world, based on within field (sub-field) land and crop management. Its impact is shown in Table 10.1.

Table 10.1 Expected impacts in land and crop management.

Improvement of ground and surface water quality	Weed detection promotes weed activated spraying as a weed control practice. It is a practice that reduces soil and ground water contamination and mobility of herbicides through targeted spraying. The reduction of water contamination by agrochemicals is in line with the European and global regulations regarding environmental protection. A redistribution of the inputs according to the real crop need will be achieved. Through this way, fertilizer leaching and runoff is minimized leading to reduced soil and ground water contamination attracting the interest to those concerned with the presence of agrochemicals in the environment
Conservation of biodiversity and wildlife	Biological benefits can derive from maintaining biodiversity and reduction of chemical use as proposed on the presented approaches. Soil biota can be negatively affected by application of pesticides and inorganic fertilizers. They transform hard organic matter in soil into a form, which can be easily used by plants. They add to soil fertility after they die. Normally, the greater the plant biodiversity - more is the animal biodiversity in the soil

Table 10.1 Expected impacts in land and crop management—cont'd

Improved human health, through the reduced release of pollutants and GHGs	The current methods adaptation will result in a reduction of fertilizer use by targeting areas of the field where pesticide and fertilizer application application, is beneficial and required. The reduced amount of fertilizers will reduce greenhouse gases' (GHG) emissions; hence, indirect energy consumption is also reduced
Increased farmers' competitiveness through the reduction of production costs	The main purpose of the presented work aims to increase farms' competitiveness by providing a system that is a support for improved spatially targeted variable rate fertilizer and pesticides applications. The integration of data on soil, crop cover and yield will furnish the farmers with ground truth decisions on applications of fertilizer and pesticides. Homogeneous application of pesticides and fertilizers is adopted by the majority of farmers worldwide. Variable rate application of input will result in increasing crop yield, at reduced inputs. Therefore, manufacturers of precision agriculture fertilizer spreaders The overall economic impact will be increased farming efficiency for farmers all over the world. The redistribution of inputs according to the real crop need will lead to improved yields (quantitative, but mostly qualitative) and reduced over-all inputs
Reduction of the negative environmental impact of crop production through more rational use of external inputs	A significant positive environmental impact is expected by reducing fertilizer use by targeting areas of the field to apply the precise fertilizer rates and optimizing spatial pesticide application. A reduction in fertilizers and pesticide use would lead to reduced soil and ground water contamination and therefore this research will be of significant interest to those concerned with the presence of agrochemicals in the environment. The reduced amount of fertilizers and pesticides will reduce GHG and GWP, which will have a positive impact on the environment

10.2 Economic benefits for agricultural production

Smart farming approaches fuse sensor data acquired by heterogeneous sensory systems, proximal and remote sensors which most farmers cannot afford due to its high cost, but it can be offered as a service at a low cost thanks to the infrastructure of Copernicus and products offered by several SME agricultural consultancy companies. By this way, the farmers are assisted to make more profitable decisions regarding crop management, without high cost services needed. Evolution supported by the proposed approaches enables farmers to assimilate actual information regarding the condition of farming assets (soil, water, crop) in an automatic way. The current techniques offer the potential of an elevating farming productivity, leading to high economic profits achieved through higher yield by employing Data Mining based approaches.

Smart farming is an enabling factor for enhancing the adoption of Best Management Practices (BMPs). The above given management examples outline the economic benefits through the proposed smart farming approaches as follows:

1. **Soil quality and fertility**. Yield variations across a field can be attributed to the variability of soil nutrient content. Decisions regarding different fertilizers applications rates of fertilizer, spatial decision for lime, mode and spatial decisions for manure deposition can be derived from these information. Soil compaction, buffer strips, sufficient soil drainage, and soil erosion can be controlled based on decision from data mining patterns, for applying topsoil depth, and soil organic matter.
2. **Pest management**. Since several types of infections, insects, and pesticides presence tend to appear in patches, the proposed technologies manage to detect effectively the possible infected filed areas at an early stage. Consequently, the farmers are able to adapt a site-specific treatment capable of minimizing the water contamination risk in a cost effective way. Moreover, the proposed methods support the precise fertilization applications, leading to the minimization of crucial environmental impact and economic loses. The proposed techniques are fundamentally based on precise and customized nutrient applications harmonized to individual crop needs and health status. Taking into consideration the proposed customized and precise application of fertilizers with the proposed methods assistance, spraying overlapping is avoided since pesticides are applied solely in the infected and nutrient stressed field areas.
3. **Decisions on planting**. The proposed methods enrich the yield potential, through assisting site-specific seeding application. The sensor driven decision element regarding the seeding location and application

rate, help to the maximization of the yield potential according to the type, morphological characteristics and nutrient content of soil.

4. **Proper data analysis**. Taking into consideration the increase of the production costs, concerning seeds, water, energy and farm machinery, resource optimization for improving the overall agricultural production is needed. Moreover, it is questionable whether the existent conventional methods concerning data acquisition in agriculture are able to perform effectively and precisely. The presented methods are capable of employing effectively heterogeneous sensory systems and advanced data mining algorithms, by yielding heterogeneous types of high volume data that are gathered, stored and fed to the proposed AI based applications for cost effective decision making

10.3 Social benefits for agricultural production

10.3.1 Environmental impact derived from the proposed methods

Considering the fact that some fertilizers are water-soluble and can remain in groundwater for decades, the addition of more nutrients over the years has an accumulative effect. Nitrogen groundwater contamination contributes further to an important environmental problem: the "dead zones". Nitrogen deposition, together with agricultural runoff, increases nutrients in water bodies, resulting in eutrophication and harmful algal blooms. The so-called "dead zones" are areas of large bodies of water—typically in the ocean—that do not have enough oxygen to support marine life. The cause of such "hypoxic" (lacking oxygen) conditions is usually eutrophication, leading to excessive blooms of algae that deplete underwater oxygen levels. Nitrogen and phosphorous from agricultural runoff are the primary culprits. This has an impact not only on the aquatic ecosystem, but on local societies which depend on food sourced from these areas.

The novel technologies that are presented, assist the reduction of extensive fertilizer applications and pesticide inputs having a significant impact on:

- Reducing nitrate loss to water from agriculture and at the same time increase the efficiency of N-use by the crops.
- Promoting farm practices necessary for the maintenance of bio-diversity and landscape.
- Helping to reduce eutrophication of fresh surface waters, thus reducing the health risk associated with toxic algae in inland waters.

- Preventing and reducing underground aquifers pollution by deep leaching of Nitrogen.
- Preserving the biodiversity in the soil, coastal and natural ecosystems
- Decreasing the use of scarce resources such as fossil fuels used in Nitrogen fertilizer
- Reducing indirect and direct energy inputs to crops leading to increased energy efficiency and reduce GHG emissions
- Increasing environment protection due to prevention of environmental damage using methods to assess the real crop requirements and covered them at the accurate amounts through the Variable Rate (VR) inputs application.

Table 10.2 demonstrates the environmental impact of the application of the primary precision agriculture approaches.

10.3.2 Standardization framework and traceability

Generally, Good Agricultural Practices (GAP) involve a variety of farming activities associated with:

- food safety indicators arising from the compliance to generic HACCP framework;
- avoiding unjustified chemical applications and decrease the presence of residues contained on food crops;
- sustainable agricultural production by minimizing its impact on the environment by controlling the harmful inputs;
- improved living conditions of rural populations, through improving occupational health and safety standards, and livestock welfare on farms;

GAP is well known and has already been applied in several countries at different levels of development by adopting sustainable practices as integrated pest and preservation agriculture. GAP takes advantage of existent best practices to handle the ambient, financial and societal impact from rural production through the food chain, aiming at offering food of high standards and nutritional value. The GAP protocol has found application in a variety of farming practices and production volumes, aiming to ensure food supply, supported by government policies and directives (FAO, 2003). A GAP framework to farming concerns the formation of best practices for throughout the production chain. This includes the compliance to these standards by farm managers and food chain stakeholders, and the promotion of these

Table 10.2 Envisaged environmental benefits by adopting PA approaches (https://publications.europa.eu/en/publication-detail/-/publication/6a75e0ac-90ae-11e9-9369-01aa75ed71a1/language-en/format-PDF/source-search).

No.	Process	Technique	Expected environmental gains
1	Timeliness of working under favorable weather conditions	Automatic machine guidance with GPS	Reduction in soil compaction Reduce carbon footprint (10% reduced fuel consumption in field operations)
2	Leave permanent vegetation on key location and at field borders	Automatic guidance and contour cultivation on hilly terrain	Reduction of erosion (from 17 to 1 tons/ha/year and perhaps lower) Reduction of runoff of surface water and reduced runoff fertilizers Reduced flood risk
3	Reduce or slow down water flow between potato/vegetable ridges to slow water	– micro-dams or micro- reservoirs made between ridges("tied ridges") – ridges along field contours	Reduced sediment runoff Reduced fertilizer runoff
4	Keep fertilizer or pesticide at recommended distances from water ways	– Automatic guidance based on geographic information – Section control of sprayers and fertilizer distribution	Avoidance/elimination of direct contamination of river water
5	Avoid overlap of pesticide or fertilizer application	– Section control of sprayers and fertilizer distribution	Reduce/avoid excessive chemical input in soil and risk of water pollution
6	Variable rate manure application	On the go manure composition sensing Depth of injection adjustment	Reduced ground water pollution Reduced ammonia emissions into the air
7	Precision irrigation	Soil texture map	Avoidance of excessive water use or water logging. Reduction of fresh water use
8	Patch herbicide spraying in field crops	Weed detection (on line/weed maps)	Reduction of herbicide use with map- based approach (in winter cereals by 6–81% for herbicides against broad leaved weeds and 20–79% for grass weed herbicides (Gerhards & Oebel, 2006).

Continued

Table 10.2 Envisaged environmental benefits by adopting PA approaches (https://publications.europa.eu/en/publication-detail/-/publication/6a75e0ac-90ae-11e9-9369-01aa75ed71a1/language-en/format-PDF/source-search)—cont'd

No.	Process	Technique	Expected environmental gains
			Reduction of 15.2–17.5% in the area applied to each field was achieved with map-based automatic boom section control versus no boom section control (Luck et al., 2010). 24.6% average herbicide savings was achieved in tramline spraying field trials
9	Early and localized pest or disease treatment	Disease detection – Multisensor optical detection – Airborne spores detection – Volatile sensors	Reduction of pesticide use with correct detection and good decision model (84.5% savings in pesticides possible (Moshou et al., 2011)
10	Orchard and vineyard precision spraying	– Tree size and architecture detection – Precision IPM	Reduction in pesticide use up to 20–30% Reduction of sprayed area of 50–80%
11	Variable rate nitrogen fertilizer application according to crop requirements and weather conditions	Crop vegetation index based on optical sensors Soil nutrient maps	Improvement of nitrogen use efficiency. Reduction of residual nitrogen in soils by 30–50%
12	Variable rate phosphorus fertilizer application according to crop requirements and weather conditions	Crop vegetation index Soil nutrient maps	Improvement of phosphorus recovery of 25%
13	Crop biomass estimation	Crop vegetation index	Adjust the fungicide dose according to crop biomass (Jensen & Jørgensen, 2016)
14	Mycotoxin reduction	Crop vegetation index and fungal disease risk	Optimization of fertilizer dose and fungicide use on the basis of higher disease risk in areas with high crop density

standards as quality labels to the wider audience so that the image of GAP compliant products become synonymous to food safety and quality (Hobbs, 2003).

FAO (2003) also associates GAP to various resources closely related to agricultural activities like pest management, crop processing, animal welfare, waste handling, and energy use optimization. Food safety is not a direct target, but it is an important issue.

The GLOBAL GAP (2016) standard concerns the entire agricultural process originating from the seed stage to the post-harvested unprocessed end product. More specifically, GAP the fruit and vegetables standard covers several areas including: food traceability, quality and safety, soil and water management, crop monitoring, fertilizer application and plant protection management.

There are also domain specific guidelines concerning livestock welfare expressed with targeted standards. The GAP standard can be applied through a guidance document that illustrates the procedure of certification to the GAP allowing the producer to attain the certification and remaining compliant. Different GAP frameworks contain similar requirements, however they might have different focus depending on the region where they were customized to specific local needs.

Qualität und Sicherheit GmbH (QS, 2016) is a separate framework provides enabling tools to producer associations by compiling practical requirements into guidelines in that can be verified by checklist compliance. Initially these procedures may look different, but it is possible to apply benchmarking evaluation or to adapt checklists so that they are mutually applicable for audits that work for different schemes simultaneously without adaptation.

10.3.2.1 Food safety

Food safety is closely associated to GAP directives, which result as a measure to protect human health. Additionally, the retail sector have the expectation that the food production actors and suppliers will have setup, compliant to food safety framework that can be examined. Private organizations including SAI-Global, and TÜV Rheinland try to establish food safety standards. The formation of those systems and the initiators of these frameworks have many differences, but they share common interest concerning GAP and food suitability. The main key players are listed in Table 10.3 as follows:

Table 10.3 Key players.

Organization	Certification standard
SAI-Global, 2016	Certification of quality management systems
FSSC/FS 22000 (Food Safety System Certification standard)	Certification scheme for food manufacturers, focused on the integration of ISO 22000:2005 Food Safety Management
Systems standard and Publicly Available Specification (PAS)	Certification promoted by FoodDrinkEurope, FS 22000, supported by the Global Food Safety Initiative (GFSI)
ISO 22000	Certification offering a standard with a wider appeal than those of food processors, directed across the whole food chain covering packaging and ingredient producers, logistics, chemical and agromachinery industries with application to farms
BRC Global Standards	Worldwide earliest GFSI-approved standard, option for retailers globally seeking assurance from food chain providers
SQF (Safe Quality Food institute)	a globally leader of food safety and quality management systems, complying to the needs of retailers and suppliers globally. It certifies that a supplier will provide a level of safety and quality meeting the global regulations
HACCP (Hazard Analysis and Critical Control Points)	a risk handling system that recognizes, evaluates, and handles food hazards throughout the food supply chain
IFS International Food Standard	quality and food safety standard for market branded food products, which promotes uniform compliance scheme that unifies the different approaches
GFSI	supported by the Global Food Safety Initiative (GFSI), seven major retailers embracing four GFSI validated food safety frameworks
PACsecure	HACCP-based certification concerning packaging industry
Global GAP	promoted by FoodPLUS GmbH, derived from GLOBALGAP, to enhance standards in the production of fresh fruit and vegetables. Certificates promote an equalized level of safety and quality, and assures that the producers follow continuously improvement process to elevate the standards

10.3.2.2 Chemical residues

Pesticide application is heavily regulated regarding the type of molecules that are permitted on certain crops and the allowed time interval between chemical application and harvest. This comes additionally to the Maximum Residue Level (MRL) on the product. Additionally there is no effective MRL regulation in the occasion that when different types pesticides are combined and applied on the same crop. The issue concerns the cumulative dose or MRL defined by molecule. It is obvious that no residues of unregistered molecules are permitted for application. Within the European areas, the Commission accesses all active compounds used for pests/plant diseases elimination (the "active substance") to prevent entering the market in a product. The MRL is applicable throughout the EU region. However, the composition of many authorized products differs a lot between different countries. In the European region, the crop trade between countries is not affected by these different compositions since the MRL's are permitted across EU. On the other hand, for international trade out of EU it is stricter for farmers and exporters, because a molecule permitted in the country import may not be accepted in the production country or the opposite and also the MRL's may have different definitions.

10.3.2.3 Microbial control

The European Community (EC) microbiological protection standards for food products have been modified and additional crucial standards have been defined through the years. EC directive No 2073/2005 on microbiological safety for food products, sets food safety thresholds concerning specific bacteria detected in food and their relevant toxins and metabolites they produce during different metabolic stages. The regulation complies fully with by food industry stakeholders when applying the appropriate hygiene measures. Moreover the microbiological protection standards meet the globally accepted directives (e.g., Codex Alimentarius).

In 1998, in USA, the regulatory bodies for Food and Drugs Administration (FDA) and the United States Department of Agriculture (USDA) introduced the "Guidance for industry—Guide to minimize microbial food safety hazards for fresh fruits and vegetables." This guide concerns detailed description of microbial food hazards and proposes agricultural BMPs for the cultivation, harvesting, post-harvest handling and transportation of most agricultural, ready for consumption products, in order to minimize microbial contamination risk (USDA, 1998, 2015). More specifically, it introduces eight principles of microbial protection standards related to the cultivation, harvesting, post-harvest handling and transportation of most agricultural products, defining the basis for adopting BMPs and minimizing the arising microbial contamination risk.

Most of the times, the majority of consumers define fresh leafy vegetables or fruit (either raw or minimally processed) as health promoting. However, recent events of foodborne hazards were linked to raw products- contamination (Food and Drug Administration, 2007). The main aim of the retailers of leafy vegetables is to minimize the risk and ultimately achieve zero-risk production chains. It has been indicated that the contamination of products is due to irrigation water, but unfortunately, there is lack of adequate information regarding the microbial load in agricultural water. A combination of several solutions including chemical or physical treatment and drip irrigation water placement are regarded crucial so as to comply to the microbial thresholds in leafy vegetables, enhancing food safety for consumers (Allende & Monaghan, 2015). Van Der Linden et al. (2014) confirmed that bacteria like *Salmonella* and *E. coli O157:H7* presence is variable between different irrigation samples.

10.3.2.4 Toxin safety

Microbial and toxin food contamination can take place at the growth stage, while also at harvest and postharvest stages, too. According to GAP regulations, this type of contamination derives from the worker personal hygiene and the repetitive sanitation of the equipment used for harvesting and transportation.

Other features of microbial and mycotoxin risks are related fungi activity on crops. Owing to their chemical makeup, cereals and soybean are vulnerable to microbial infections, particularly by filamentous fungi. These crops can be infected by fungi, either during growth or during post-harvest. Fungi affecting cereal seeds and oilseeds are regarded crucial for the assessment of the potential risk of mycotoxin presence. Mycotoxins belong to fungal secondary metabolites, carrying high toxicity for vertebrate animals even in low concentrations either orally or inhaled. Field transmission of fungi to cereals in the field, occurs though soil, birds, animals, insects, organic fertilization, and from other types of vegetation in the field area. Fungal growth is affected by a complex relationship between several environmental parameters including pH, temperature, humidity, nutrient content, structural damage and the microbial load. Low hygiene standards, temperature level fluctuations and moisture distribution after harvesting, post-harvesting, processing and handling is likely to result to higher food contamination.

10.3.2.5 Relation of mycotoxin transmission to crop management practices

Mycotoxin content in the field area can be affected by the adaptation of the suitable BMPs like crop rotation, soil cultivation, proper irrigation, and fertilizers application. Crop rotation is regarded important, limiting the growth of contagious material through wheat/legume crop rotations. The need of the soil regarding the fertilization or soil conditioners has to be evaluated in order to guarantee the level of soil pH and the level of plant nutrients to prevent crop stress, especially at growth stage. As an example fertilizer application may affect *Fusarium* infection probability and severity either through changing the residue speed of decomposition, by inducing stress on the host plant or by changing the crop architecture. In some occasions, the increase of the nitrogen-rate concentration leads to the wider spread of *Fusarium* occurrence in wheat, barley, and triticale crops. Moreover, ambient parameters including relative humidity and temperature have crucial effect on crop health status.

Water stressed crops are more vulnerable to several infections, so crop planting should be constructed in such a way so to avoid that stress during development maturation in order to prevent heat and water stress during growth. Another factor which decrease resistance to toxigenic mold invasion is injuries by insects, birds, or rodents. Insect damage and fungal infection have to be eradicated in the region of the crop though applying registered chemicals, and adopting other suitable practices offered by an integrated pest management control.

As a conclusion the compliance to good agricultural practices, will alleviate the occurrence of mycotoxins in the food chain. The variability of field conditions increase the risk for infection, when environmental parameters are suitable, therefore the site specific management can control this risk factor. The management practices have to be supported by suitable data acquisition during plant growth and also information about the previous field interventions to enable decision making on treatments or harvesting operations.

The early crop disease identification and their removal achieved by the presented techniques in previous chapters encourages the design and equipment engineering sectors to focus more on food safety oriented design by adopting suitable sanitation processes. The equipment used at the food chain sectors that comes in contact with food should be augmented microorganism or volatile sensors so that any microbiological hazard or toxin contamination can be detected in real time. These sensors will be able to alert for possible problems and will trigger the proper intervention. The information concerning the interventions along the food chain together with contamination triggers should be recorded with times stamps and location information in the traceability system.

10.3.2.6 Weed control

GAP compliance criteria concern the minimal herbicides application reporting of time location and quantity of chemical products use. It has been indicated that the presented weed detection mapping methods can enable the site specific use of chemicals only on where they are needed. These methods offer new insights on weed management practices, posing new challenges for the investigation of new solution in the near future given as follows:

- Several techniques for the discrimination between weeds and the main cultivated crop have utilized spectral signatures and/or image recognition (Slaughter, Giles, & Downey, 2008; Vrindts, De Baerdemaeker, & Ramon, 2002). However, when the operations take place in field conditions there are several factors that impose challenges to the correct acquisition. Some examples of these challenges concern the mutual leaf coverage of weed and crop, their variability due to the growth stage and the variable illumination conditions during the day, unless artificial lighting is provided;
- Contemporary weed management approaches are not accurate and precise enough to avoid side effects without harmful consequences to the crop productivity. The problems of mechanization in agricultural operation have led to the development of systems that can actually can be adapted to tasks of mechanical weeding by using sensor assisted weed management, enabling the use of cutting edge technology, meeting their needs and raising their profits (Young, Meyer, & Woldt, 2014);
- A further challenge concerns the ineffectiveness of single strategies. Therefore an integrated weed management approach is capable of offering enhanced weed control. There is a need for additional research to assess the feasibility of these approaches and to enhance their applicability in a such a way that it can satisfy both technical and cultural criteria. These methods optimize crop management to meet the needs of sustainable and conservation techniques (Bajwa, Mahajan, & Chauhan, 2015);
- Adoption by growers non chemical weed control as well as approaches that augment other systems with herbicide use, has been minimal. Future weed management by growers will be more knowledge intensive, will require

more effective planning and it will be more time consuming and expensive compared to the past (Shaner & Beckie, 2014);

– A precision agriculture approach can be combined with a model describing the weed dynamics. A weed dynamics model can predict the evolution of weed establishment though the years, as a function of the cropping system and parameters of soil and weather. This weed dynamics model is useful for the definition of managemet practices concerning weed-related biodiversity (weed species abundance and equitability, weed-based food source for birds, insects and pollinators) and weed borne damage (yield destruction, inclusion of weeds during harvesting, weed dispersion in the field, propagation of crop disease from the weeds) (Mézière, Petit, Granger, Biju-Duval, & Colbach, 2015). The weeds dynamics models are crucial for future approaches concerning weed management and decision making in the framework of IPM.

10.3.3 Employment

Through the development and the adaptation presented technologies for precision farming which optimizes soil nutrient and pesticide management, a contribution for a sustain and increase agricultural income will be achieved. As a consequence jobs can be saved or even new jobs can be created, both in agriculture and in related businesses. Savings on fertilizer costs will have reduced production costs to the farmers and allowing extra investment of precision agriculture enabled equipment including a more accurate new VR fertilizer applications for farmers. In addition, the development of more precision agriculture using such novel technologies, will create new jobs for people working in agriculture machinery manufacturing sector. Therefore, manufacturers of agricultural machinery and related devices will benefit from the proposed methods as well. New machinery will be developed or already existing will be adapted to meet the VR application of the inputs. Suppliers to these manufacturers (e.g., the sensing and wireless communication/networking electronics and I.T. sector) will gain more work as well and specific components in particular.

The impact of the expected spin-offs of using these technologies, i.e., the development of precision farming applications, will be substantial. The newly adopted approaches including hardware and software will open a new business opportunity to commercialize the results, by partnering with the partner SMEs or by creating new businesses in the agriculture sector.

More employment in a new generation of fertilizer and pesticides technology to meet these needs is therefore expected. In addition, there will be a need for more working positions creation, concerning crop monitoring, aiming to ensure that the needs of sustainable agriculture and water quality are being met. These new technologies may also attract young people and improve precision agriculture's potential and prospects.

10.3.4 Food quality, geo-traceability and authenticity

Food quality and authenticity can also be a potential domain concerning the application of the proposed technologies. As the food processing industry is a direct successor of the agriculture domain it has the strongest link of all to the above named markets. The transfer of automation and autonomous monitoring technology into the agriculture domain might lead to more efficient agriculture processes with, for example, new timing constraints. The food processing industry also has to adopt these processes by responding to changes in farm practice brought about by increased use of automation and autonomous monitoring technology.

By using such monitoring technologies, the traceability of food production will be more efficient since the recorded information will enable automatic compliance to food safety standards and all information on treatments and origin will be available for use. Through PF, almost all (if not all) data and activities are digitally geo referenced. Consequently, it can be rather straightforward to ensure geo-traceability of farm products (e.g., farm to fork, cradle to gate) ensuring quick and accurate trace-back and recall when necessary or providing information on agricultural products provenance to the public. This is a growing requirement from food safety agencies, certification bodies, but also from the European consumer.

10.3.5 Health

Groundwater and food contamination has been linked to gastric cancer, goitre, birth malformations, and hypertension, testicular cancer and stomach cancer. Excessive airborne fertilizer and pesticides components have been linked to irritation of eyes, nose, mouth and skin. Air- and water-borne nitrogen from fertilizers and pesticides can cause respiratory and cardiac disease, several types of cancer, decelerate children's growth and "affect the dynamics of several vector-borne diseases, including West Nile virus, malaria, and cholera"(http://www.sustainablebabysteps.com/effects-of-chemical-fertilizers.html). Fertilizers can induce even dermatological problems, if the skin comes in contact with contaminated soil or airborne fertilizers. When on the skin, they may draw out moisture and cause skin burns. Liquid anhydrous ammonia is used for injecting nitrogen into the soil, where it expands into a gas and combines with moisture. If the liquid or gas contacts the body, cell destruction and burns can occur, and permanent injury can result without immediate treatment (http://www.ilo.org). Maybe the most serious health effect of fertilizers is a disorder called "methaemoglobinemia" or "Blue Baby Syndrome". It is characterized by reduced ability of the blood to carry oxygen because of reduced levels of normal haemoglobin. Infants are most often affected, and may seem healthy, but show signs of blueness around the mouth, hands, and feet. The most common cause of methemoglobinemia, which can be even fatal, is high levels of nitrates in drinking-water. The risk most often occurs when infants are given formula reconstituted with nitrate contaminated water.

The adoption of the presented approaches will help to the elimination of health risks posed by the excessive use of chemicals in crop production. Using remote sensing and field sensors to optimize both fertilizer and pest management, potential losses of fertilizers through soil, water and atmosphere can be promptly detected and managed, so that the related stresses on human health can be avoided.

10.3.6 Development of skill/education and training

The introduction and implementation of such technologies will require further education and training at different levels. Education and training at academic level (both agricultural as well as technical colleges) is needed to safeguard further research and development. To successfully implement the technology on farm level it will be imperative that cross-area training will be carried out. Extension services and farming machines suppliers need the education and training to demonstrate the applications. The consortium implementing this project represents the educational (internet-based e-learning) and research network very well and has good access to extension services and applied research and demonstration projects.

10.3.7 Other socio-economic impacts

Furthermore, the impact the presented techniques on socio -economic problems includes but is not limited to the following:

- Increased food safety due to prevention of food poisoning by using the appropriate amounts of fertilizers and by reducing chemicals application through the reduction of pesticides use in crop production.
- The proposed methods are able to form new products of existing agriculture machinery as well as sensor and IoT manufacturers.

10.4 Prospects for further development of the proposed sensing techniques in the near future

The current approaches demonstrated in the presented case studies, have the potential of imposing a significant impact on further applications where the specific algorithms can be employed. Table 10.4 summarizes the future perspectives of the presented approaches into pertinent scientific regions regarding the theoretical expansion and applicability in Biosystems Engineering.

Table 10.4 Synopsis of the utilized methods for every application of the presented tutorials and the relevant potentials of evolution and expansion of each application into other scientific fields.

Method	Application	Evolution potential of the application in the near future	Expansion of the application in other scientific fields
Hierarchical SOMs	Weed species detection	Scale up of the algorithm that can recognize x10 weed species	Phenotype recognition for species discrimination in crop genetics applications
LSSVM	Identification of *Septoria* infection	Detection & Discrimination of fungal infections	Identification and discrimination of more types fungal infections, viral infections and nutrient stresses
Hierarchical SOMs	Yellow Rust detection	Detection & Discrimination of fungal infections	Recognition and discrimination of more types of viral infections and nutrient stresses
SVM combined with Conflict resolution	Recognition of three different types of infection in 25 different crops	Scale up of the algorithm that can recognize and discriminate x10 different types of infections	Recognition and discrimination of different weed species

Table 10.4 Synopsis of the utilized methods for every application of the presented tutorials and the relevant potentials of evolution and expansion of each application into other scientific fields—cont'd

Method	Application	Evolution potential of the application in the near future	Expansion of the application in other scientific fields
Hierarchical SOMs	Yield prediction in wheat crop	Expansion of the method aiming to predict the yield in other crops	Prediction and quality assessment – expansion of the algorithm so as to be embedded in traceability models
Hierarchical SOMs	Classification of different harvesting stages in lettuce plants	Harvesting stage assessment in other horticultural products	Quality prediction of horticultural products in association with nutrition elements (phenols, vitamins, sugars, proteins, etc.)

10.4.1 Impacts of the projected methods to the surroundings

The next shown PA methods are thought-about to own the best beneficial environmental impacts:

- Patch deposition with chemicals in field crops, supported financial and crop yield issues of weed establishment;
- In orchards and vineyards: implement and high heat weed management inside and among rows;
- Use of PA methods together with digital elevation maps and erosion prevention to cut back erosion dangers;
- Introduction of a toolbox for stress and disease management in orchards that embody tree design, tree volume and form within the proximal future this might be combined with disease recognition;
- Use of unmanned airborne vehicles (UAV) recognition of infections and weeds as an additional versatile tool. If appropriate image processing is applied this technique may be a valuable tool for crop monitoring and early alarm of future hazards;

- Employment of reduced rates of pesticides by higher risk-based decisions and temporally exact interventions;
- Monitor insects that will harm grain kernels that successively will cause reduced plant growth and poisonous substance production. Once the insect number is immense then insect powdering is also required.

Alternatively, site specific antifungal application is also needed on areas wherever fungi are active so as to cut back the occurrence of mycotoxins;

- Exploit the connection between nitrogen-fertilizer rate, timing of application, crop and condition and also the risk for plant growth for site specific variable rate (VR) use of fertilizers;
- Focus in additional detail and employ a more flexible discovery of variability and accessibility of phosphorus within the soils, in reference to plant needs;
- Use of sensing systems for chemical element and phosphorus in water and also the relation with temporal schedule and rate of chemical application additionally as climate and soil maps (or soil spatial variations). Analyze this info over time in season and over many season) to draw conclusions for higher management;
- Avoid irrigation induced contaminants by specific arrangement of drip irrigation, watching the system so that water doesn't interfere with foliate vegetables;
- Utilize GPS, weather conditions and wind status to avoid spray drift which will harm proximal crops or make them improper for consumers. Similar management is required for manure application on fields;
- Utilize PA for alleviating risks by microorganism contamination or poisonous substance contamination that would have an effect on consumer health;
- Check out potential branching effects of various native, national or European policy frameworks that will cause negative environmental impact once applying PA;
- Low price environmental watching tools for nitrates, or alternative chemical is that the emptying or runoff water from fields may be a stimulant as these might show the environmental impacts of PA and use a reward scheme. Nowadays, the regional nitrate watching supplemented with management rules is demonstrating its result on drinking water quality.
- Information technology in farming will support and boost the environmental impacts of Precision agriculture in many alternative ways;

10.4.2 Foresights of the projected techniques integrated within the industry

The farming friendly Fourth Industrial Revolution can increase the market appeal of agriculture in numerous fields, like culture, welfare, and health promotion in terms of product orientation. As shown in Fig. 10.1, the 4IR can cause a bigger quantity of communal and freelance farming through cultural activities, like associating farming with recreational activities, human health and wellbeing agriculture within the health claim realm for elderly, and farming activity with crops and livestock (Food and Drug Administration, 2007). The enlargement of farming in the context of the 4IR is anticipated to be very variable within the growth sector, logistics, and retail-consumer levels.

Adaptations of farming activities within the 4IR can appear mainly in farming installations with smart farming techniques. In advanced installations, the control of the growth can increase the worth of farming products. In Korea, three stages should be operational so as to push smart farming. The primary stage, completed before 2017, is that the convenience enhancement stage. During this stage, facilities received upgrades to permit farmers to ascertain the expansion standing of farming through mobile devices. Thus, farmers don't got to trip farms for simple tasks like temperature management. The second stage, that is anticipated planned to be ready until 2020, is

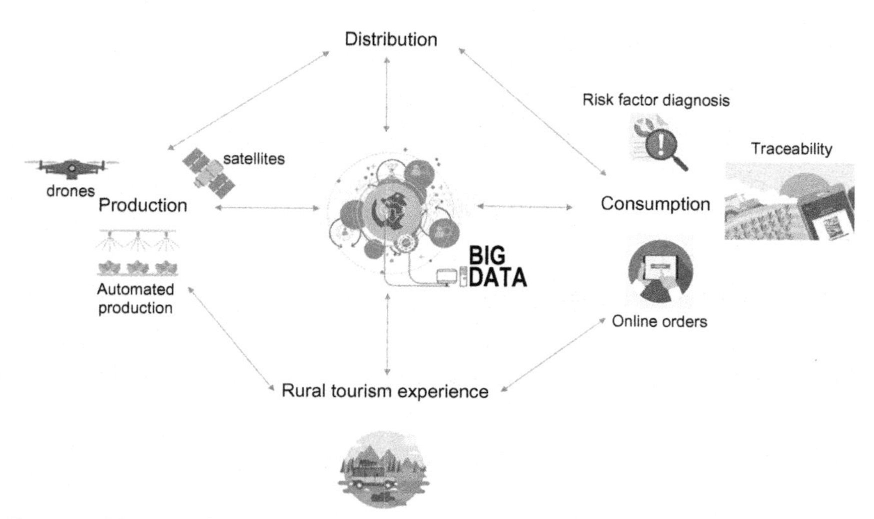

Fig. 10.1 The Fourth Industrial Revolution and its impacts in different fields of the market.

productivity enhancement. During this stage, profits are accumulated through precise management and optimum management of farming. The third stage is that the completion stage, during which all of the processes are automatically adjusted consistent with the growth plan of the crop primarily based on the crop prediction of growth and targets set. The Korea Rural Development Administration has set up and providing numerous sensors and technologies in smart farms, so as to assist farmers speedily and with efficiency pass through the three stages.

As depicted in Fig. 10.2, the 4IR also will build a giant distinction in open-field agriculture. There are three stages during which this technology may be used: monitoring the field for crop growth, analyzing knowledge within the decision-making phase, and employing variable rate treatments by smart farm machinery. Monitoring crop growth status encompasses not solely the health condition of crops however additionally environmental condition info, environmental info, and growth info, and it's apace evolving simultaneously in large-scale industrialized agriculture, as within the USA, and intensive agriculture, as in Korea.

it's attainable to maximize production volume and minimize the likelihood of failure because of natural hazards, system failures, and alternative factors by getting knowledge on growth, the weather, and farming instrumentation. Analyzing knowledge within the decision-making phase concerns data mining coming from sensing stage and plan farming tasks that are needed. During this stage, collected knowledge is stored, processed,

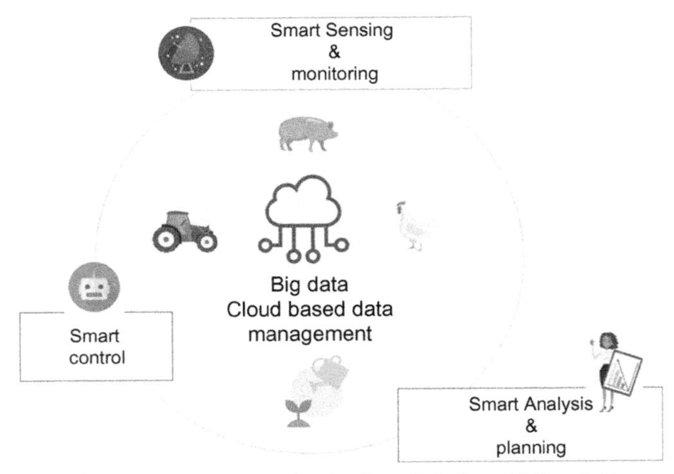

Fig. 10.2 Smart farming components in the Fourth Industrial Revolution context 4IR.

and analyzed as Big Data. Then, economical and precise decisions regarding the data are created during a process that evolves beyond human intelligence, wisdom, and skill. What is more, it's attainable to gather environmental knowledge on farming practices through a farming service platform that enables the exploitation of big data. This knowledge may be exploited to predict market sale trends consistent with market preference analysis, so the info (the cultivation surroundings, stress and hazards data, climate and weather info, soil fertility, geographical relevance, etc.) may be supplied to producers to optimize farming processes. In recent years, huge knowledge and computer science are accustomed to maximally expand the fields of biotechnology with relation to agriculture and stock. Inside the realm of resolution laws, rules, and moral problems, it'll be attainable to produce edible crops and biology crops that grow in extreme climatic conditions or dry areas. It'll even be attainable to rework animals' genes so as to create additional economical and appropriate genotypes for native environments. Variable Rate Application exploitation in smart farming machines is the third stage of this method. Within the previous stage, the optimum plan was chosen for every location. During this stage, it's needed to administer the proper material that appropriate for the specific location. In intensive agriculture a fleet of tractors are going to be able to accomplish identical tasks (i.e., weed killer spraying) at totally different positions (i.e., variable rate application) by following fixed intervals.

At night, once the farmer is dormant, robotic vehicles might be guided via GPS and electronic maps, enter the farm, end any needed farming tasks, and come to back earlier than dawn. This dream is going to be a reality within the near future. It'll be introduced by the Fourth Industrial Revolution.

10.4.2.1 Preparing for the fourth industrial revolution

In the era of the Fourth Industrial Revolution, new technological paradigms and novel business models that can't be predicted by the current legal and systems framework are going to be developed. The positive regulation methodology of regulating gene expressions in genetic therapies is presently not legal. So as to employ positive regulation, businesses intending to incorporate novel techniques and services have to look forward to laws to be enacted, which permit the employment of positive regulation.

To allow the 4IR to be embedded in agriculture, it's needed to support the protection of farming work and rural livelihood and to establish a convenient condition for cyber techniques and cloud infrastructures. This will protect against health related and sociologically negative situations in rural regions. Wearable IoT and mobile devices are concrete strategies based on which we will apply farming tasks safely, cyber physics systems (CPS), remote medicine practice, cyber cultural life, and elderly farmer life support systems for safe and healthy living by feeding streams of data into a Big Data framework.

The conceptual framework of precision farming is one during which agriculture work isn't really becoming additionally precise, however in it the farming systems as an entire move from a statistically monitored mentality to a precise sensing, monitoring and control framework. Hence, it's not considered an exaggeration to mention that the aim of precision agriculture is the entire agricultural system to be optimized. As a farming framework, three frameworks of technology should be used so as to totally develop precision agriculture (Fig. 10.3).

The first framework is that the acquisition of data associated with the surroundings wherever crops will be grown, like crop growth standing, soil fertility, and climate by location. The suggests that of getting such data by sensing nodes placed at every location, which might monitor a variety of conditions together with the yield of crops, soil moisture concentration, soil nutrients, water stress, and also the occurrence of pests or weeds. These sensor systems don't just acquire data to be later analyzed during a laboratory;

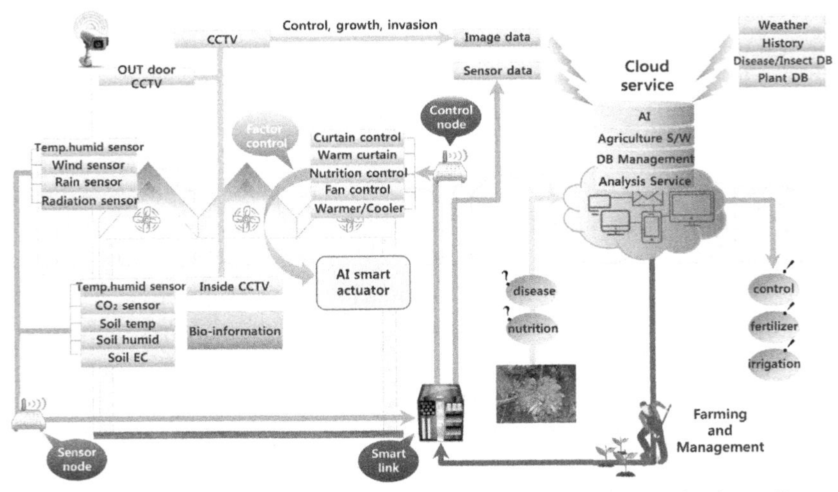

Fig. 10.3 Crop production framework through Sensing and Artificial Intelligence (Sung, 2018).

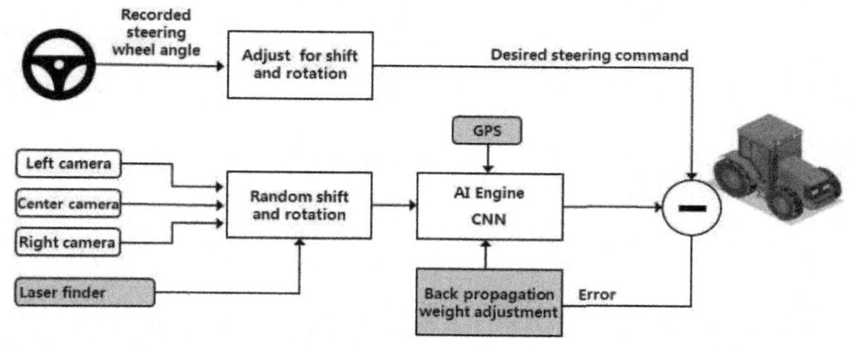

Fig. 10.4 Artificial Intelligence based autonomous routing of unmanned farm vehicles (Sung, 2018).

they're able to instantaneously process and save datasets of farming sensor data in real time.

The second framework is that the application of needed farming intervention material into the crops primarily based on decision-support-systems and best management practices, machines will unleash seeds, nutrients, and phytochemicals to the crops.

The third framework is that the process of processed geographical info and databases at the side of the farmers' prescribed inputs so as to drive the management systems of assorted farm platforms. Although the primary two frameworks are in an advanced level of development, it's troublesome to hold out precision farming if the third decision support framework methodology is not equally ready to implement (see Fig. 10.4).

As one sector of farming is evolving and the farmer operations have become technically advanced during this manner, it doesn't mean that precision farming is operational. That is, precision agriculture doesn't evolve farm by farm. Precision agriculture isn't a word concerning one technology, instead **it encompasses an overall conception of evolution in agriculture.**

References

Allende, A., & Monaghan, J. (2015). Irrigation water quality for leafy crops: A perspective of risks and potential solutions. *Int J Environ Res Public Health, 12*, 7457–7477.

Bajwa, A. A., Mahajan, G., & Chauhan, B. S. (2015). Nonconventional weed management strategies for modern agriculture. *Weed Science, 63*(4), 723–747.

FAO, 2003. Development of a Framework for Good Agricultural Practices. http://www.fao.org/docrep/MEETING/006/Y8704e.HTM; accessed 31/03/2016.

Food and Drug Administration, FDA Finalizes Report on 2006 Spinach Outbreak, 27 March 2007, Available from: http://www.fda.gov/NewsEvents/Newsroom/PressAnnouncements/2007/ucm108873.htm

Gerhards, R., & Oebel, H. (2006). Practical experiences with a system for site-specific weed control in arable crops using real-time image analysis and GPS-controlled patch spraying. *Weed Research*, *46*(3), 185–193.

GLOBAL G.A.P (2016). http://www.globalgap.org/export/sites/default/.content/.galleries/documents/160202_FV_Booklet_en.pdf; accessed 31/03/2016.

Hobbs, J., 2003. Incentives for the adoption of Good Agricultural Practices: Background paper for the FAO Expert Consultation on a Good Agricultural Practice approach; http://www.fao.org/prods/gap/Docs/PDF/3-IncentiveAdoptionGoodAgrEXTERNAL.pdf, accessed 3/04/2016.

Jensen, P. K., & Jørgensen, L. N. (2016). Interactions between crop biomass and development of foliar diseases in winter wheat and the potential to graduate the fungicide dose according to crop biomass. *Crop Protection*, *81*, 92–98.

Luck, J. D., Pitla, S. K., Shearer, S. A., Mueller, T. G., Dillon, C. R., Fulton, J. P., et al. (2010). Potential for pesticide and nutrient savings via map-based automatic boom section control of spray nozzles. *Computers and Electronics in Agriculture*, *70*(1), 19–26.

Mézière, D., Petit, S., Granger, S., Biju-Duval, L., & Colbach, N. (2015). Developing a set of simulation-based indicators to assess harmfulness and contribution to biodiversity of weed communities in cropping systems. *Ecological Indicators*, *48*, 157–170.

Moshou, D., Bravo, C., Oberti, R., West, J. S., Ramon, H., Vougioukas, S., et al. (2011). Intelligent multi-sensor system for the detection and treatment of fungal diseases in arable crops. *Biosystems Engineering*, *108*(4), 311–321.

QS (2016): Guideline Production Fruit, vegetables, potatoes. https://www.qs.de/documentcenter/dc-production.html, accessed 3/04/2016

Shaner, D. L., & Beckie, H. J. (2014). The future for weed control and technology. *Pest Management Science*, *70*(9), 1329–1339.

Slaughter, D. C., Giles, D. K., & Downey, D. (2008). Autonomous robotic weed control systems: A review. *Computers and Electronics in Agriculture*, *61*, 63–78.

Sung, J. (2018). The fourth industrial revolution and precision agriculture. *Automation in Agriculture: Securing Food Supplies for Future Generations*, *1*, .

USDA (1998). http://www.fda.gov/downloads/Food/GuidanceRegulation/UCM169112.pdf (web archive link, 03 April 2016) http://www.fda.gov/Food/GuidanceRegulation/GuidanceDocumentsRegulatoryInformation/ProducePlantProducts/ucm187676.htm accessed 3/04/2016.

USDA, N. (2015). The PLANTS Database (http://plants.usda.gov, 1 May 2018). National Plant Data Team, Greensboro.

Van Der Linden, I., Cottyn, B., Uyttendaele, M., Berkvens, N., Vlaemynck, G., Heyndrickx, M., et al. (2014). Enteric pathogen survival varies substantially in irrigation water from Belgian Lettuce Producers. *Int J Environ Res Public Health*, *11*, 10105–10124.

Vrindts, E., De Baerdemaeker, J., & Ramon, H. (2002). Weed detection using canopy reflection. *Precision Agriculture*, *3*, 63–80.

Young, S. L., Meyer, G. E., & Woldt, W. E. (2014). Future directions for automated weed management in precision agriculture. In S. L. Young & F. J. Pierce (Eds.), *Automation: The future of weed control in cropping systems* (pp. 249–259). Dordrecht, The Netherlands: Springer.

Further reading

Dammer, K. H., & Wartenberg, G. (2007). Sensor-based weed detection and application of variable herbicide rates in real time. *Crop Protection*, *26*(3), 270–277.

Mézière, D., Colbach, N., Dessaint, F., & Granger, S. (2015). Which cropping systems to reconcile weed related biodiversity and crop production in arable crops? An approach with simulation-based indicators. *Eur J Agron*, *68*, 22–37.

Index

Note: Page numbers followed by *f* indicate figures and *t* indicate tables.

I

K

L

M

N

T

U

V

W

X

Y

Z

Printed in the United States
By Bookmasters